The Chemical Industry

M000237706

Also edited by Alan Heaton

An Introduction to Industrial Chemistry
Second edition
(a complementary volume)

Contents: Editorial introduction *C.A. Heaton*. Introduction *C.A. Heaton*. Sources of chemicals *C.A. Heaton*. The world's major chemical industries *C.A. Heaton*. Organization and finance *D.G. Bew*. Technological economics *D.G. Bew*. Chemical engineering *R. Szczepanski*. Energy *J. McIntyre*. Environmental pollution control *K.V. Scott*. Chlor-alkali products *S.F. Kelham*. Catalysts and catalysis *J. Pennington*. Petrochemicals *J. Pennington*. Index.

The Chemical Industry

Second edition

Edited by

Alan Heaton

School of Chemical and Physical Sciences
Liverpool John Moores University

WARNER MEMORIAL LIBRARY
EASTERN UNIVERSITY
ST. DAVIDS, PA 19087-3696

BLACKIE ACADEMIC & PROFESSIONAL
An Imprint of Chapman & Hall

London · Glasgow · New York · Tokyo · Melbourne · Madras

1-25-8

Published by
Blackie Academic & Professional, an imprint of Chapman & Hall
Wester Cleddens Road, Bishopbriggs, Glasgow G64 2NZ

Chapman & Hall, 2–6 Boundary Row, London SE1 8HN, UK

Blackie Academic & Professional, Wester Cleddens Road, Bishopbriggs, Glasgow G64 2NZ, UK

Chapman & Hall Inc., One Penn Plaza, 41st Floor, New York, NY 10119, USA

Chapman & Hall Japan, Thomson Publishing Japan, Hirakawacho Nemoto Building, 6F, 1-7-11 Hirakawa-cho, Chiyoda-ku, Tokyo 102, Japan

DA Book (Aust.) Pty Ltd, 648 Whitehorse Road, Mitcham 3132, Victoria, Australia

Chapman & Hall India, R. Seshadri, 32 Second Main Road, CIT East, Madras 600 035, India

First edition 1986

Second edition 1994

© 1994 Chapman & Hall

Typeset in 10/12 pt Times by Thomson Press (India) Ltd, New Delhi
Printed in Great Britain by the Alden Press, Oxford

ISBN   0 7514 0018 1 (PB)

Apart from any fair dealing for the purposes of research or private study, or criticism or review, as permitted under the UK Copyright Designs and Patents Act, 1988, this publication may not be reproduced, stored, or transmitted, in any form or by any means, without the prior permission in writing of the publishers, or in the case of reprographic reproduction only in accordance with the terms of the licences issued by the Copyright Licensing Agency in the UK, or in accordance with the terms of licences issued by the appropriate Reproduction Rights Organization outside the UK. Enquiries concerning reproduction outside the terms stated here should be sent to the publishers at the Glasgow address printed on this page.

The publisher makes no representation, express or implied, with regard to the accuracy of the information contained in this book and cannot accept any legal responsibility or liability for any errors or omissions that may be made.

A catalogue record for this book is available from the British Library

Library of Congress Cataloging-in-Publication data

The Chemical industry / edited by Alan Heaton. -- 2nd ed.
    p.   cm.
  Includes bibliographical references and index.
  ISBN 0-7514-0018-1
  1. Chemical industry.   2. Pharmaceutical industry.   3. Chemical
engineering.   I. Heaton, C.A.
HD9650.5.C522   1994
338.4′766--dc20
                                                93-28656
                                                CIP

⊚ Printed on acid-free text paper, manufactured in accordance with ANSI/NISO Z39.48-1992 and ANSI Z39.48-1984 (Permanence of Paper).

HD 9650.5 .C522 1994

The Chemical industry

Contributors

Dr Stephen Black   Research and Technology Department, ICI Chemicals and Polymers plc, The Heath, Runcorn, Cheshire WA7 4QD

Dr Jack Candlin   Department of Science and Technology, Teesside University, Middlesbrough, Cleveland TS1 3BA

Dr Peter Gregory   Specialties Research Centre, Zeneca Specialties, Blackley, Manchester M9 3DA

Dr Alan Heaton   School of Chemical and Physical Sciences, Liverpool John Moores University, Byrom Street, Liverpool L3 3AF

Mrs Christine Ryan   Educational Consultant, Udine, Quarry Lane, Kelsall, Cheshire CW6 OPD

Dr Tony Ryan   Research and Technology Department, ICI Chemicals and Polymers plc, The Heath, Runcorn, Cheshire WA7 4QD

Dr Craig Thornber   Zeneca Pharmaceuticals, Mereside, Alderley Park, Macclesfield SK10 4TG

Professor Mike Turner   The Advanced Centre for Biochemical Engineering, Department of Chemical and Biochemical Engineering, University College of London, Torrington Place, London WC1E 7JE

For Liam

Contents

6 The sulphur, phosphorus, nitrogen and chlor-alkali industries   189
Stephen Black

Editorial Introduction
to the First Edition

Background

This book provides an introduction to the main sectors of the chemical industry, and complements *An Introduction to Industrial Chemistry* (subsequently referred to as Volume 1) which covers the physicochemical principles of the subject, as well as introductory technical economics and chemical engineering.

Processes considered include the large-scale production of polymers (up to 1 000 tonnes per day for a single plant); the chlor-alkali, nitrogen, sulphur and phosphorus industries; and the production, on a smaller scale, of dyestuffs, pharmaceuticals and agrochemicals. The rapidly developing area of biotechnology is dealt with under biological catalysis. The consequences of scale of operation are also highlighted in the chapter on The Future.

Each chapter includes common themes, such as brief history, present position, major products and the future. The final chapter links together the predictions made for the future of each sector, to give an overall projection for the whole chemical industry; the quadrupling of oil prices in 1974 and the widespread recession at the beginning of the 1980s provide a salutary lesson about the difficulty of such projections.

Due to the difficulty of comparing statistics, it is emphasised that the reader should regard all figures not as absolute or currently accurate (even though the most up-to-date statistics available have been used) but rather as being indicative only of orders of magnitude, trends or relative positions. It is the conclusions drawn from the statistics which are important. Readers requiring current figures are directed to journals such as *Chemical and Engineering News, Chemical Marketing Reporter* and *European Chemical News*, and to the specific sources detailed at the end of each chapter.

Nomenclature and units

The importance of the use of trivial names in industry will be apparent from the examples given; indeed, they are used throughout the book. This should not present any difficulty to readers who are conversant with systematic (IUPAC) names, since chemical structures are also usually given.

We have also, as in Volume 1, standardised on the following units: tonnes for weight, °C for temperature, and US$ and UK£ for monetary values. Conversion to other units can be made using the table in Volume 1, p. viii.

I wish to acknowledge the co-operation of all contributors and to stress that they are expressing their own views, which are not necessarily those of their respective companies. We thank our colleagues and families for their help and support in this venture and the publishers for their advice.

Acknowledgements   We wish to thank the following for the use of previously published material. Imperial Chemical Industries PLC (Figures 3.6, 3.7, 3.8, redrawn from *Steam*, **2**, 4–5; Figures 3.11, 3.13, 3.14, redrawn from *Steam*, **1**, 22–23); The McGraw-Hill Book Co. (Figures 3.5, 3.9, redrawn from *Chemical Process Industries*, 4th edn., R.N. Shreve and J.A. Brink, Figures 13.2 and 18.8); Van Nostrand Reinhold Co. (Figure 3.16, redrawn from *Reigel's Handbook of Industrial Chemistry*, 8th edn., J.A. Kent, p. 546); The Royal Society of Chemistry (Table 3.13, from *The Modern Inorganic Chemicals Industry*, R. Thompson, p. 196; Figure 3.10, redrawn from W.J. Bland, *Educ. Chem.* **21**, 9); The American Chemical Society (Figures 3.2, 3.3 and 3.4, redrawn from S. Venkatash and B.V. Tilak, *J. Chem. Educ.* **6**, 276); The Longman Group Ltd (Figures 3.1, 3.12, redrawn from *Waste Recovery and Pollution Control Handbook*, A.V. Bridgewater and C.J. Mumford).

Preface
to the Second Edition

In this second edition the original chapters have all been revised and up-dated, with major revisions to the chapters on Dyestuffs and the Sulphur, Phosphorus, Nitrogen and Chlor-Alkali Industries, because of the different approach adopted by the new authors.

We have taken the opportunity in this second edition to add three complete-ly new chapters, as follows: Chapter 1, Introduction, provides an overview of the chemical industry and also acts as a lead into the chapters which follow. The growing importance of, on the one hand Quality and Safety and, on the other Environmental Issues, are recognised by devoting chapters 2 and 3, respectively, to these topics. This has resulted in the original chapters being re-numbered.

Almost all statistics and tables have been brought completely up-to-date and we hope that this plus the increased range of topics will widen the readership of the book. It is also worth emphasising here that a second edition of the companion volume—*An Introduction to Industrial Chemistry* by C.A. Heaton—was published in 1991 and it is to that edition (referred to as Volume 1) that references in this book refer.

On a personal note I would like to thank all contributors for their co-opera-tion and stress that they are expressing their own views and not necessarily those of their respective companies. I also wish to thank Phil Hughes of Zeneca Agrochemicals and Dr Albert Percival of Schering Agrochemicals for advice and information on chapter 7. Thanks are also due to Mrs Margaret Glynn and Mrs Sue Abraham for typing my own contributions, and to my wife Joy for her patience and support during the writing of this book.

The team of authors hope you enjoy reading this new edition and find it both informative and interesting. I would be pleased to receive comments on the book.

Alan Heaton

Introduction 1
Alan Heaton

The aims of this chapter are twofold. Firstly, it will give an overview of that diverse part of manufacturing industry which is called the chemical industry. In doing so a number of topics are introduced briefly which will be discussed in greater detail in later chapters in this book. It will give the reader an idea of what the industry comprises, what it does, how it is organised and sub-divided, and how its products are converted into useful end or consumer products. The second aim of this chapter is to serve as a lead into the specific topics which are covered in more detail in the following chapters.

Throughout the book frequent reference will be made to a complementary volume—*An Introduction to Industrial Chemistry*, 2nd edition, edited by C.A. Heaton, Blackie—which will be referred to as Volume 1. The two volumes are designed to be used as a two-book set which gives a comprehensive picture of the chemical industry.

Although the use of chemicals dates back to the ancient civilisations, the evolution of what we know as the modern chemical industry started much more recently. It may be considered to have begun during the Industrial Revolution, about 1800, and developed to provide chemicals for use by other industries. Examples are alkali for soapmaking, bleaching powder for cotton, and silica and sodium carbonate for glassmaking. It will be noted that these are all inorganic chemicals. The organic chemicals industry started in the 1860s with the exploitation of William Henry Perkin's discovery of the first synthetic dyestuff—mauve. At the start of the twentieth century the emphasis on research on the applied aspects of chemistry in Germany had paid off handsomely, and by 1914 had resulted in the German chemical industry having 75% of the world market in chemicals. This was based on the discovery of new dyestuffs plus the development of both the contact process for sulphuric acid and the Haber process for ammonia. The latter required a major technological breakthrough—that of being able to carry out chemical reactions under conditions of very high pressure for the first time. The experience gained with this was to stand Germany in good stead, particularly with the rapidly increased demand for nitrogen-based compounds (ammonium salts for fertilisers and nitric acid for explosives manufacture) with the outbreak of World

1.1
Development of the chemical industry

War I in 1914. This initiated profound changes which continued during the inter-war years (1918–1939). Further details are given in section 3.1 of Volume 1.

Since 1940 the chemical industry has grown at a remarkable rate, although this has slowed significantly in recent years. The lion's share of this growth has been in the organic chemicals sector due to the development and growth of the petrochemicals area since 1950. The explosive growth in petrochemicals in the 1960s and 1970s was largely due to the enormous increase in demand for synthetic polymers such as polyethylene, polypropylene, nylon, polyesters and epoxy resins. Further details are given in section 3.1.4 of Volume 1 and also in chapter 4 of this book.

The chemical industry today is a very diverse sector of manufacturing industry, within which it plays a central role. It is one of the largest parts of manufacturing industry and in the UK it usually occupies third or fourth place—based on the value of the products made—with annual sales of the order of £29 000 million! It makes thousands of different chemicals which the general public only usually encounter as end or consumer products. These products are purchased because they have the required properties which make them suitable for some particular application, e.g. a non-stick coating for pans or a weedkiller. Thus chemicals are ultimately sold for the effects that they produce.

It is important to recognise that, in addition to the considerable efforts of the research chemists in synthesising new molecules and discovering their useful properties, a considerable debt is also owed to the technologists in ensuring that the useful properties can be harnessed and applied. Two examples will serve to illustrate this point. Firstly, although the remarkable properties of PTFE or polytetrafluoroethylene (very low coefficient of friction, great chemical inertness, etc.) were soon recognised, these could not be exploited for several years because the material resisted processing by conventional techniques. It had to await the development of powder metallurgy techniques. A second example was the initial difficulty in satisfactorily dyeing synthetic fibres such as nylons and polyesters. Existing dyestuffs at the time would not satisfactorily adhere to these synthetics, although they had worked perfectly well with natural fibres like cotton. The combined efforts of dyestuff chemists and technologists were required to overcome the problem.

1.2 What the chemical industry does

The chemical industry takes a relatively small number of natural raw materials such as oil and natural gas, limestone and salt and converts them (by chemical processing or chemical reactions) into several thousand chemical intermediates. As we have already seen, these are then converted into end or consumer products. It is important to note that value is added at every stage in this process and the final product may have a value many times that of the raw materials which were used at the beginning. For example:

	Relative value
Crude oil	1
Fuel	2
Typical petrochemical	10
Typical consumer product	50

Clearly the increase in value added at each stage must exceed the processing costs if the company is to realise a profit on its activities.

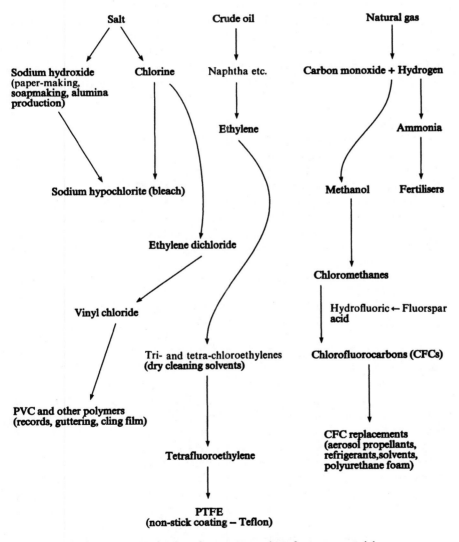

Scheme 1.1   Production of consumer products from raw materials.

Scheme 1.1 illustrates one small area of the chemical industry's activities in converting raw materials via intermediates to end or consumer products. Note the inter-relationships between many of the key chemicals appearing early on in the scheme. This point is further emphasised and dealt with in more detail in section 6.1, and also in section 6.6.2 which specifically considers the manufacture of methyl methacrylate.

**1.3
Characteristics of the
industry**

In what we regard as the developed world (Europe, USA, Japan, etc.) the chemical industry may be regarded as having become a mature manufacturing industry, following its explosive growth in the 1960s and 1970s. Although its rate of growth has slowed in the last decade it still exceeds that of manufacturing industry as a whole by a factor of 1.5–2 times. Many of the basic processes for producing key intermediates are mature, since they have been operated for more than 20 years, and this has also meant that their patent protection has expired. An example of this is the production of ethylene from crude oil fractions by cracking. This has also meant that countries wishing to venture into this area can now buy such plants 'off the shelf' and can soon have them successfully operating even though they previously had very little expertise or experience in these areas. As a result countries with reserves of crude oil, such as Korea, Mexico and Saudi Arabia, have entered and rapidly expanded their production of the basic alkenic and aromatic petrochemical intermediates. With their cheap and readily available raw material supplies, they have competed very well in the world markets for these products, resulting in the growth rates for their chemical industries being very much higher than those of the countries of the developed part of the world. There is also a growing shift in emphasis on the markets for chemical products, and although North America, Europe and Japan remain pre-eminent, opportunities are growing very rapidly in the Pacific Rim countries of south-east Asia. Clearly this is a consequence of both the rapidly growing population and the industrial development of these countries. At the dawn of the next century there is no doubt that China, with its enormous population, will become both a major market and a major player in chemicals production.

One of the main reasons for the rapid growth of the chemical industry in the developed world has been its great commitment to, and investment in, research and development (R&D). A typical figure is 8% of *sales income*, with this figure being almost doubled for the most research intensive sector, pharmaceuticals. It is important to emphasise that we are quoting percentages here not of profits but of sales income, i.e. the total money received, which has to pay for raw materials, overheads, staff salaries, etc., as well. For comparison, in the UK a figure for the engineering industry would be 1%. In the past this tremendous investment has paid off well, leading to many useful and valuable products being introduced to the market. Examples include synthetic polymers like nylons and polyesters, and drugs and pesticides. Although the number of new products introduced to the market has declined significantly in recent years, and in times of recession the research department is usually one of the first to suffer cutbacks, the commitment to R&D remains at a very high level. It is discussed in more detail in the chapters on the agrochemical and pharmaceutical sectors in sections 7.2.2 and 8.2 respectively.

With the increasing competition from the developing nations, mentioned above, in producing basic chemicals such as ethylene and benzene, there has been a trend for the chemical companies of the developed nations to shift their emphasis to downstream products, i.e. fine chemicals like drugs and plant-protection agents, and speciality polymers such as poly(aryl ether ether ketone) (PEEK) rather than commodity polymers like polythene (see Table 4.9 and section 4.4.1). It is much harder for developing countries to acquire the research skills and experience needed to successfully develop these high-tech products. In addition their advantage in having cheap raw material supplies (crude oil) is considerably smaller than it is with basic chemicals or intermediates. This is because the further downstream that the products are from the starting crude oil, the smaller the influence of its price on the cost of producing the final product.

The chemical industry is a very high technology industry which takes full advantage of the latest advances in electronics and engineering. Computers are very widely used for all sorts of applications, from automatic control of chemical plants, to molecular modelling of structures of new compounds, to the control of analytical instruments in the laboratory.

Individual manufacturing plants have capacities ranging from just a few tonnes per year in the fine chemicals area to the real giants in the fertiliser and petrochemical sectors which range up to 500 000 tonnes. The latter requires enormous capital investment, since a single plant of this size can now cost £150 million! This, coupled with the widespread use of automatic control equipment already mentioned, helps to explain why the chemical industry is capital- rather than labour-intensive.

Although companies specialising in pharmaceuticals and pesticides have always had to take a worldwide view on production and marketing, all the major multinational companies are also developing global strategies. This may involve either setting up manufacturing and marketing operations in other countries or acquiring companies which are already operating there. As an example, the UK's largest chemical company, ICI, now has over 60% of its manufacturing operations based outside the UK. Political developments such as the formation of the EC have encouraged this. Hence even more companies from the USA have set up both research and production units in the UK in order to serve this large European market. In addition, products requiring enormous R&D costs, e.g. drugs and pesticides, cannot possibly recoup these in their home market. They must therefore necessarily be sold worldwide. Another factor is the new markets which are growing very rapidly in importance; for example, the Pacific Rim in south-east Asia has already been mentioned.

1.4
Sectors of the industry

The major sectors of the chemical industry are those which form most of the chapter headings in this book plus chapters 9 and 11 in volume 1. Thus they are:

Petrochemicals	Chlor-alkali products
Polymers	Sulphuric acid (sulphur industry)
Dyestuffs	Ammonia and fertilisers
Agrochemicals	(nitrogen industry)
Pharmaceuticals	Phosphoric acid and phosphates
	(phosphorus industry)

An alternative categorisation, shown in Table 1.1, is based on the end uses of the chemicals. Here the basis of the comparison is value added, i.e. roughly the difference between the selling price and the price of the raw material plus processing costs. This tends to balance high-tonnage but relatively low-priced products like basic petrochemicals against low-tonnage but very high-value products like pharmaceuticals, although, as Table 1.1 shows, the latter come out very much on top.

The petrochemicals sector provides the key intermediates or building blocks (derived from oil and natural gas) such as ethylene, propylene, benzene and toluene. These are the starting points for the synthesis of an enormous range of industrial organic chemicals, which are produced in the downstream processing of the key intermediates in some of the other sectors listed.

The polymers sector is the major user of petrochemical intermediates and consumes over half of the total output of organic chemicals. A brief inspection of chapter 4 indicates a large number of high-tonnage commodity polymers, e.g. polythene, polystyrene and poly(vinyl chloride) (PVC), and demonstrates why such large quantities of ethylene, benzene, etc., are required.

Although the dyestuffs sector is much smaller than the previous two, it has strong links with them. This arose because the traditional dyestuffs, which were fine for natural fibres like cotton and wool, were totally unsuitable for the

Table 1.1 Sectors of the UK chemical industry, 1990 (gross value added)

	% Share
Pharmaceuticals	33
Specialised chemical products (industrial/agricultural)	18
Organics	12
Soap and toilet preparations	10
Paints, varnishes and printing inks	6
Plastics, synthetic resins/rubbers	6
Dyestuffs and pigments	5
Inorganics	4
Specialised chemical products (mainly for household and office use)	4
Fertilisers	4

new synthetic fibres like nylons and polyesters. A great deal of research and technological effort within the sector has resulted in the amazingly wide range of colours in which modern clothing is available. More recent high-tech applications of dyestuffs, such as optical data storage, are covered in section 5.8.2.1. Along with drugs and plant-protection agents (pesticides), dyestuffs are examples of fine chemicals, i.e. chemicals produced in relatively small tonnages which are of high purity and high value per unit weight.

Agrochemicals (pesticides) are covered in chapter 7 and in recent years, along with the pharmaceuticals sector, have constituted a blue-chip sector of the chemical industry, i.e. a very profitable one for those companies which can continue to operate in it.

Pharmaceuticals have been the glamorous sector of the industry for some years now because of their high levels of profitability—although this often seems to attract criticism that profits on some drugs are too high. The latter comments have to be set against the very high risks and costs (in excess of £100 million per marketable compound) of discovering and developing new products (see section 8.3). This sector is unique in both its relationship with its customers (see section 8.1) and the fact that demand for its products is little affected by economic recessions, in great contrast to all the other sectors.

The chlor-alkali products sector produces principally the very high-tonnage chemicals chlorine and sodium hydroxide, and comprises the subject matter of chapter 9 in Volume 1 and also section 6.5 of this book. It is an area which demonstrates very nicely the influence of new technology and energy costs on chemicals production.

Sulphuric acid (sulphur industry) is the most important chemical of all in tonnage terms. It is discussed in detail in section 6.2.

Ammonia and fertilisers (nitrogen industry) are an area where it has been very difficult to balance capacity with demand, and this has led to cost-cutting and losses for many companies. It centres around the very energy-demanding Haber process, and is discussed in section 6.4.

Phosphoric acid and phosphates form the subject matter for section 6.3, and their major uses are in phosphate and compound fertilisers, detergents and plasticisers.

1.4.1 High-volume and low-volume sectors

We can divide the various sectors of the chemical industry into these two types. In the high-volume sector, individual chemicals are typically produced on the tens to hundreds of thousands of tonnes per annum scale. As a result, the plants used are dedicated to the single product, operate in a continuous manner and are highly automated, including computer control. Sectors categorised as high-volume are therefore those covered in chapter 6, plus petrochemicals and commodity polymers such as polythene. With the exception of the latter, they are key intermediates, or base chemicals, which are feedstocks for the production of a wide range of other chemicals, many of which are also required in large quantities.

In contrast, low-volume sectors are largely involved in fine-chemicals manufacture, and individual products are produced only on the tens of tonnes to possibly a few thousand tonnes scale. However, they have a very high value per unit weight, in contrast to high-volume products. Actual figures are given on p. 190 which illustrate this point very well. Fine chemicals are usually produced in plants operating in a batch manner and the plants may be multiproduct ones.

Thus, low-volume sectors are agrochemicals, dyestuffs, pharmaceuticals, and speciality polymers such as PEEK.

1.5
Conversion of chemicals into consumer products

As has already been stated, the chemical industry produces chemicals which are then, in one or more steps (usually the latter), converted into consumer or end products. These consumer products are purchased by the public to satisfy a particular need or because they have the required properties. Thus the chemicals are bought for the effects they produce. To illustrate this, an insecticide is bought because it will kill insects that are causing problems, and a polyurethane varnish is purchased because it will impart a tough, protective, and probably decorative, finish to wood. Members of the public buying these products would certainly not think of them as chemicals, but rather just as insecticides or paints.

It is interesting to trace the steps involved in producing the polyurethane varnish from the starting crude oil or natural gas to its sale in the paint shop. This starts with the reforming of a suitable petroleum fraction over a catalyst such as platinum to produce a mixture of aromatic compounds, principally benzene, toluene and xylenes. Toluene is separated from this mixture by fractional distillation. Nitration of the toluene under suitable conditions gives mainly 2,4-dinitrotoluene together with a little 2,6-dinitrotoluene. Reduction of the nitro groups followed by reaction of the resulting diamine with phosgene yields tolylene di-isocyanate (TDI), which is

Finally, reaction of TDI with linseed oil, a glyceride of linoleic acid, produces a polyurethane-linked polymer. Suitable formulation of this in a solvent, etc., produces the polyurethane varnish. Considering all the steps that are involved, it is a tribute to the efficiency with which these chemical processes are carried out that the selling price of the varnish is so relatively low.

1.6.1 Its importance

The central role of the chemical industry within manufacturing industry has already been mentioned, and it is true to say that all other sectors require the products of the chemical industry to a greater or lesser degree. Whereas the contribution to computer manufacture might be relatively small, consisting only of the supply of inert solvents for cleaning purposes (if we disregard preparation of the pure silicon chips), a much larger contribution is evident in the manufacture of packaging materials, with the chemical industry supplying the polymer raw materials which are used. Many sectors require chemicals for the treatment of effluent and also for the analysis of raw materials and finished products. The major contribution of the chemical industry to catering for our major needs will be apparent from reading in particular chapters 7, 8 and 9 in this book. This topic is also discussed in section 3.2.2 of Volume 1.

1.6.2 Major chemicals-producing countries

As one might expect, the major chemicals-producing countries are to be found in North America, Europe and Japan. The league table in 1991 was:

	Sales ($ thousand million)
USA	285
Japan	180
Germany	100
France	65
UK	50
Italy	50

Accurate statistics on countries such as China and Russia are difficult to obtain, but by the end of this century both of them may well figure prominently in this list. Further details are given in section 3.3.2 of Volume 1.

1.6.3 Multinational chemical companies

The largest multinational chemical companies are listed in rank order in Table 1.2, together with their countries of origin. Two points should be noted. Firstly, although the country of origin is given, all these companies operate throughout the world. For example, the major UK company, ICI, prior to its split into ICI and Zeneca in 1993, had over 60% of its manufacturing capacity located abroad and operated in over 140 countries. Secondly, the dominance at the top of the list of the three major German companies and, in contrast, the presence of only two Japanese companies, both well down the list, are notable. This is despite the fact that, as already seen, the value of Japan's production of chemicals is almost twice that of Germany.

Table 1.2   World's largest chemical companies, by sales proceeds ($ million)

		Country of origin	1991	1990	1989	1988	1987
1.	Hoechst	Germany	31 147	30 017	27 162	23 096	23 545
2.	BASF	Germany	30 778	31 195	28 180	24 733	25 636
3.	Bayer	Germany	27 989	27 863	25 624	22 816	23 664
4.	ICI	UK	23 346	24 909	21 258	21 158	20 989
5.	Dow	USA	18 807	19 773	17 600	16 682	13 377
6.	Rhone-Poulenc	France	18 111	15 483	12 648	10 783	10 564
7.	Du Pont[1,2]	USA	17 941	22 286	22 104	19 608	17 601
8.	Ciba-Geigy[3]	Switzerland	15 538	15 459	13 364	11 753	12 422
9.	Elf Aquitaine[1,2]	France	14 035	14 323	9 855	8 089	7 961
10.	EniChem	Italy	11 699	13 363	12 120	5 616	5 324
11.	Shell[1,2]	UK/Holland	11 208	12 703	11 075	11 838	11 717
12.	Sandoz	Switzerland	9 911	9 703	8 104	6 761	7 075
13.	Akzo	Holland	9 871	10 229	9 820	8 259	8 804
14.	Exxon[1,2]	USA	9 171	8 591	9 210	8 797	7 177
15.	Mitsubishi Kasei[4]	Japan	9 161	8 949	7 297	5 049	7 965
16.	Monsanto	USA	8 864	8 995	8 681	8 293	7 639
17.	Sumitomo Chem[4]	Japan	8 696	7 867	6 547	7 179	4 262
18.	Merck & Co	USA	8 603	7 672	6 551	5 940	5 061
19.	Roche	Switzerland	8 442	7 587	6 364	5 788	6 071
20.	Solvay	Belgium	8 146	8 265	7 244	6 815	6 784

[1] Chemicals only
[2] Excludes inter-company transfers
[3] Data based on current values
[4] Year end 31 March
Adapted from *Chemical Insight*

The scale of operation of these companies is truly enormous, and the value of sales of the larger companies exceeds the gross domestic product of some of the world's smaller countries!

1.7
Quality and safety

1.7.1   Quality

During the past decade, the recognition of the importance of quality in manufacturing industry has increased enormously. This was initiated by the success of Japanese manufacturing industry which, with its emphasis on quality products, achieved dominance in several sectors, perhaps the best examples being motor cycles and cars, and electrical goods such as television sets. Although this development has been going on for at least two decades, it is only during the last decade, and particularly the latter part of it, that the large chemical firms have acted to educate their staff on the importance of quality and developing a company environment conducive to a quality operation. An important stimulus has been the increasingly competitive nature of the international chemicals business, coupled with the worldwide recession of the late 1980s and early 1990s. Many of the companies have spent over

£10 million in putting in place a Total Quality Management System and in educating and training staff to be able to contribute to it and use it. As the case histories in section 2.1.6 demonstrate, it not only makes for a more efficient system but it actually can save the company money in the long run.

1.7.2   Safety

In chapter 3, we trace the awakening and then growing concern over the general public's attitude to the safety of chemicals in the environment. This rapidly extended to areas such as chemicals in food, solvents in paint and glue, etc., and the safe operation of chemical plant. Add to this the occasional accident involving a road tanker which resulted in spillage of chemicals on to the highway, which provided another opportunity for the media to criticise the chemical industry, and it is easy to see why public pressure on the companies has increased greatly in recent years. This pressure has resulted in stricter regulations to ensure that plants are operated safely, so that hopefully we will never have another disaster such as Flixborough (section 3.2.3) or Bhopal (section 3.2.5), and even the occasional leak of hazardous chemicals will be less likely to occur. There has also been insistence on much more extensive testing to ensure that compounds are as safe as possible before they can be used as pesticides, drugs, etc. Perusal of Table 7.4 and sections 7.3 and 8.3 will demonstrate just how comprehensive, time-consuming and expensive this is. Note that we, the general public, ultimately end up paying for the cost of this in the increased price of the chemical which we purchase as a consumer product. Although we all accept the importance and necessity for this, it must be pointed out that however much testing is carried out no compound can ever be proved safe. Water is non-toxic and essential for life but it can still kill people by drowning! It is salutary to note that if we applied the same testing and safety standards to aspirin (and water) that we currently apply to new compounds being developed as drugs they would almost certainly be banned from reaching the market! This testing programme and submission of the results to the appropriate regulatory authorities for approval now take at least seven years, and some of the biologically active compounds start to lose their effectiveness within four to five years of first being marketed because of pest resistance by, for example, insects or bacteria. My personal opinion is that we must now reach a balance and taper off our calls for more and more testing, since the additional tests will probably give us little or no additional safeguard on the safety of the product. There is, however, scope for making the tests more efficient and developing alternatives, for example in the area of animal testing.

Safe working practices and the good health of employees are closely linked. Partly perhaps because of trade-union pressure, triggered by incidents such as some workers on plants using naphthylamines (for dyestuffs) and vinyl chloride monomer (for PVC) developing rare forms of cancer, and partly due to a more enlightened and open approach by the companies, both safety and health and hygiene of workers now command a much higher profile. Cynics might say a lot of this was forced on the companies because of the poor public image of the chemical industry, but this is not justified. As we show in a number of places in this book the public perception of aspects of the chemical industry (usually very negative) can be at variance with the facts. For example, the one or two disasters in the industry (see Table 2.4) have led the public to view the industry as a particularly dangerous sector of manufacturing industry. The figures in Table 2.3 tell a different story and show a very interesting comparison with some other occupations.

It is encouraging that the industry is now taking a much more positive and open approach and bringing the facts on what it is doing, and also often the problems it still faces, to the attention of the public. Schools–industry projects such as the one directed by the author[1,2] provide factual material for schoolchildren and their teachers and allow them to undertake works visits, enabling them to acquire a much more balanced view of the industry, but still allowing them to make up their own minds about the industry and its operations. Their major concern and interest relate to environmental matters, which are discussed in chapter 3.

We recognise the importance of health and safety by devoting the second half of chapter 2 to this topic.

1.8 Environmental aspects

Rachel Carson in her book *Silent Spring*, published in 1962, first drew public attention to the possible harmful effects of some chemicals on the environment. She could not have imagined in her wildest dreams the interest, emotion and controversy in this topic which would ensue, and the remarkable growth in the size of the environmentalist movement. It has led to the birth of organisations like Greenpeace and 'Green' political parties, and a proliferation of environmental-type degree courses, the latter in response to the great interest in the topic amongst young people.

Chapter 3 is devoted to this topic, but it is perhaps worth stating here that it is an area where emotions have run high and sometimes facts have been ignored in presenting views, and points have been taken out of context— which can convey a quite different impression.

1.9 The importance of team-work

One of the most striking differences which a science graduate going to work in the chemical industry, and particularly in a research department, will encounter, is the importance of team-work. In the academic environment, practical work is invariably carried out by oneself, with perhaps just occasional working in pairs. Research projects in industry are carried out by teams of scientists and perhaps engineers and others. It is therefore essential that, as well as being good at their science, graduates are able to fit into a team. This in turn means that they must develop more general skills, such as becoming good at verbal and written communication, and being able to get on with other people.

Another important general skill—problem-solving—is developed in all science courses. Many universities have recognised the growing importance of these general skills—particularly when science graduates are in competition with arts graduates for the same job—and now include formal courses on them within the science degree. The Royal Society of Chemistry recently emphasised their importance in its *Review of* (chemistry) *Degree Courses.*[3]

References

1. Alan Heaton, 'People and processes in the chemical industry,' *Chem. & Ind.,* 1985, 372.
2. Alan Heaton, 'People and processes in the chemical industry—The continuing story,' *Chem. & Ind.,* 1988, 265.
3. 'Degree Courses Review,' Royal Society of Chemistry, London, 1992.

2 Quality and safety issues
Christine Ryan and Tony Ryan

2.1.1 The quality revolution

What is meant by the word 'quality'? In the modern context quality is not merely concerned with the quality of a particular product or service but rather the quality of the business as a whole, embracing everything that happens in the company. For this reason it is often referred to as 'total quality' or 'total quality management' (TQM).

The concept of quality with which we are familiar today was probably first established in Japan, having been marketed there, most notably, by two American academics, Edwards Deming and Joseph Juran. After seeing the success that resulted from the adoption of the quality system, previously sceptical American organisations and latterly, in the 1980s, the UK and other European nations then took it up. No doubt in the 1990s we can expect the developing countries, such as Taiwan, to follow suit.

Since its introduction, when total quality was concentrated on quality control in the manufacturing sector, quality has spread to embrace the service industries and the professions. The original concepts are still applied but have since developed and expanded to suit new applications and have moved away from narrow quality-control ideas to broader concepts that place the satisfaction of the customer as a top priority.

Quality is about the total way in which a company does business, and it is based on establishment of the principle of giving complete customer satisfaction, at the most favourable cost to all concerned, by getting the job right first time and every time. Everyone involved in the business should be treated as a customer whether outside or inside the organisation. For a laboratory technician the primary 'customer' may be the project or laboratory manager, whereas the managing director's primary 'customers' might be the shareholders of the company.

Over recent years the quality concept has become increasingly adopted by commercial organisations including multinational chemical companies such as BP, BASF, DuPont, Exxon, Glaxo, ICI and Shell, to name but a few. Their interest has been driven by the need to satisfy their customers in an increasingly competitive world. Failure to do so would invite the obvious consequence of lost business.

There are many publications dealing with specific quality issues and numerous books and courses are now available which detail the intricacies of training and implementation of quality principles, but these are outside the scope of this book.[1-5] Instead, we are interested here in giving an overview of how quality is applied in the chemical industry and illustrating what this means to the individual working in such an environment.

Three of the most renowned quality specialists have each devised their own quality programme. However, there are many areas of general agreement between them and the key features of each are summarised below.

Crosby
- Importance of senior management commitment.
- Education programme aimed at the individual and all aspects of work.
- Simple problem-solving techniques.
- Attitude changes and a common quality vocabulary.
- Continuous improvement.

Deming
- Importance of senior management commitment.
- Statistical approach to quality control.
- Continuous improvement.
- Delighting the customer.

Juran
- Importance of senior management commitment.
- Problem-solving techniques and remedial action.
- Greatest emphasis on management action, rather than the individual and attitude changes.
- Annual improvement.

Having briefly established what quality is, how then do we identify a 'quality product'? The first thing to remember is that there is no absolute definition of a quality product other than to say that it is 'that which meets the customer's requirements'. A 'quality car', for example, is not necessarily a Rolls-Royce. If the customer has only a small garage, and income to match, then his or her idea of a quality car might be a Mini or Compact. However, it should be borne in mind that what satisfies the customer now is almost certain not to do so in the future, and a successful company will be one that has foreseen its customer's future demands. What matters is the needs, and the anticipated needs, of the customer and this should provide the direction for an organisation's change and development.

Failure to recognise the need for change in the workplace, in terms of the quality of a business, has led to the decline of many industries. One very striking example is that of the television manufacturing industry. In 1978 a popular UK consumers' magazine recommended the purchase of Japanese-made television sets in preference to those made in the UK.[6] Analysis had shown that the average time between breakdowns for the best Japanese-manufactured models was ten times greater than for the best British-made sets. This sorry state of affairs had arisen because, unlike their British counterparts, the Japanese (faced with a multipart product) had concentrated on reducing the tolerances on every component they produced. The result was that overall the Japanese set was much less likely to fail. In other words, the Japanese manufacturers, unlike their competitors, had a policy of continuous improvement i.e. a quality approach. Although British television sets are now considered to be equal in reliability to the Japanese product the improvement has come much too late for many manufacturers to hold their previous market share and, for them, business has been lost.

Those who remember the 1950s will recall that Japanese-manufactured goods were generally regarded as shoddy and offering poor value for money. Since then, Japanese industry has provided fierce competition worldwide (sometimes resulting in domination) in many industries, for example, the motorcycle, electronics, steel and shipbuilding industries. By adopting quality principles we can hope to ensure that chemicals are not added to this list.

So, the need for quality is clear. It is all about survival—of the industry, the company, and of the individual's job. TQM can provide the techniques needed for improving the overall performance of an organisation. There are many companies, such as Rank Xerox and British Telecom, that can point to tangible results, but it should also be remembered that total quality is not a substitute for good research and development, although it does have a very important role to play.

2.1.2   The quality philosophy

Although there are different versions of the quality concept on offer they all have certain basic essentials in common, even if the way in which they are expressed within a particular company varies. These essentials have been defined by one quality expert, Philip Crosby,[1] as the 'four absolutes of quality' and they may be summarised as follows.

- Definition—what is quality?
- Systems—how can quality be implemented?
- Performance standard—what standard is the aim?
- Measurement—how can the success of quality be demonstrated?

These four absolutes should be evident in any good chemicals company by the way in which it operates for both its customers and its employees.

Definition.   An outline of what is meant by quality has already been given but more specifically the definition of quality is 'conformance to requirements'. This can apply to all types of job as they all have requirements that must be met. If the employer or customer fails to communicate these requirements clearly then quality will be impossible to achieve.

The importance of good communications between customer and supplier can be illustrated by the story, no doubt apocryphal, of the Japanese company that received an order from a major computer manufacturer to supply certain parts to a standard of 'not more than five defective parts out of 1000'. The order was filled as requested and was dispatched to the customer. However, when the consignment arrived the customer was puzzled to find, along with the order, a smaller parcel containing the specially produced defective parts as requested! This last example shows a lack of understanding between customer and supplier not only with respect to the requirements but also with the expected performance standard for a quality product—but more about that later.

A process model, Figure 2.1, is sometimes used as a tool to identify the requirements, that is to ensure quality, in any job or process under consideration. The terms used in the process model are defined in Table 2.1.

The use of this simple type of model provides a starting point for implementing a quality approach and helps in the identification of those factors that will have most influence on the outcome.

Systems. Conventional systems use a 'quality control' or appraisal system which relies upon finding errors or defects in a product or sample and then

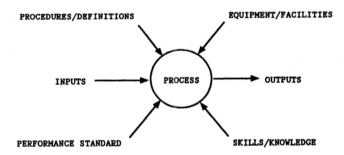

Figure 2.1 A process model.

Table 2.1 Definition of terms used in the process model

Term	Definition
Outputs	What is agreed with the customer
Inputs	The 'things' being provided to the process
Equipment/facilities	The items and amenities needed to do the job
Training/knowledge	The skills and understanding needed to ensure a conforming output
Procedures/definitions	Instructions, process manuals, analytical methods, etc.
Performance standard	The standard to which the output is required

correcting, recycling, or disposing, as appropriate. In other words the product is inspected after the error has been made (rather than checking the process before it can happen). This is an expensive way to do business as it still allows errors or non-conformances to occur. Such a conventional approach to production will never bring about any significant improvement. In order to cause real quality changes a system of defect prevention, not appraisal, is needed. Improvement through defect prevention means first identifying those areas of a process that lead to non-conformances occurring so that preventive measures can be built into the system. The consequence of this approach is the removal of those problems that stimulate customer dissatisfaction and the removal of the costs involved in correcting mistakes.[1] The relationship between defects and costs in a prevention system compared to an appraisal system can be illustrated graphically (Figure 2.2).

Performance standard.   It is essential to establish what is meant by a quality standard in an organisation if that standard is to be reached on every occasion. The quality performance standard can be defined as 'zero defects'. This does not mean perfection, but rather the attitude that non-conformances are unacceptable.

Establishing zero defects as the performance standard for every task removes the chances for misunderstanding. There can be no mistaking how often the requirements should be met, because by definition it is every time. Traditional standards such as targets, yields, faults per batch and so on can be misunderstood because, by their nature, they imply that mistakes are acceptable. However, we can all think of examples where there are no acceptable levels of error, such as in matters of health and safety. Within a quality culture, where there is a proper policy of education, errors are no longer inevitable.

A classic example of failure to understand the real concept of quality could be seen in the performance of a particular national postal service that, at the time, did not employ a standard of zero defects. Upon introducing their 'first

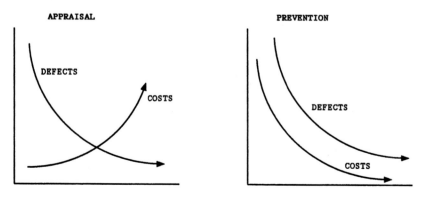

Figure 2.2   Cost comparison: appraisal versus prevention.

class' service of next-day delivery, for which the customer paid a premium, they made the proud boast that their aim was to get 90% of the letters delivered on time. To some this would seem an admirable achievement but an alternative way of looking at this is that this organisation intends to fail 10% of the time. They would not meet the requirements of the customers who had paid for first class service. The fault here lies in their stated aim of 90%, because it allows for relaxation once this target is reached; there is no incentive for further improvement. To introduce a target at anything less than 100% reliability automatically builds failure into the system. 'Zero defects' is really an attitude of mind that says that non-conformances are not acceptable.

At first the notion of zero defects may seem an impossible standard to achieve, and yet there are many instances in our personal lives when this is precisely what we expect. When taking a flight across the Atlantic, for example, we would find a stated failure rate of 10% to be completely unacceptable and assume that there are preventive measures in place which ensure that we make the crossing safely. It is apparent that if a requirement is taken seriously enough then errors can be prevented.

Introducing a quality culture into an organisation and ensuring implementation by the individuals within it should help to prevent unsatisfactory situations from occurring. Quality will provide a climate where all operations are done right the first time, where it is realised that ultimately everyone in an organisation is working for the customer, even if their tasks seem far removed from that contact.

Measurement.   The importance of measurement, in the quality context, is as a positive process aimed at causing improvement by eliminating non-conformances. At first glance, getting a quantitative measurement of the success of a process or operation can sometimes seem impossible, but quality can be calculated, very precisely, by knowing what it costs when things are done wrongly. In quality terms this is known as the price of non-conformance (PONC).

The PONC, for example, may be the money wasted on reprocessing, warranty, breakdowns, dealing with customer complaints, downtime, excess stock, unproductive meetings, accidents and absences. In research, it might be working on the wrong projects. It has been estimated that most organisations spend between 20 and 40% of their operating costs on these types of waste.[1,2] If an effective prevention system is put in place many of these costs can be removed and although it may require some initial financial investment, sometimes called the price of conformance (POC) (usually about 5%), there are clearly large savings to be made. This state of affairs can be dramatically represented by an iceberg where the apparent costs are readily seen above sea level but the much larger hidden costs reside beneath the surface (Figure 2.3).

The present situation is that all the larger chemical companies and many smaller ones now have some type of quality system in operation. Such

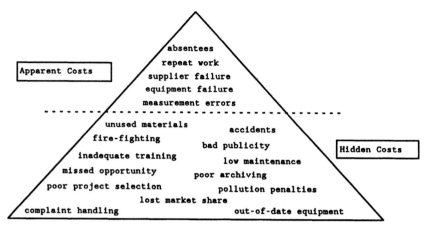

Figure 2.3   Hidden costs of non-conformance.

companies are committed to quality improvement and demonstrate this by the involvement of the most senior managers as well as the general workforce in the quality process.

2.1.3   Quality accreditation

Quality assurance.   The purpose of quality assurance in the chemicals industry is to ensure that the materials or services provided are fit for the tasks specified by the customer. Indeed, many customers will now only purchase from organisations that have achieved a recognised quality standard such as the British Standard, BS 5750.[7] The chemical company also benefits from implementing an effective quality structure through increased efficiency and lower costs, a reduction in the amount of downgraded materials formed and a more consistent product or service at the end of the process. In addition to these obvious benefits they have an improved business image, with the opportunities that this may bring in terms of opening new markets by developing an edge over the competition.

International quality assurance standards.   It is apparent that both customers and suppliers may determine what the quality standards are for a particular product or service but, increasingly, the setting of standards has become the work of independent standards authorities. This development has led to the formation of an international standards body, made up of the standards authorities from different countries, called the International Standards Organisation (ISO). In practice this means that those companies that have been inspected and that are found to comply with the required quality assurance standards may then be issued with a 'certificate' by the standards authority.

The standards for quality assurance upon which the certification is based are laid down in the ISO 9000 series.[8-11]

ISO 9000 certification indicates that the company is committed to quality and is able to supply its customers to a particular quality standard. It provides a framework for a quality management system and has a twenty-point checklist that top management can use to secure the involvement of staff at every level. This information forms the basis of the company's own 'quality system manual' with clearly laid-out quality assurance documents so that they can demonstrate the continuing application of the appropriate standards.[8-13].

Organisations that are committed to quality will provide extensive training of all personnel and will demonstrate their ongoing commitment by performing regular quality audits. Indeed, for those organisations that are quality certified, regular internal checks or audits must be carried out or the quality registration will be lost.

2.1.4 Quality structure

The policy statement.   Companies vary in the details of their quality management structure, but typically it will consist of a quality manager, a clearly documented organisation chart and a written commitment to a stated quality policy. A good example here is the quality policy statement made by ICI Chemicals and Polymers Ltd, which is based upon the following principles to reinforce the company's quality improvement process.

- Continuous improvement of everything we do, which will take us towards the standard of best imaginable.
- Systems for meeting requirements, for preventing errors and for continuous improvement must be established and maintained.
- Measurement is the only certain test of performance and improvement.
- Everyone has customers. Delighting them is everyone's business.
- Every member of C & P staff has an important contribution to make.

Not all policy statements are this detailed. A simple, but memorable, example given in one company is 'To provide goods that don't come back to customers that do'. The important thing in both of these examples is that there is clarity of purpose and a definite commitment to providing quality output. Policy statements should be carefully formulated not only for the company in its widest sense but also for the divisions and departments within it. Thus, the company's safety department, intellectual property department, training function, research groups, product groups, businesses and so on, should all be clear in what they are trying to achieve and this should be reflected in the respective policy statements.

The customer/supplier relationship.   Another key requirement for a successful quality organisation is that there should be a good communications system in place for ensuring that suppliers meet the company's specified quality requirements. All tasks are supplier-related (whether goods or services); if the supplier provides non-conforming inputs then we cannot expect a conforming output (see the process model, Figure 2.1). True quality improvement in any business is impossible without the full participation of its suppliers.

Traditionally, suppliers were kept at arm's length and attempted to meet the customer's needs without proper discussion. The tendency was also to regard suppliers as dispensable, such that if another supplier emerged with goods or services offered at a lower price than the existing one, then the current supplier was discarded; loyalty was neither given nor expected. In addition, it was considered prudent to keep the supplier in the dark about the processes they were servicing for fear of them passing on confidential or proprietary information.

The quality supplier–customer relationship is very different from this traditional model. It involves close collaboration, with the supplier regarded as an extension of the customer's business. It is now believed that a full exchange of information on processes and procedures is often desirable if the supplier is to provide the most conforming inputs. For this reason suppliers must be chosen carefully with the result that, on introduction of a quality system, there is often a great reduction in the number of suppliers used with the corresponding improvement in efficiency and reliability that this brings. An example, *par excellence*, is that of one chemicals company which has its supplies of liquid air delivered without ever having to place an order. The manufacturing supplier monitors the stock-tank levels telemetrically at all times and knows precisely when to top up. This type of arrangement saves time and money for both customer and supplier. Whilst the benefit to the customer is clear, the benefit to the supplier is that it is locked into its customer's business and that it is able to run its own business more efficiently.

Teamwork. Most of the problems encountered in the chemical industry are highly complex and, as such, the solutions are generally beyond the power of any one individual to solve. For this reason the synergistic power of the team can be applied. Although the psychology and dynamics of teamwork are beyond the scope of this work, the teamwork approach probably contributes one of the biggest changes in moving from an undergraduate or graduate research environment at university to work in industry.

The quality improvement team (QIT) is responsible for the planning and administration of the quality improvement process (QIP) in any particular function such as research, distribution and so on. It has to provide resources and support to the process, but more than anything else the QIT is charged with the task of creating the environment under which the quality improvement process can flourish. In large organisations there may be many QITs, but their common role is to ensure the implementation of the improvement process through the leadership and guidance of the quality manager. This involves the induction and education of all employees in quality, the formulation of a quality policy statement, measurement of the costs of quality, the formal establishment of a corrective action system, and continuous reinforcement of the quality message with formal recognition of success.

Each QIT within the organisation may be further subdivided into teams for each group or department, who are responsible for the implementation within their own area. These teams will define the problems occurring and select the appropriate individuals to form or comprise the corrective action team (CAT) (see case history no. 3)

The CAT is at the heart of the problem-solving activities of the quality process. Having defined the non-conformance, they will identify the root cause of the problem and formulate a suitable corrective action. The action taken should always be followed up and evaluated to ensure that the non-conformance has been permanently eliminated. Wherever possible the correction should be mistake-proofed.

Not all problems can be solved straight away and it may be necessary to use a temporary quick fix of the problem until a permanent solution can be put into place. For example, a defective output must be eliminated for the company to stay in business whether the elimination is by scrapping, reworking or by returning defective inputs. However, it must be recognised that 'fixing' costs profit. Management must realise that 'fixing' only appears to cost less and that ultimately (because it fails to deal with the root cause of the problem) it will cost more than the permanent solution.

A process for the elimination of non-conformances (referred to as DICE) is summarised in Figure 2.4.

It is important that individuals within the company realise that anyone can initiate a corrective action request (CAR) and feel confident that their concerns will be acted upon. It is solving all of the 'small' problems in the organisation that results in overall massive improvement.

2.1.5   Techniques of quality implementation

There is no shortage of techniques that can be employed in the measurement or implementation of quality. DICE and the process model have already been discussed. Broadly, the techniques can be split into statistical and non-statistical methods, and a few examples of the more useful techniques are given in this section.

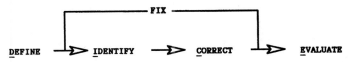

Figure 2.4   Process for the removal of non-conformance.

Statistical methods.[14]   Statistical quality control (SQC) and statistical process control (SPC) are essentially a collection of statistical techniques used to analyse the variables in a process that affect its capability to produce material consistently that is in specification.

SPC may involve the use of control charts to analyse the variables which affect the performance of a process in order to maintain or improve its capability to produce consistent outputs. When applied to the measure of product quality, SPC should be interpreted as statistical product control or statistical quality control (SQC). The success of the technique lies in the fact that measurements are made during the operation of the process and not at the end. Upper and lower limits, based on statistical criteria, are calculated and the process is monitored continuously so that the trends can be measured and action taken to control the process when these limits are approached, thus avoiding an undesirable result. A typical control chart, used to monitor the purity of a product during its manufacture, is illustrated in Figure 2.5.

Other common statistical techniques such as frequency distribution graphs and Pareto charts may also be adopted to aid the improvement process. A Pareto chart is a bar-diagram used to rank events or problems in order of their importance so that the most frequent or most costly can be dealt with first. Experience indicates that the removal or correction of the biggest problem gives the greatest return for the least effort. This is sometimes known as the Pareto principle or 80:20 rule which, in a quality context, refers to the phenomenon that about 80% of the non-conformances in an organisation or system tend to arise from only about 20% of the potential causes. An example of a Pareto chart is shown in Figure 2.6.

ISO 9000 contains clauses that require producers to give due attention to the application of SPC where appropriate.[8-11,15]

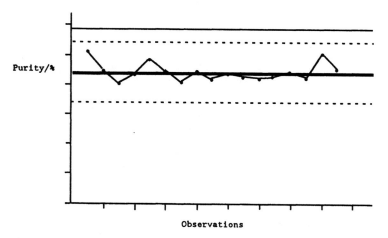

Figure 2.5   Statistical process control: (——) mean; (---) upper and lower limits.

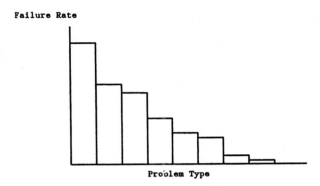

Figure 2.6 Pareto chart.

Non-statistical methods.   Despite its unattractive and uninformative title, a particularly useful technique for providing a systematic approach for the translation of the customer's requirements into a conforming product or process is known as quality function deployment (QFD).[16] This quality tool has recently emerged from Japan, where it has been successfully applied to the car industry, with real examples of customers' wishes (such as the design of car doors) being translated into new models with remarkable speed. The basic philosophy in QFD is that time spent in planning and preventing non-conformances yields a better return than fire-fighting and problem-solving once manufacture has begun. The methodology is therefore consistent with the first and second absolutes of quality defined in section 2.1.2 (conforming to customer's requirements; prevention systems) although it requires a different, unconventional, cultural approach for its implementation.

QFD is essentially a planning tool that begins with an understanding of the customer's (often vague) perceived requirements and translates them, quantitatively, into product design requirements. These product requirements are then refined into the material or other component elements needed to provide these properties, and the refinement is continued until every item is quantified and actionable. The final plan is captured on to a single chart with the main objective of the exercise kept in mind: to satisfy the customer's requirements. This technique has been examined in ICI for the design of a new barrier coating for packaging film and for the design of a fluorination catalyst for the manufacture of ozone-friendly fluorocarbon refrigerants.

The 'fishbone' (cause and effect) diagram is a systematic, graphical means of investigating the root causes of a problem. As such, it is essentially a structured aid to brainstorming for the elimination of non-conformance. It is particularly useful when there are a wide variety of opportunities for error and has been applied, for example, to problems regarding the incorrect dispatch of chemical drums (Figure 2.7).

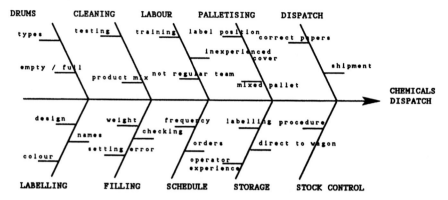

Figure 2.7   Example of 'fishbone' diagram.

Basically, the effect (i.e. the problem) is put at the fish head with the ribs representing the possible causes. The main ribs are frequently labelled as materials, manpower, methods and machinery ('the four Ms'), but whatever makes sense can be used. When carrying out the analysis of root causes, it is often enlightening to refer to operating manuals and procedural instructions, just to see how operations are supposed to be executed.

2.1.6   Quality case histories

This section presents a series of case histories where the application of quality principles has resulted in improved performance, often with significant cost savings. Looking at case histories serves to demonstrate the practical applications of quality, where the general techniques used may stimulate their application in new areas and hopefully, by providing specific examples, will help to prevent similar mistakes from being repeated in the future.

The first example chosen illustrates the principle of using a simple Pareto analysis, the second shows the importance of the customer/supplier relationship, the third describes the use of a corrective action team to solve a multipart problem, and the fourth demonstrates the importance of the quality education process and the vital role of management support.

1.   Customer complaints—how to remove them.   Poly(methyl methacrylate), better known as Perspex®, is a widely used material which accounts for a significant portion of the acrylics industry. A major manufacturer of poly-(methyl methacrylate) noted that the largest number of customer complaints originated from the handling of the product after manufacture. The company therefore set itself the task of reducing the number of complaints by doing whatever was necessary to remove the root cause of the problem. By using a simple Pareto analysis they were able to identify the vital few non-conformances that were leading to greatest customer dissatisfaction. The complaints concerned matters such as poor packaging, late deliveries and delivery of incorrect grades of material.

The Pareto analysis showed that one of the more important errors was that incorrect product was being loaded at the distribution site. The problem was tackled at source, first by identifying the opportunities for error, for example by making sure that pallets were correctly located and that the movement of pallets within the warehouse was kept to a minimum, so reducing the potential for paperwork mistakes. Secondly, a monitoring system, involving all members of the department, was introduced for monthly random sampling of pallet locations.

The result of these quality actions was that location errors fell dramatically and that distribution complaints, from all causes, were reduced by 65% in one year.

2. Dry ice losses—benefits of customer/supplier co-operation. A long-standing problem associated with the supply of dry ice is the difficulty of maintaining the very low temperatures necessary to prevent sublimation during packing, transport and delivery. The usual method of packing solid CO_2 is to place it in cool boxes made of fibre-glass or similar material. A large supply organisation found that a recurring complaint from customers was that the deliveries of dry ice were underweight owing to losses from sublimation of the solid during transit.

Discussion with the customer led to their suggestion of placing a thin layer of scrap dry ice pellets on the top of each box to inhibit the evaporation. The supplier was happy to adopt this solution and, as a result of no longer having to compensate customers for underweight deliveries, made a net saving of £26 000 per annum.

3. Corrective action team—how to make permanent improvements. Methylene chloride (dichloromethane) is a chemical that is in demand worldwide and as such, like many chemical exports, presents particular demands for specifications that meet international standards. A major UK manufacturer of methylene chloride found that it had an on-going problem with the containment drum suppliers. The manufacturer exported their methylene chloride to over 92 countries and required from the suppliers several different drum types, often at short notice. However, they found that frequently there were errors in the types of drums supplied and it was obvious that the root cause of the problem was the instantaneous demand for a wide range of drum styles.

A CAT was formed and it was quickly decided that the ideal corrective action was to rationalise to one drum and one fill weight. The team found it necessary to enlist the help of other departments and of the supplier in order to acquire the skills necessary for their new initiative, and eventually a totally new drum was designed. The new drum design not only successfully replaced the existing wide range of different styles but was also made to take account of newer and more stringent international regulations.

The benefits for both the supplier and the customer were numerous, such as productivity improvement and cost savings in excess of £100 000, a reduction in end-user complaints and additional business of approximately 95 000 drums per annum. The improvements came about as the result of finding the correct blend of skills amongst the team members, including those of the supplier as well as the customer, and then implementing the corrective action that they recommended.

4. *Quality education—measurement, communications and recognition.* A plant team set about improving an environmental problem associated with unpleasant smells. A solution to the problem was found which not only stopped the emissions but also increased production by 20% and provided more business for their internal feedstock supplier. The material, trimethylamine hydrochloride, is produced by the addition of trimethylamine (TMA) to hydrochloric acid. When an excess of the amine is used in the reaction a powerful fishy smell is produced. Complaints were made by residents from as far as six miles from the location of the plant. The problem was identified to be the pH meter used for the process control. A team of process operators collected plant data around the clock and looked at the relevant measurement control. It was found that the TMA was being added too fast for the pH meter to register the resulting change in acidity. The fix was to add the amine slowly at the back end of the process. Although this approach resolved the environmental problem, the batch times were slowed and the subsequent production rates were cut. By using simple brainstorming techniques, the team conceived the idea of adding the TMA at an initially faster rate over the first few hours, followed by a trickle addition of the remaining batch. By operating the process in this way batch times were cut and the smells, which were the initial cause of the problem, were completely eliminated. This is an example where a problem is turned into an opportunity. The quality improvement arose from the education of the process operators in the techniques of quality measurement and communication and the empowerment of the team by their management. The team responsible for this initiative were recognised by receiving a prestigious award from the Chief Executive.

2.1.7   Conclusion

Quality in the chemical industry is essential to prevent the overtaking of the industry by foreign competition. It can only be accomplished by cultural change, moving from a fixing–inspecting–fire-fighting mentality to one of prevention, continuous improvement and teamwork. Quality is about people as much as systems and depends more than anything else on a change in each individual's attitude.

The Quality process is a strategic process for improving everything that an organisation does—to the benefit not only of themselves but to their customers and the community in which they operate.

2.2.1 *Introduction*

The chemicals industry is an extremely diverse business covering everything from food additives to fluorochemicals. Yet all these many manufacturing processes encounter common problems of personal and public health and safety.

Health and safety at work has in the past been regarded by some employers and employees as a peripheral chore divorced from the real business of the company. Thankfully, this view has now been largely replaced by the realisation that it forms the foundation of an efficient, as well as a caring, industry. Apart from the emotional aspects, in today's harsh business climate the bottom line for any chemical company is that if they allow unsafe practices to occur then there may be direct financial consequences, both in terms of fines through prosecutions and also through production time lost when key workers are absent or when plant is out of commission. In effect, good safety provisions contribute to the overall profitability of any company or organisation.

The cornerstone of good company policy on health and safety, as with all quality issues, is full management support. Indeed, an assessment of a company's attitude to health and safety can be used as a fairly reliable guide to its general ethos on matters such as employee care, customer commitment and community relations.

There are many hundreds of publications available on specific issues of health and safety within the chemical industry, and it is not the intention here to give detailed or specific information.[17] Instead, it is hoped to provide an overview of the general types of hazards that particularly concern the chemical industry and the sorts of legislation and management operations that affect both the company and the individual worker.

It is a little-known fact that chemical industries in general have often been at the forefront of improvements in the safety culture of the workplace, and it is worth remembering at this stage that the chemical industry has an excellent safety record when compared to industry in general.[18,19] Some safety statistics are given on pages 62–64 of Volume 1.

Unfortunately, there is a popular mythology that the industry has a cavalier attitude to safety, a view no doubt fuelled by a very small number of tragic incidents such as the one at Bhopal, central India, in 1984, discussed in section 3.2.5. It is certainly true that because of the nature of the work done in the chemical industry there is often the potential for serious accidents, but the fact remains that the most common problems that affect workers in this industry are the same as those found in any workplace, and the highest fatalities are those, as in society in general, associated with driving a motor vehicle.

Nonetheless, it should be recognised that the chemical manufacturing industry is involved with many potentially hazardous processes, as well as the additional problems associated with pilot plant and scale-up. When major accidents do occur then the resulting bad publicity reflects, in the public mind, on the whole industry. It is therefore in everyone's best interest i.e. the public, the chemical industry and its workers, to ensure the safest working environment possible.

2.2.2   Safety organisation

The basic pattern for developing safe practice applies equally to an individual working on one specific task as to the safety structure for an entire multi-national company.

The starting point for a successful health and safety policy is to recognise the hazards present. This will be relatively easy in some instances but less clear in others. Once a hazard has been identified then the risk involved must be assessed, and finally those practices that will ensure prevention of accidents must be put into place.

Safety starts at the top with the commitment of the most senior people. A safety culture that is openly and fully supported by management is the key to good practice; without it there is an in-built failure system.

All reputable chemical companies now have established workplace-orientated safety schemes, often as part of a comprehensive quality structure. A good example of such a scheme is the 'Unsafe Acts Prevention Scheme' instigated by DuPont and now adopted by many other companies. This scheme is designed to highlight the importance of the individual in maintaining a safe working environment whilst also providing, for the workforce, a full management-supported referral structure. A regular maintenance schedule should also form an integral part of any scheme and it should include all plant, even those items that are rarely used, such as the diesel safety generators that act as back-up during power failures. Machinery like this may go unused for years at a time but its maintenance should not be neglected.

Senior staff must show that they believe accidents can be prevented and be prepared to enforce safety discipline throughout the company. They must ensure that safe operating procedures are actually practised on the work site and not simply paid lip-service. This type of self-regulation, with spot checks from regulatory authorities, should provide for a quality health and safety system.

2.2.3 Legislation

There are considerable variations in the legislation in different countries, but the USA is probably the most regulated with over 16 acts specifically affecting the chemical industry. Legislation in the USA may originate with both state and federal agencies operating to national standards. The National Institute for Occupational Safety and Health (NIOSH) is the independent standards authority charged with developing and establishing the national standards. The Occupational Safety and Health Act, 1970, covers much of workplace health and safety matters and, like its European counterparts, lays down duties on both employers and employees, determines standards and allows for the inspection and enforcement of the legislation.

One of the most comprehensive acts is the Toxic Substances Control Act (TSCA), 1976, which covers the regulation of all existing and new chemicals used and manufactured. The act is administered by the Environmental Protection Agency (EPA), and under the legislation it has the power to delay or ban the manufacture of chemicals.

In Europe the European Community is currently in the process of harmonising legislation for all its member countries and has recently produced six European Community Directives, developed under Article 118A of the Treaty of Rome, on health and safety at work. The regulations, which replace many old pieces of legislation, cover: health and safety management,[20] equipment safety,[21] manual handling of loads,[22] workplace conditions,[23] personal protective equipment,[24] and display screen equipment.[25]

Health and safety management.   These regulations endeavour to encourage a systematic and thorough approach to matters of health and safety. The regulations require employers to:

1. Assess the risks to the health and safety of employees and others affected by the work activity. Assessment findings must be recorded and employers with more than five employees must produce a written health and safety policy.
2. Implement the measures that follow from the risk assessment. Planning, organisation, control, monitoring and review must all be covered.
3. Provide appropriate health surveillance for employees where the risk assessment shows it to be necessary.
4. Appoint competent people to help in the compliance of the duties under health and safety law.
5. Set up emergency procedures.
6. Provide employees with understandable information on matters of health and safety.
7. Co-operate with other employers sharing the work site.
8. Ensure adequate health and safety training for employees so that they are capable enough at their jobs to avoid risks.

9. Provide temporary workers with health and safety information to meet special needs.
10. Place duties on employees to follow health and safety instructions and report dangers.
11. Consult employees' safety representatives and provide facilities for them.

Equipment safety. These regulations are designed to give a clear direction on the use of equipment at work and are aimed at the protection of workers. In this instance 'equipment' may mean anything and everything from simple hand tools to a complete chemical plant, and 'use' includes starting, stopping, repairing, modifying, installing, dismantling, programming, setting, transporting, maintaining, servicing and cleaning. The duties specified are:

1. Make sure that the equipment is suitable for its use.
2. Select equipment suitable for the working conditions and hazards in the workplace.
3. Ensure equipment is used only for those operations and conditions for which it is suitable.
4. Ensure that equipment is maintained in efficient working order and in good repair.
5. Give adequate training, instruction and information.
6. Provide equipment that conforms with EC product safety directives.

Other specified requirements cover:

- Guarding of dangerous parts.
- Protection against specified hazards such as rupture of equipment parts, fire, unintended discharge of substances, explosion.
- Equipment and substances at high and very low temperatures.
- Control systems and control devices.
- Isolation of equipment from energy sources.
- Stability of equipment.
- Lighting.
- Maintenance.
- Warnings and markings.

Manual handling of loads. This legislation is aimed at reducing the large numbers of injuries caused by the incorrect handling of loads and applies to any manual handling operations which may cause injury at work. The operations will have been identified by the risk assessment carried out under the Management of Health and Safety at Work Regulations, 1992. Employers will be expected to have taken three precautions:

1. Avoidance of hazardous operations where practicable and where possible to substitute mechanical handling for manual handling.

2. Make a full assessment of any hazardous operation.
3. Reduce the risk of injury as far as reasonably practicable, remembering that additional training may be required.

Workplace conditions. These regulations cover many aspects of health, safety and welfare in the workplace and apply to all places of work except: means of transport; construction sites; and sites where extraction of mineral resources or exploration for them is carried out. The regulations govern four broad areas:

Working environment—temperature; ventilation; lighting; room dimensions and space; and suitability of workstations and seating.
Safety—safe passage of pedestrians and vehicles; windows and skylights; use of safety material, safety devices and marking for doors, partitions, gates and escalators; floors (construction, maintenance and hazards); falling into dangerous substances; falling objects.
Facilities—toilets; washing, eating and changing facilities; clothing storage; drinking water; rest areas.
Housekeeping—maintenance of workplace, equipment and facilities; cleanliness; removal of waste materials.

Personal protective equipment. Personal protective equipment (PPE) should be relied upon only where the risks cannot be controlled by other means. The employer then has a duty to provide, free of charge, suitable PPE appropriate for the risks and working conditions. The employer also has duties to:

1. Assess the risks and PPE to ensure that it is suitable.
2. Maintain, clean and replace PPE.
3. Provide storage for PPE when not being used.
4. Ensure PPE is properly used.
5. Give training, information and instruction to employees on its use and maintenance.

Display screen equipment. These regulations cover a new area of work activity and cover any employee who habitually uses display screen equipment as a significant part of normal work. There are also some duties towards the self-employed using display screen equipment in their undertakings. The regulations cover equipment used for the display of text, numbers and graphics. Employers' duties include:

1. Assessment of display screen equipment workstations and reduction of risks discovered.
2. Ensure that workstations satisfy minimum requirements for the screen itself, keyboard, desk and chair, working environment and task design and software.

3. Plan work so that there are breaks or changes of activity.
4. Provide information and training for display screen equipment users.

Users will also be entitled to appropriate eye tests, and special spectacles if required.

All the directives required implementation by January 1993 and aim to make clear to employers and employees alike what is expected of them.[21-28] Essentially, employers must provide a safe place of work, information, training, proper plant and appliances and safe systems. In the EC, as in most countries with extensive health and safety legislation, the regulations are enforced by inspectors who have a wide range of powers under the law. They can, for example: carry out examinations and inspections; take possession of anything which they think is dangerous or arrange for it to be tested; and inspect and copy documents. In cases where there is significant risk to health and safety they also have the powers to prosecute people and companies or to revoke licences (e.g. nuclear installations). In the UK, however, the Health and Safety Executive (HSE) is unlikely to consider formal enforcement unless:

a. the risks to health and safety are evident and immediate; or
b. what needs to be done is not new (i.e. existing duties transposed into new legislation); or
c. employers appear deliberately obdurate and unwilling to recognise their responsibilities to ensure the long-term health, safety and welfare of employees and others affected by their activities.[26]

In the UK, companies that are currently complying with the Health and Safety at Work Act, 1974, should find little difficulty in meeting the new requirements. The main change embodied in the regulations is one of approach. The emphasis is placed on management and on broad duties of risk assessment and prevention, as is demonstrated by, for example, the Control of Substances Hazardous to Health Regulations (COSHH), 1988.[27,28] Non-EC

Table 2.2   Sources of Health and Safety Information, USA and EC

United States of America
Chemical Manufacturers Association, Washington, DC, USA
Hazardous Substances Data Bank (HSDS)
National Library of Medicine, Bethesda, MD, USA
National Institute of Occupational Safety and Health (NIOSH)

European Community
Chemical Manufacturing; list of publications May 1992.
Health and Safety Commission, London, UK.
HSELINE—a computer database of bibliographic references to published documents on health and safety at work. It contains over 140 000 references, and 12 500 additions are made each year, and is available through the European Space Agency Information Retrieval Service (ESA-IRS) Dialtech.
British Library, London, UK.
Health and Safety Executive, Bootle, Merseyside, UK.

---

countries, such as Japan and Australia, have also enacted similar legislation to that found in the USA and Europe.

Sources of specific information on health and safety matters for the USA and the EC are given in Table 2.2. This list is by no means exhaustive, and readers should keep up to date with the legislation from relevant bodies.

2.2.4   Hazard and risk

The public's perception of risk, e.g. the risk posed by living close to a chemical plant, is often very different from the mathematical reality. It is a subjective view. Unfortunately, there have been instances when misleading statements have been made or details concealed when accidents have occurred and this has produced a public mistrust of the industry. This situation can only be improved by a more open and honest approach when accidents do happen.

Before going into detail about specific dangers it is useful to look at the facts for fatalities in the chemical industry when compared with other types of workplace. Table 2.3 shows examples of typical Fatal Accident Frequency Rates (FAFR) (i.e. roughly the number of fatalities in a group of 1000 people over their working lifetimes) for a variety of occupations.

In addition, if we compare the FAFR for activities such as travelling by car (FAFR, 57), we can see that working in the chemical industry is much less hazardous and is only slightly more dangerous than staying at home (FAFR, 3).[18,19]

The terms 'hazard' and 'risk' are frequently misused and so for clarity here we will use the following definitions.

Hazard—a physical situation with a potential for human injury, damage to property, damage to the environment or some combination of these.

Risk—The likelihood of a specified undesired event occurring within a specified period or in specified circumstances.[29,30]

Hazard identification can be carried out in three stages: firstly, what can go wrong (qualitative); secondly, how often—i.e. frequency (quantitative); and thirdly, what are the effects and consequences of the hazard (quantitative).

Risk analysis and quantification of hazard are much more complex procedures than qualitative hazard identification and are best regarded as a

Table 2.3   Fatal accident frequency rates

Occupation	Fatalities per 100 million hours
Chemical industry	3.5
British industry overall	4.0
Coal mining	40.0
Flat racing jockeys	50 000.0

specialist job. For those readers seeking more detailed information a useful summary may be found in the book by Marshall.[29]

When risk assessments are carried out the low probability event (LPE) and the low probability, high consequence event (LPHC) must be looked at as well as the more obvious risks. It should also be remembered that chemicals do not exist in isolation but interact with other hazards such as electricity, machines and inclement weather. However, although major chemicals usage and manufacture can represent high levels of hazard they are also characterised by a low level of risk.[29]

In terms of the modern chemical industry, there are some hazards that arise from the inherent nature of the materials or substances being handled, for example they may be toxic, flammable, corrosive, explosive, or even a mixture of all these. In addition it is known that prolonged or intermittent exposure to some chemicals can cause certain occupational diseases, such as silicosis, dermatitis or various cancers. Clearly, therefore, it is essential that safety is the first consideration whenever chemicals are being manufactured, used, transported or disposed.

Unfortunately, the hazards that may be presented by the chemicals industry are brought to public attention in a very dramatic way when there are major accidents involving people outside the industry itself or affecting the environment outside the chemical plant. Examples of this type of incident are given in Table 2.4.

Although such major disasters make headline news they are in fact rare events. It is the individual workers in the industry who are potentially the most exposed, but as we have already noted, the chemical industry has a very good record in terms of the health and safety of its workers. Accident records show that the great majority of employees who are killed or injured in chemical and process plants suffer their injuries through personal accidents. A third or less of

Table 2.4   Accidents in the chemical industry[19,29]

Type of incident/material	Location	Date
Explosion and fire		
Explosion in caprolactam plant	Flixborough, UK	1974
Propylene gas explosion	San Carlos, Spain	1978
Explosion in LNG storage tank	San Juanico, Mexico	1984
Toxic release		
Phosgene	Hamburg, Germany	1928
Chlorine	Rauma, Finland	1947
Ammonia	Potchefstroom, South Africa	1973
Dioxin	Seveso, Italy	1976
Methyl isocyanate	Bhopal, India	1984
Chronic poisoning		
Thalidomide	Various	1961
Mercury via dimethyl mercury	Minamata Bay, Japan	1965
Dioxin	Seveso, Italy	1976

these are from the specific technology, whilst the majority are due to events such as falls, collisions, electric shock and the use of machinery or hand tools. A smaller proportion may suffer from recognised occupational diseases in addition to those affected by such things as stress, noise or even poor lighting.[29,31]

2.2.5   Conclusion

It is essential that the chemical industry does not place its employees, customers and the general public at unnecessary or unacceptable risk from its activities. All accidents are preventable (zero defects) and it is imperative that the industry continues to work towards this goal through continuous improvement such as is dictated by the quality initiative. Both safety and quality make good business sense.

Acknowledgements

The authors are grateful to Terry Tribbeck, Hugo Steven, Roger Pybus and Ian Grieve for their guidance and valuable comments and to the Directors of ICI Chemicals and Polymers Ltd for their permission to publish this chapter.

References

1. 'Quality Improvement through Defect Prevention,' Philip Crosby Associates Inc., Winter Park, Florida, 1985.
2. 'Quality is free,' P.B. Crosby, McGraw-Hill, New York, 1979.
3. 'In Search of Excellence: Lessons from America's Best Run Companies,' T.J. Peters and R.H. Waterman, Harper & Row, New York, 1982.
4. 'Managing Quality,' D. Garvin, Free Press Macmillan Inc., New York, 1988.
5. 'Quality Control Handbook,' J.M. Juran, 3rd edn, McGraw-Hill, New York., 1979.
6. *Which?*, Consumers' Association, London, January 9, 1978.
7. British Standards Institution BS 5750; Part 1, London, 1987.
8. Comité Européen de Normalisation, ISO 9000, 1987.
9. Comité Européen de Normalisation, ISO 9001, 1987.
10. Comité Européen de Normalisation, ISO 9002, 1987.
11. Comité Européen de Normalisation, ISO 9003, 1987.
12. *BSI News*, August 1987, 11.
13. D. Johnston, *Chemistry and Industry*, 1988, 365.
14. 'SPC and Continuous Improvement,' M. Owen, IFS Publications, 1989.
15. 'Total Quality Management,' A.R. Tenner and I.J. DeToro, Addison-Wesley, New York, 1989.
16. 'House of Quality,' J. R. Hauser and D. Clausing, Harvard Business Review, **66**, 1988, 63.
17. 'Chemical Manufacturing: List of HSC/HSE Publications,' Health and Safety Executive, 1992.
18. 'Safety in Industry,' B.F. Street, paper presented at the 121st Annual General Meeting of the Institution of Gas Engineers, 22 May 1984.
19. 'Safety in the Chemical Industry,' O.P. Kharbanda and E.A. Stallworthy, Heinemann, London, 1988.
20. 'The Management of Health and Safety at Work Regulations 1992,' HMSO, London, 1992.
21. 'Provision and Use of Work Equipment Regulations 1992,' HMSO, London, 1992.
22. 'Manual Handling Operations Regulations 1992,' HMSO, London, 1992.
23. Workplace (Health, Safety and Welfare) Regulations 1992,' HMSO, London, 1992.
24. 'Personal Protective Equipment at Work (PPE) Regulations 1992,' HMSO, London, 1992.

25. 'Health and Safety (Display Screen Equipment) Regulations 1992,' HMSO, London, 1992.
26. 'HSE26 10/92,' Health and Safety Executive, HMSO, London, 1992.
27. 'Health and Safety at Work Act 1974 (HSW),' HMSO, London, 1974.
28. 'Control of Substances Hazardous to Health Regulations, 1988 (COSHH),' HMSO, London, 1988.
29. 'Major Chemical Hazards,' V.C. Marshall, Ellis Horwood, Chichester, 1987.
30. 'Nomenclature for Hazard and Risk Assessment in the Process Industries,' Institution of Chemical Engineers, Rugby, 1985.
31. 'The Safe Handling of Chemicals in Industry,' Vol. 1, P.A. Carson and C.J. Mumford, Longman Scientific and Technical, New York, 1988.

Environmental issues 3

Alan Heaton

Up until about 1960 the general public was not really aware of chemicals as such. They met them only as consumer or end products such as insecticides like DDT, the penicillin antibiotics, and new fibres such as nylons and polyesters. These were all clearly desirable products which were very beneficial, and in some cases life-saving. Thus the public's concept of chemicals, although admittedly based on a limited understanding of them, was a very favourable one. The industry which produced them, the chemical industry, was therefore clearly needed. It produced these by now essential products and provided employment for a large number of people. Furthermore it was expanding rapidly, introducing many important new products, particularly several polymers and plastics, and was also very profitable. A major reason for this was that it was the era of very cheap raw materials, e.g. crude oil. Of course, this was well before the shock of OPEC's quadrupling of crude oil prices in 1973. The expansion of the industry created more opportunities for employment and in some areas one or two large chemical companies dominated the local economy and community. An example was ICI on Teesside, which had an enormous local impact through both the tens of thousands of people whom it employed and the amount of money which it injected into the economy through the services that it purchased and the salaries that its employees spent.

The consequence of all this was that the chemical companies had a great deal of freedom to operate very much as they wished at that time. It should be noted also that there was only a very limited awareness and concern for pollution, which was attributed largely to the smoke produced in iron and steel works and also by coal-fired power stations. Very little blame or complaint was laid at the door of the chemical industry.

Two events in the early 1960s initiated a major transformation in the general public's awareness of, and particularly attitudes to, chemicals and their possible effects on the environment. These were the thalidomide tragedy (1961) and the publication of the book *Silent Spring* by an American marine biologist Rachel Carson (1962).

Thalidomide was first made in 1953 and was first marketed in West Germany in 1957 and in the UK in 1958. It was initially developed as a sedative and hypnotic, being a non-addictive alternative to the barbiturates. The latter were the major cause of drug-induced deaths at this time. It proved particularly beneficial in treating geriatric patients. However, problems occurred when it was prescribed as a sedative and anti-emetic for women in the early stages of pregnancy. As the drug became more widespread medical practitioners became aware of an increasing frequency of foetal abnormalities. A link between the drug and damage to the foetus was suspected. Eventually, in December 1961, the Distillers Company withdrew the drug from the market, and the originators of the product, Chemie Grünenthal, later withdrew it from the West German market. Sadly, before this could happen, some 10 000 children were born with physical abnormalities—in some cases very severe ones.

There are a number of points that have to be considered in making a judgement on the thalidomide tragedy. Firstly, the company carried out all the tests that would have been reasonably expected at the time during the research and development of the drug. Secondly, in the 1950s it was not considered likely that any drug could pass from the mother to the foetus in the womb and therefore put the developing child at risk. The company therefore appears to have done what was required at that time in checking the safety of the drug. However, it has been suggested[1] that the companies were aware of the dangers of thalidomide to the central nervous system before it was withdrawn.

Thirdly, it is now clear that the product should have been withdrawn from the market much earlier than it was. Although one can appreciate that a company would be reluctant to withdraw a product which they had spent tens of millions of pounds developing until its link with causing problems was clearly established, in this case the company was rightly strongly criticised for not acting more quickly. Sadly, the company's reluctance to accept responsibility and pay compensation to the victims led to protracted and distressing litigation. Final settlement only took place in 1972.

What lessons were learned from this tragedy? It led to much stricter and more extensive testing being required for both drugs and pesticides (see sections 7.3 and 8.3), including tests on at least two species of mammals for teratogenicity, the ability to cause malformed foetuses to develop. The latter also includes following the test through several generations of the mammal.

Thalidomide has the structure:

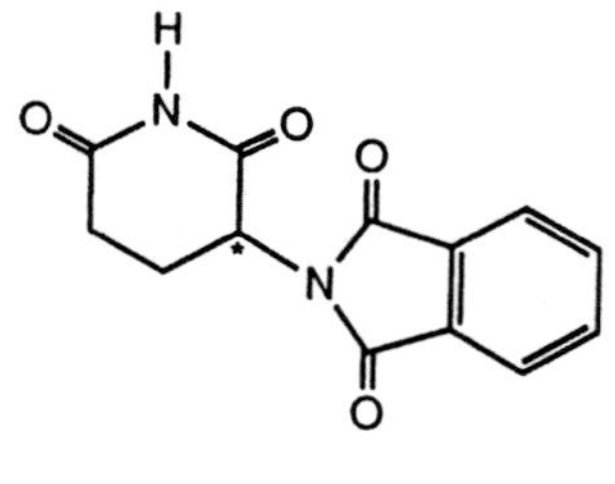

and this has one chiral centre, marked *. It can therefore exist in two enantiomeric forms. The product that was marketed in the late 1950s was the racemic mixture of these two. Research over the years has established not only the link between the drug and its teratogenic effects, but also that, as expected, only one enantiomer is active as the drug—the R-enantiomer. It is the other, inactive enantiomer that is responsible for the teratogenic effects. This has lent support to the trend in the agrochemical and pharmaceutical sectors nowadays to market only the active stereo-isomer rather than a mixture of stereo-isomers. Despite the additional processing costs involved this has environmental advantages. The pesticide Fusilade, discussed in section 7.5.8, illustrates this.

Prior to the publication of *Silent Spring* by Rachel Carson in 1962, no-one really thought very much, or worried, about the effects of chemicals on the environment. She, however, suggested that pesticides like DDT were having an adverse effect on the environment, and began a campaign to have them banned. Not surprisingly, she attracted vigorous opposition from the agrochemical companies who were making such products. A long, and sometimes bitter, campaign ensued, punctuated occasionally by dramatic events. For example, the representatives of the companies proved that DDT had a very low toxicity to humans by taking it daily during several months of litigation, without suffering any ill effects. Nevertheless, the outcome was that the use of DDT as an insecticide in the USA was banned from 1973. Most other developed nations have followed in also banning DDT. However, because of its low cost and many plus points as an insecticide (see section 7.6.3.1), it is still used in developing countries.

The major consequence of these two events, and particularly the latter, was to awaken a public concern with chemicals—in food, in the environment, as drugs, and as pollutants. As we shall see, this has led to a very adverse or negative attitude towards chemicals and the chemical industry. We will look into the reasons for this, but will also see why this point of view is not really tenable when we compare the pluses and minuses of chemicals.

In the next section, five environmental disasters which made the headlines between 1961 and 1984 will be considered, and we will then try to analyse why they happened and the lessons to be learned from them. Even here it is very important to place these in context, by bearing in mind that there are many tens of thousands of chemical plants operating round the clock throughout the world, yet we do not hear about them operating safely, effectively and without causing adverse environmental problems. Clearly some of them do occasionally cause problems and, of course, it is these that the media seize on. It is these problems and only these, therefore, that the general public hears about, since safely operating plants are not newsworthy. Clearly, from the other side, the chemical industry must put a lot more effort and resources into making the public aware of the good things that they are doing, so that people have a more balanced view. In the past the industry only tended to speak very publicly when it had to respond to some pollution incident. In recent years there has been a much more positive attitude towards educating the public and, as we will see in section 3.4, a much more open relationship with the general public.

After looking back at examples of environmental disasters we will trace the effect that these have had in raising the level of people's awareness and concern for the environment. Then, following a look at the chemical industry's response to these changing public attitudes towards it and its products, and its current approach to environmental issues (and a consideration of CFCs), we will try to come to some conclusions regarding the chemical industry and the environment.

3.2 Some environmental disasters

3.2.1 DDT

DDT is the abbreviated name for p,p-dichlorodiphenyltrichloroethane and, although this is considered in detail in section 7.3.1, it is still worth considering briefly here because, as we have already seen, it was this compound above all others that first sparked off concern for possible adverse effects of chemicals on the environment.

Despite its apparently being an ideal insecticide, one of its desirable properties—its great chemical stability—proved to be its undoing. As a result of this it got into the food chain unchanged, and as it passed up it became concentrated in fatty tissues, reaching levels in birds, and particularly birds of prey like peregrine falcons, where it caused serious problems and even death. Consequently, it has now been banned and replaced by other more acceptable insecticides throughout the developed world.

Against these adverse environmental effects must be balanced the many millions of lives that DDT saved in worldwide malaria eradication programmes in the 1950s and 1960s.

3.2.2 Minamata Bay

In 1965 in the area of Minamata Bay in Japan a number of people were affected by a strange disease, which also affected birds and animals. Its symptoms included lack of co-ordination, problems with eyesight and even death. Eventually almost fifty people died and over one hundred became seriously ill with this disease. The source of the problem was finally traced to the eating of contaminated fish. Since the bay was a fishing area, seafood played a major part in the diet of the local population.

What was the contaminant and where had it come from? Consideration of the symptoms of the disease and analysis of the contaminated fish showed that the culprit was the highly toxic dimethyl mercury, $(CH_3)_2Hg$.

Its source was traced to a plastics factory situated a little way up-river from the bay. This was discharging mercury in its effluent into the river. If the mercury had just passed into the sea unchanged then there would have been no problem. However, certain micro-organisms which live in the muds and silt at the bottom of the river and estuary convert the mercury into dimethyl mercury. This then enters the food chain and concentrates as it passes up it—note the parallel with DDT.

As a result, Japan now has very strict and low limits of about 0.01 ppm of mercury in effluent.

3.2.3 Flixborough

This major disaster occurred in 1974 at the Flixborough works of Nypro (UK) Ltd on Humberside in the UK. The plant involved was one used for the manufacture of caprolactam (which is the precursor for nylon-6) and in particular for oxidation of cyclohexane to cyclohexanol plus cyclohexanone. One of a series of reactors had been removed for repair and temporarily replaced by a large-diameter pipe which was inadequately supported. A crack developed in the pipe through which the cyclohexane, under pressure, leaked, forming a massive cloud of highly flammable material. This ignited with devastating consequences. The resulting explosion caused 28 deaths plus almost 100 injuries, as well as damaging or destroying some 2 000 houses, shops and factories. It is fortunate that the explosion occurred on a Saturday when relatively few people were working at the factory, otherwise the loss of life would have been very much greater.

The inquiry into the disaster noted that there was no plant engineer because the post was vacant, and that no design study had been carried out.[2] Note that this, the UK chemical industry's blackest day, was the result of human error.

3.2.4 Seveso

The town of Seveso, situated near Milan in Northern Italy, was probably only known to relatively few people outside Italy, but in the summer of 1976 it became known throughout Europe and beyond. The reason was the explosion that occurred at the Givaudan plant there. This company was a subsidiary of the Swiss pharmaceutical giant, Hoffman La Roche (now simply Roche), which is perhaps best known for its tranquillisers Librium and Valium.

The Seveso plant was making the herbicide 2,4,5-T (2,4,5-trichloro-phenoxyacetic acid) by the route shown on page 242, and the problems concerned the second stage. The reaction had been completed and water cooling of the reactor continued. It is believed, however, that it was not continued for the full prescribed length of time. Whatever the reason, there was a rapid, unexpected rise in temperature, causing a rapid increase in pressure and resulting in a safety valve blowing and an explosion. This released a cloud of chemicals over the surrounding area. The result would have been bad but not disastrous except that, as shown on page 242, Dioxin is produced to a minor extent in a side reaction. Dioxin is one of the most poisonous substances known and its teratogenic properties had been discovered after the use of

2,4,5-T and 2,4-D (2,4-dichlorophenoxyacetic acid) as the defoliant Agent Orange in Vietnam. Dioxin can also damage several organs and cause cancers as well as being teratogenic and can produce these effects at concentrations of only a few tens of ppm. It is estimated that 2 kg of Dioxin were in the cloud released over Seveso. Not surprisingly the reaction to the explosion was therefore much more emotional than it would have been in the absence of the Vietnam experience.

The result was that the immediate area was sealed off and the population of a large surrounding area was evacuated and given health checks. Locally grown food was banned for several months and several inches of top soil were removed and incinerated in order to destroy the Dioxin. It was some weeks before the evacuees were allowed to return to their homes. Despite all the worries it appears that to date only one person has died from liver cancer, although there has been some skin disease (chloracne). No employees of the company became ill or suffered damage to their organs. The greatest worry at the time was the possible effect on pregnant women. Although some women had abortions, amongst the rest the incidence of miscarriage or the birth of malformed children has been no higher than would have been expected in the absence of the explosion.

Although we cannot be certain it again looks as if human error was the cause.

3.2.5   Bhopal

In 1984 the chemical industry's worst ever disaster occurred at the Union Carbide works in the Indian city of Bhopal.

The catastrophe happened at a plant that was used for making the insecticide Carbaryl (or Sevin), which is active against a very wide range of insect pests and is therefore used to protect many fruit and vegetable crops. The company also manufactured this product at its plant in the USA but different chemistry was involved (Scheme 3.1).

Both of the routes shown in Scheme 3.1 are potentially quite hazardous since they each involve a highly toxic gas, i.e. phosgene and methyl isocyanate, respectively. The decision to use a different route in the Indian plant would seem a logical and commercially sensible one, since if we assume that the total cost of chemicals is very similar then the Bhopal route has the advantage of being a single-step process compared to the two steps of the other route.

Be that as it may, on the fateful day there was a release of a deadly cloud of methyl isocyanate over the plant and the surrounding area. The exact sequence of events that led to this may never be known, but what clearly happened was that the pressure in the methyl isocyanate storage tank rapidly built up causing a safety disc to burst. This would normally have been a matter of some, but not major, concern because gas would only be released into the atmosphere after passing through caustic scrubbers. The sodium hydroxide

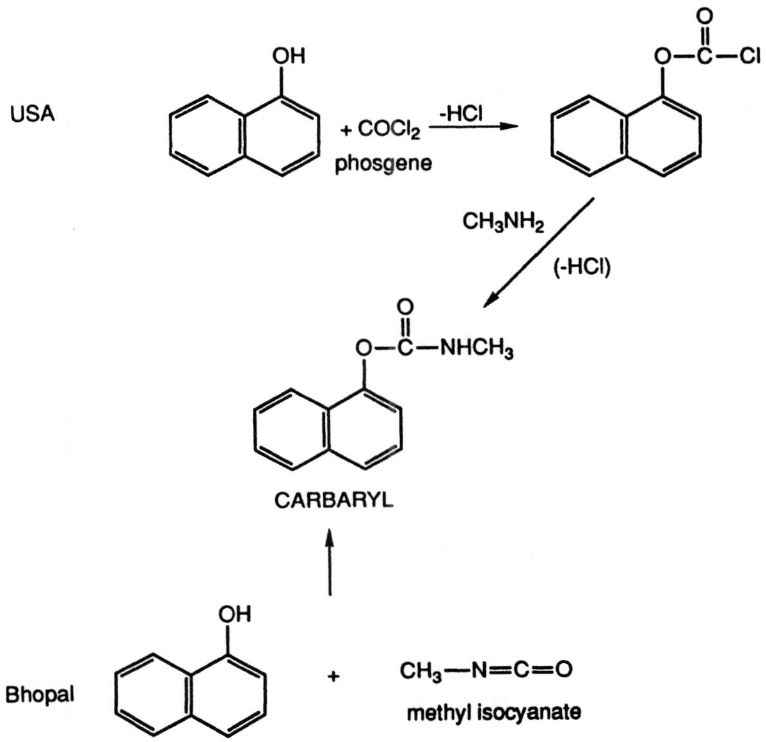

Scheme 3.1.

would react with the methyl isocyanate and render it relatively harmless. Tragically the scrubbers were not working so the toxic gas was released unchanged.

As a consequence some 3 000 people were to lose their lives and many more thousands were injured, often seriously. This toll was so high for two reasons. Firstly, many houses had been constructed right next to the factory. Secondly, company officials initially assured medical teams treating the victims that there was no problem with the material released, and they did not reveal its identity until a few days later. Thus vital time was lost in which correct treatment could have alleviated the suffering and mortality to some extent.

One suggestion was that the cause was an operator opening the wrong valve and allowing water to be pumped into the methyl isocyanate tank. This would cause an exothermic reaction which would generate carbon dioxide gas (Scheme 3.2), thereby causing the rapid increase in pressure. This, plus the fact that the scrubbers had not been repaired and made operational again, shows that human error was the cause yet again.

Protracted negotiations for compensation were eventually concluded when the Indian Government, on behalf of the victims, accepted Union Carbide's

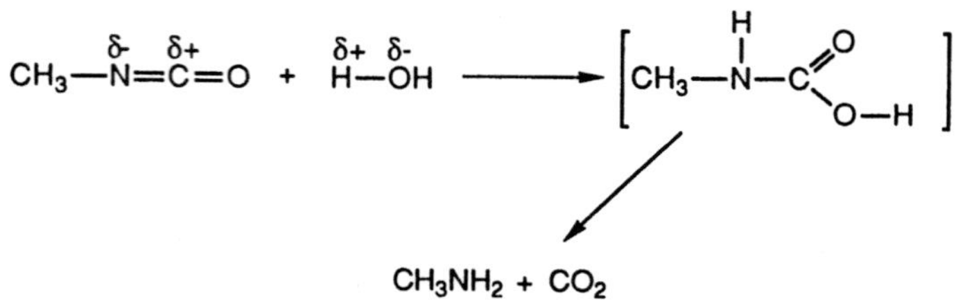

$$CH_3{-}\overset{\delta-}{N}{=}\overset{\delta+}{C}{=}O \ + \ \overset{\delta+}{H}{-}\overset{\delta-}{OH} \longrightarrow \left[CH_3{-}\overset{\overset{\displaystyle H}{|}}{N}{-}C\overset{\displaystyle O}{\underset{\displaystyle O{-}H}{}}\right]$$

$$CH_3NH_2 \ + \ CO_2$$

Scheme 3.2.

offer of $470 million in 1989. Critics have suggested this was too low, since if the claims had been upheld in courts in the USA the award would have totalled several billion dollars.

In April 1993 it was reported that several staff of the plant had been summoned to appear before the Indian courts to face charges arising from the incident.

3.2.6   Other examples

Table 2.4 lists a few other examples of disasters in the chemical industry as well as those already discussed.

It is important to place these in context and note that over the last thirty years only these dozen or so serious incidents/disasters have occurred world-wide. Whilst none of them should have happened—since as we have seen human error was the cause in most cases—we should acknowledge the many tens of thousands of potentially very hazardous processes that are operated day in, day out, round the clock in a safe manner. However, there is always room for improvement, particularly in reducing the number of minor pollution incidents.

3.3
Public response to the disasters

It is perhaps appropriate here to consider the response in general terms and in a collective manner.

There is no doubt that from *Silent Spring* onwards, and increasingly as each one of the disasters which we have considered happened, the general public has become more aware of, and more concerned for, chemicals in the environment, in its broadest sense. This ranges from gaseous pollutants such as sulphur dioxide causing acid rain—although the chemical industry makes only a small contribution to the total problem—to water pollutants of various sorts (e.g. PCBs (polychlorobiphenyls), mercury and heavy metals). Another area which generates a great deal of debate (and research) is additives in food and drink. Views range from 'it is wrong to add anything at all' to 'additives are essential'. A case in point is the addition of fluoride to water. Research has clearly established that doing this improves resistance to tooth decay, but some would argue that even at the low levels added there could be some long term adverse effects—although there does not appear to be any scientific evidence in support of this. At one extreme, people would see the addition of fluoride to water as an infringement of civil liberty since they would expect to have their own choice on whether to add it or not. They and we should note, however, that there can surely be no argument over adding some things (e.g. chlorine or hypochlorite) to water to kill bacteria to ensure that it is safe to drink.

Food additives are another controversial area which is often linked with the good or bad effects on health. Thus debates have ensued on whether tartrazine should be banned as a colouring additive (as it now has been) and also monosodium glutamate as a flavour enhancer and sodium cylclamate as a sweetener. In the last-named case, as more and more research work has been reported the evidence has swung both for and against it. Another area is the trend towards softer margarines and low-fat butters which are higher in polyunsaturated compounds. Medical opinion suggests that these cause less cholesterol to be deposited on artery walls than products which are lower in polyunsaturates and therefore higher in saturated compounds (e.g. butter and animal fats). However, recent reports suggest that the situation is not so clear-cut and is much more complicated than this, being related to amounts of low-density versus high-density lipoproteins.

The overall result has been the growing numbers and strength of consumer and environmental groups and their increasing influence at the decision-making and political levels. Examples are Greenpeace, Friends of the Earth and in Germany the Green political party, some of whose candidates have been elected.

What effect has this had on the chemical industry? It has certainly forced it to look more closely at its processes and products in terms both of safety and absence of adverse effects on the environment. This has in part been forced on the industry by the pressures just discussed, which have resulted in much tighter laws on the control of effluent disposal and testing to ensure that products are as safe as possible, particularly those that are biologically active. We have recognised the importance of these issues by devoting a full chapter in Volume 1 (chapter 8) to environmental pollution control. Testing to ensure the safety of chemicals crops up in a number of places in this book, particularly sections 7.3.1, 8.3.3 and 8.3.4.

Surveys have also shown that the state of our environment is an item of major concern amongst young people. One result of this, which has already been mentioned, is proliferation of environmental courses in academic establishments.

What has been the chemical industry's response to the increasing public pressure discussed in the previous section and the tighter pollution control legislation and measures that have resulted? Partly as a result of the increased penalties for causing pollution and breaking the appropriate statutes, and

3.4
The chemical companies' response

partly as a result of a conscious decision by companies to adopt a better, more open and more enlightened approach, there is no doubt that there has been a considerable improvement in performance. There has also been an increasing realisation that the company and its employees are part of the community and they need to live in harmony with other members of it. Thus a more open approach to discussing what companies are doing to minimise pollution, as well as sharing knowledge of the difficulties which they still face, has been apparent in the last five years. Another driving force has been the need to improve the industry's poor public image.

A great deal of money has had to be spent on measures to reduce pollution caused by existing processes. This is often difficult because of physical constraints imposed by the fixed nature of existing plant. Much effort is therefore devoted to anti-pollution measures at the design and planning stages of new plants. The aim now is to devise processes that do not produce by-products (which can become pollutants) in the first place, rather than having to develop containment and treatment methods for the by-product—which is the situation pertaining to most current processes.[3] It also makes economic sense, since unwanted by-products push up costs, and therefore reduce the profitability of processes in three ways. Firstly, they represent the loss of some feedstock. Secondly, there will be the cost of separating the by-product(s) from the desired product. If this is very energy demanding, like distillation for example, it can add significantly to the overall costs of the process. Finally, there is probably the cost of treating the by-product(s) so that they can be safely discharged as effluent.

A simple example is the process for the electrolysis of brine to give sodium hydroxide and chlorine. Older plants use mercury cells (section 6.4.2) which inevitably results in some loss of mercury in the effluent (about 10 g per tonne of chlorine produced) which is costly to replace. The possible environmental consequences were clearly spelled out in section 3.2.2, about the Minamata Bay incident, if levels of mercury become too high. As we see in chapter 6, the solution to the problem was to develop first diaphragm cells and now membrane cells. An additional attraction is their lower energy (electricity) consumption leading to reduced costs.

Let us conclude this section by looking at one company's efforts in limiting toxic chemical emissions into the environment and then considering what some national industry organisations are doing to encourage member companies to all work towards achieving a cleaner environment.

Monsanto claims to be on target to reducing air emissions of toxic chemicals by 90% in 1992 both in the USA and world-wide plants (based on emissions reported in 1987).[4] Another aim, to reduce toxic chemical releases into air, land and water underground injection wells by 70% by the end of 1995 (compared with 1990 releases) is also on target. Ironically, this has been in part by the enforced closure of one plant in Wales because of the world recession! Further plant shutdowns, to increase competitiveness, will also play a significant part. Another irony is that 25 of the 130 members of the environmental safety and health department will lose their jobs as part of these cutbacks! Monsanto have spent $100 million over the past four years on capital projects in the USA to reduce toxic air emissions. Although some processes gained economically, the overall result was that it had little or no effect on the company's profitability.

Interestingly, DuPont's chairman, Edgar S. Woolard, and Dow's chairman, Frank Popoff, produced a report suggesting that more efficient use of resources and elimination of wastes would actually promote economic development and environmental protection simultaneously 'because a business that is not sustainable will be destroying the very basis of its existence'.[5]

Turning to national schemes, the Dutch chemical industry and government have agreed on environmental measures to help implement the Dutch National Environment Plan, which will cut a wide range of emissions.[6] The agreement is a voluntary covenant, but back-up legislation is being prepared just in case industry does not deliver the goods. It includes the reduction of emissions, energy savings, reducing waste, soil clean-up and health and safety. Each company will draw up an individual environment plan showing how it is going to achieve its share of the cuts in emissions. Particular items included are sulphur dioxide, oxides of nitrogen and waste metals. It is planned to cut their emission by 70% (based on 1985 levels) by the year 2000.

Finally in the UK the trade body, the CIA (Chemical Industries Association), of which all the major companies are members, has a Responsible Care Programme and any new company wishing to join the association must agree to comply with this.

3.5 The way forward?

Most of the large chemical companies have published their environmental policies and strategies and are increasingly turning to environmental auditing to help them to fulfil their responsibilities. In the UK, Norsk Hydro broke new ground when it published an externally verified environmental report on its operations in 1990.

As we have already mentioned a major target with new processes is to 'design out' pollution by trying to ensure that unwanted products are not produced in the first place, thus improving the economic viability of the process.

The development of CFC replacements is worth considering briefly because it demonstrates just how much can be achieved if there is a strong enough will to succeed.

CFCs (chlorofluorocarbons) were hailed as wonder products, following Midgley's discovery of the first one, CFC 12 (CF_2Cl_2) in 1930, with their special properties such as chemical inertness and non-flammability. Large-scale uses for them as safe refrigerants, aerosol propellants and foam blowing agents (in polyurethane foam) soon followed.

However, in 1974 Sherwood Roland and Mario Molina first suggested that they might be responsible for destroying the ozone layer which surrounds the earth high up in the stratosphere, and which screens us from the sun's harmful UV-B radiation. Since the chemical reactions involved were taking place some 20–40 km above the earth's surface it is not easy to prove or disprove this theory. The situation is also made very complex because variations in ozone concentrations (up to 25%) occur naturally as a result of climatic changes.

After almost 30 years of extensive investigations and research a link between CFCs and the depletion of the ozone layer has been established, and it is, ironically, the great chemical inertness of the CFCs that is ultimately the cause of the problem (cf. DDT). Because of this they have very long lifetimes and therefore pass unchanged up into the stratosphere. There they are exposed to the short-wavelength UV-B radiation (hardly present at the earth's surface) which causes photolytic decomposition of the CFC molecule. In particular, chlorine atoms or radicals are produced, and it is these which attack and destroy the ozone molecules. Even worse, one chlorine atom can destroy many ozone molecules. It is important to note that some 300 million tonnes of ozone are created and destroyed by natural processes each day, but these largely balance each other. What the CFCs are doing is to tip the balance, only slightly but significantly each year, in favour of greater destruction. Our concern therefore is that diminution of the ozone shield will allow more dangerous UV-B radiation to reach the earth, causing a considerable increase in the incidence of skin cancers.

Concern for the problem was so great that it quickly led to Canada and Sweden banning the use of CFCs as propellants in non-essential aerosols, even though there were no data available to support Roland and Molina's hypothesis at the time. This was followed by an unprecedented event in 1987 with the first ever international agreement on an environmental issue, the signing of the Montreal Protocol by more than 50 nations, who between them produced 80% of the world's CFCs. They pledged to cap production and consumption at 1986 levels, and to reduce this by 20% in 1993 and 50% by 1998. Nations then vied with each other to do even better, resulting in the signing of a revised agreement by over 70 nations in 1990. This set the year 2000 as the deadline for the complete phasing out of CFCs and related chemicals, e.g. CCl_4.

The willingness to do even better resulted in the United Nations Ozone Layer Conference in Copenhagen in November 1992, at which 80 nations pledged essentially to complete the phase-out by 1996, although there were some essential-use exemptions. Friends of the Earth have criticised the latter as 'a loophole that could allow production to continue indefinitely'.

Since CFCs have such important uses, phasing them out by an agreed timetable was only one side of the coin, because this would clearly depend upon the availability of suitable replacements. Aerosol propellant replacements were soon available, e.g. butane, but it should be noted that the latter is inflammable, so one should take care when spraying hairspray! Suppliers have jumped on the Green bandwagon by labelling these products 'ozone friendly'. Finding alternatives for use as foam blowing agents and refrigerants has proved much more challenging. All the major manufacturers e.g. DuPont,

Hoechst and ICI, soon came up with the same list of possible compounds, so the challenge was to develop the most economic route as quickly as possible. These compounds are HCFCs (hydrochlorofluorocarbons), but since this name is close to that of CFCs the alternative HFAs (hydrofluoroalkanes) has now been adopted since it is far better from a public-relations point of view. It should be noted that a whole family of HFAs will be needed to encompass all the uses of CFCs, but initial efforts seem to have focused on HFA 134a (CF_3CH_2F) as a refrigerant. The normal timescale for such a project would be upwards of seven years. Remarkably, with the pressures to move very quickly resulting in the making available of enormous manpower (over 100 scientists and engineers in one company's team) and equipment resources at a cost in excess of £100 million, HFA 134a was being produced on a pilot plant within two to three years. There is an excellent account of what was involved in STEAM.[7] It required a new approach to R&D, in that engineers were involved with chemists almost from the start of the project, rather than being brought in only when the chemists had almost completed their work.

Both DuPont and ICI have plants producing HFA 134a in 1993, and a major factor in reaching this stage in such a short time was the co-operation on toxicity and ozone depletion tests. The companies jointly funded independent tests of this sort.

It will be noted that HFAs do not contain chlorine and their ozone depletion potential is therefore zero, i.e. they will not destroy ozone molecules. Also, the introduction of some hydrogen atoms into the molecules (CFCs have none) causes them to have a much shorter lifetime, because they will decompose in the earth's atmosphere before they reach the stratosphere.

3.6 Conclusions

We have looked in some detail at the public's concern over the environment and the consequences for the chemical industry plus its response. This has focused on the developed world, where the vast majority of the population receives an adequate supply of food and does not have to worry about starving. Also in a material sense they are adequately provided for, and in many cases well off. We can therefore turn our attention to thinking about and wanting a clean environment. In other words, we have a choice. However, it is worth emphasising again that there is a price to be paid. To illustrate this consider the growing of food. We have, quite rightly, banned DDT because of its adverse environmental effects. The more acceptable alternative insecticides which the farmer uses, e.g. organophosphates, are at least three times as expensive. This will be reflected in the higher price that we pay for our food.

Picture, in contrast, the poor farmer in Africa trying desperately to produce enough food to feed his family by protecting crops from the ravages of insect pests such as locusts. He certainly could not afford organophosphates, but if he could afford DDT he would use it if it could help him to produce enough food. Concern for its effect on the environment would not enter his thoughts—i.e. he has no choice. Thus our views are affected very much by our circumstances and situation.

The greater willingness of companies to improve their environmental performance and adopt a more positive and open approach with the public over issues raised is to be applauded and welcomed. There still remain many areas where improvement is needed, but we would hope for less problems with plants built in the future, since we have noted the attempts to 'design out' potential pollutants.

It is to be hoped that the type of environmental disasters of the past that were discussed in section 3.2 will not occur in the future because of the lessons learned and the stricter precautions which should be in place.

Finally, it is important to emphasise yet again the need to place the problems in context against all the many processes that are operated daily without causing problems since, as we have seen already, overemphasising just a few disasters leads to a very negative image. The total facts produce a much more positive picture—but with some blemishes.

The remarkable inter-company, national and international co-operation which permeates the development of CFC alternatives shows the way forward and provides optimism for a significant contribution by the chemical industry to improving the quality of the environment.

References

1. 'Thalidomide and the Power of the Drug Companies,' H. Sjostrom and R. Nilsson, Penguin, 1972.
2. 'The Flixborough Disaster: Report of the Court of Inquiry,' R.J. Parker *et al.*, HMSO, 1975.
3. D. Rotman, 'Successes Emerge in Search for Cleaner Processes,' *Chemical Week*, 9 Dec. 1992, 66.
4. M. Reisch, 'Monsanto's Environmental Progress Comes at High Cost,' *Chem. & Eng. News*. 14 Dec. 1992, 15.
5. *Chem. & Eng. News*, 27 April, 1993, 28.
6. 'Pollution Covenant Agreed,' *Chem. & Ind.*, 1 March, 1992, 135; see also *Chem. & Ind.*, 1989, 622.
7. 'CFCs—The Search for Alternatives,' *STEAM* (ICI Science Teachers' Magazine), No. 12, January 1990, 14.

Bibliography

'The Poison that Fell from the Sky,' J.G. Fuller, Random House, 1978.
'The Chemical Scythe,' Alastair Hay, Plenum Press, 1982.
'Thalidomide and the Power of the Drug Companies,' H. Sjostrom and R. Nisson, Penguin 1972.
'The Flixborough Disaster: Report of the Court of Inquiry,' R.J. Parker *et al.*, HMSO, 1975.
'The Chemistry of the Atmosphere,' M.J. McEwan and L.F. Phillips, Edward Arnold, 1975.

Polymers 4

Jack Candlin

Polymers are all around us. They are the main components of food (starch, protein), clothes (silk, cotton, polyester, nylon), dwellings (wood—cellulose, paints) and also our bodies (nucleic acids, polysaccharides, proteins). No distinction is made between biopolymers and synthetic polymers. Indeed many of the early synthetic polymers were based upon naturally occurring polymers, e.g. celluloid (cellulose nitrate), vulcanisation of rubber, rayon (cellulose acetate). These materials will not be discussed here.

Polymers are constructed from monomer units, connected by covalent bonds. The definition of a polymer is (Carothers 1929):

'a substance, —R—R—R—R— (or, in general —$[R]_n$—, where R is a bifunctional entity (or bivalent radical) which is not capable of a separate existence'

where n is the *degree of polymerisation*, DP_n. This definition excludes simple organic and inorganic compounds, e.g. CH_4, NaCl, and also excludes materials like diamond, silica and metals which appear to have the properties of polymers, but are capable of being vaporised into monomer units.

The molecular weight (MW) (strictly relative molecular mass) can be obtained from the MW of the monomer (or repeat unit) multiplied by n. Thus the MW of CH_4 or NaCl is 18 or 58.5 respectively, whereas the MW of a polymer can be $> 10^6$. When the value of n is small, say 2–20, the substances are called *oligomers*; often these oligomers are capable of further polymerisations and are then referred to as macromers.

By definition, 1 mole of a polymer contains 6×10^{23} polymer molecules and therefore 1 g mole = MW of the polymer in grams, which, in theory, can be $> 10^6$ g. However, by convention, 1 g mole usually refers to the MW of the repeat unit; thus 1 g mole of polyethylene —$(CH_2)_n$— is taken as 14 g (the end groups, being negligible, are ignored).

A polymer with a MW of 10^7, if fully extended, should have a length of ~ 1 mm and a diameter of ~ 0.5 nm. This is equivalent in size to uncooked spaghetti ~ 2 km in length. However, in reality, in bulk polymers the chain is never fully extended—a random coil configuration is adopted sweeping out a space of diameter ~ 200 nm. It therefore has the appearance of cooked spaghetti or worms (or more correctly, worms of different length). The movements of these polymer chains are determined by several factors, such as:

1. Temperature
2. Chemical make-up of the backbone —C—C—C— chain; whether the chain is flexible (aliphatic structure) or rigid (aromatic)
3. The presence or absence of side-chains on the backbone
4. The *inter*-polymer chain attraction (weak—dipole/dipole, H-bonding—or strong—covalent bonds, cross-linking)
5. The MW and molecular weight distribution (MWD) of the polymer.

Nearly all of the properties of polymers can be predicted if the above factors are known, e.g. whether the polymer is amorphous or partially crystalline; the melting temperature of the crystalline phase (T_m) (actually it is more of a softening temperature over several degrees); is the polymer brittle or tough; its rigidity or stiffness (called modulus), whether the polymer dissolves in solvents, etc.

Polymers are really effect chemicals in that they are used as materials, e.g. plastics, fibres, films, elastomers, adhesives, paints, etc., with each application requiring different polymer properties. Many of the initial uses of plastics were inappropriate, which led to the belief that plastics were 'cheap and nasty'. However, recent legislation on product liability and a better understanding of the advantages and disadvantages of plastics have changed this position.

Economics, that is the cost of making and fabricating the polymer is of prime importance. This has led to a rough grouping of polymers into commodity polymers, engineering polymers, and advanced polymeric materials.

4.1.1   Commodity polymers

Examples of these are:

Polyethylene $\begin{cases} \text{low density polyethylene (LDPE)} \\ \text{high density polyethylene (HDPE)} \\ \text{linear low density polyethylene (LLDPE)} \end{cases}$

Polypropylene (PP)
Poly(vinyl chloride) (PVC)
Polystyrene (PS)

Each of these is prepared on the 10 million tonnes/year scale. The price is < £1000/tonne.

4.1.2   Engineering polymers

These materials have enjoyed the highest percentage growth of any polymers over the last ten years and are principally used as replacements for metals under moderate temperature ($<150\,°C$) and environmental conditions, or they may have outstanding chemical inertness and/or special properties, e.g. low friction—polytetrafluoroethylene (PTFE). These engineering polymers include:

Acetal (or polyoxymethylene, POM)
Nylons (polyamides)
Polyethylene or polybutylene terephthalate (PET or PBT)
Polycarbonate (of bisphenol A) (PC)
Polyphenylene oxide (PPO) (usually blended with styrene).

The prices are £2000–£10 000/tonne

4.1.3 Advanced polymeric materials

These have very good temperature stability (many hours/days at 250–300 °C) and when reinforced with fibres (e.g. glass, carbon or Aramid (Kevlar) fibres), i.e. composites, they are stronger than most metals on a weight/weight basis. They are usually only used sparingly, often in critical parts of a structure. Their price can be as high as £100 000/tonne.

4.2 Molecular weight of polymers

A polymer is made up of a mixture of molecules of different MWs. It is possible to fractionate polymers (e.g. by dissolving in a solvent and cooling, precipitating the highest-MW material first) and measure the MW (e.g. by any colligative property such as osmotic pressure). When the various weight fractions are plotted against the MW of these fractions, the plot shown in Figure 4.1 is obtained.

This used to be a laborious process, but it did give an idea of the MW spread or distribution (MWD). Also included in the above graph are the values:

$\bar{M}_n$ number average MW
$\bar{M}_v$ viscosity average MW
$\bar{M}_w$ weight average MW
$\bar{M}_z$ 'z' average MW.

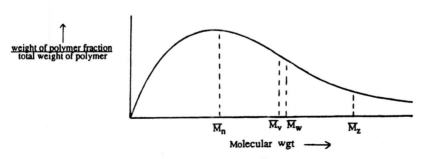

Figure 4.1   Typical molecular weight plot of a polymer.

These are the values for MW if they were carried out on the whole polymer (i.e. without fractionation). Not all of the methods used for determining MW gave the same average MW. Thus it was found that:

MW obtained from end-group analysis or any colligative method (e.g. osmometry, b.p. elevation, f.p. depression) is $\bar{M}_n$ (number average)

MW obtained from solution viscosity (which was very close to $\bar{M}_w$) is $\bar{M}_v$ (viscosity average)

MW obtained from light scattering is $\bar{M}_w$ (weight average).

The ultracentrifuge technique gives both $\bar{M}_w$ and $\bar{M}_z$. $\bar{M}_z$ gives a bias towards the highest MW fractions making up the polymer. $\bar{M}_z$ is useful because some mechanical properties of the polymer can be correlated more with $\bar{M}_z$ than with $\bar{M}_n$ or $\bar{M}_w$.

4.2.1   Gel permeation chromatography (GPC)

In recent years the situation has been transformed by the introduction of *gel permeation chromatography* or *size exclusion chromatography*, where separation–fractionation takes place in a chromatographic column filled with beads of a rigid porous gel, e.g. cross-linked polystyrene, glass or silica. The pores in the gels are of the same spatial size range as the polymer molecules in the sample being analysed.

With an analytical instrument (where a distribution curve is produced without isolating individual polymer fractions) a sample (1 cm^3) of a dilute solution of the polymer ($\sim 0.5\%$) is injected into a solvent stream flowing through the column. Polymer molecules that have a spatial size greater than the maximum pore size in the gel pass straight through the column via the interstitial volume at the same speed as the solvent flow. Smaller polymer molecules will have access to pores in the gel and will become distributed between the mobile solvent phase and the stationary solvent phase trapped in the pores of the gel. The smaller the polymer molecule the greater the volume of stationary solvent phase available to it and the longer the molecule spends on the column, i.e. the greater its retention time or elution volume. A suitable detector system (e.g. a differential refractometer) connected to the column outlet indicates when polymer is present in the eluate and its concentration.

For a particular solvent, column and temperature, polymer molecules of a given spatial size will leave the column at a fixed retention/elution volume, provided the polymer chains are flexible. Thus, the plot of detector response is a measure of the MW distribution of the sample.

Before this plot can be converted into a differential weight distribution curve, the column must be calibrated. Ideally, this is done by passing through the column an ascending series of near-monodisperse fractions of the polymer under test. If these are not available, then standard and near-homogeneous fractions of polystyrene, which are readily available, may be used and the data converted into a calibration curve for the polymer of interest.

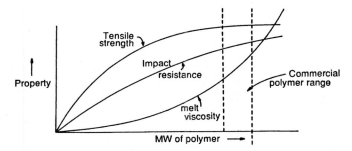

Figure 4.2   Dependence of properties on molecular weight.

It is now possible to obtain instruments where the detector system contains a low-angle laser light scattering photometer (LALLS) and, since this is capable of determining the MWs of the polymer species leaving the column directly, the need for column calibration does not arise.

As in all chromatographic processes, the band of solute leaving the column at a particular moment has been broadened by various processes. Corrections for zone broadening may be made but are relatively unimportant for polymer samples where $\bar{M}_w/\bar{M}_n > 2$. The technique has a wide range of applicability, and preparative instruments, where fractions of polymer can be isolated, are available. With cross-linked polystyrene gels a relatively non-polar solvent must be used, and tetrahydrofuran is the most favoured one. The results of GPC compare very favourably with those from more long-standing techniques. GPC gives $\bar{M}_n$, $\bar{M}_w$ and the MWD curve (Figure 4.1). If only the $\bar{M}_n$ and $\bar{M}_w$ values are known, then the ratio $\bar{M}_w/\bar{M}_n$ gives the *polydispersity* of the polymer, and the value gives an indication of the broadness of MWD; e.g. if $\bar{M}_w/\bar{M}_n = 1$, this corresponds to a perfect hypothetical monodisperse polymer, but ratios of 20–50 mean that the polymer has a very broad MWD. It cannot be stated whether large or small values of $\bar{M}_w/\bar{M}_n$ are good or bad—it depends on the particular application of the polymer. However, the absolute value of $\bar{M}_n$ or $\bar{M}_w$ is important because the polymer has to reach a certain MW before good mechanical properties become manifest. (see Figure 4.2).

**4.3
Chemistry of
polymerisation**

There are two different modes of polymerisation, the distinction being based on the kinetics of polymerisation. They are step-growth polymerisation and chain-growth polymerisation.

4.3.1   Step-growth polymerisation

This was originally called condensation polymerisation. It involves the reaction of di- (or more) functional monomers, often via a condensation reaction, e.g.

$$\text{HOOC}-\langle\bigcirc\rangle-\text{COOH} + \text{HOCH}_2\text{CH}_2\text{OH} \xrightarrow[\text{GeO}_2]{-\text{H}_2\text{O}} \left[\text{OC}-\langle\bigcirc\rangle-\text{COOCH}_2\text{CH}_2\text{O}\right]_n$$

terephthalic acid   polyethylene terephthalate
                                                                    (PET)

In actual fact it is only recently that terephthalic acid has been made pure enough to be used as a monomer. Originally, an ester exchange reaction involving dimethylterephthalate (DMT) and ethylene glycol to give di(hydroxethyl)terephthalate (DHT) was used to make PET.

$$\text{MeOOC}-\langle\bigcirc\rangle-\text{COOMe} + \text{HOCH}_2\text{CH}_2\text{OH}\,| \xrightarrow{\text{Ca(OAc)}_2} \text{HOCH}_2\text{CH}_2\text{OOC}-\langle\bigcirc\rangle-\text{COOCH}_2\text{CH}_2\text{OH}$$

DMT

$$\downarrow \text{DHT}$$

PET

Other step-growth polymerisations include nylon 6,6 (polyamide) production.

$$\text{HOOC(CH}_2)_4\text{COOH} + \text{H}_2\text{N(CH}_2)_6\text{NH}_2 \xrightarrow{-\text{H}_2\text{O}}$$

$$\left[\text{OC(CH}_2)_4\text{CONH(CH}_2)_6\text{NH}\right]_n$$
Nylon 6,6

There are quite a few nylons, e.g. 6,10; 4,6, the numbers representing the number of carbon atoms in the diamine and diacid respectively. Because in step-growth polymerisations absolute 1:1 stoichiometry of the monomers is essential to obtain a high-MW polymer, in commercial practice the nylon salt is made in water, the pH of which (7.6) guarantees a 1:1 ratio.

$$[\text{H}_3\text{N(CH}_2)_6\text{NH}_3]^{2+}\,[\text{OOC(CH}_2)_4\text{COO}]^{2-} \xrightarrow[-\text{H}_2\text{O}]{\sim 280\,^{\circ}\text{C}} \text{Nylon 6,6}$$
Nylon salt

Nylon 6 is made from caprolactam, not by a step-growth, but by a chain-growth mechanism (see section 4.3.2).

$$\begin{matrix}\text{NHCO}\\(\text{CH}_2)_5\end{matrix} \xrightarrow[\text{catalyst}]{\text{anionic}} \left[(\text{CH}_2)_5\text{CONH}\right]_n$$
Nylon 6)

Many of the advanced polymers are made by step-growth polymerisations by elimination of an inorganic salt.

$$\text{Cl}-\langle\bigcirc\rangle-\text{SO}_2-\langle\bigcirc\rangle-\text{Cl} + \text{KO}-\langle\bigcirc\rangle-\text{OK} \xrightarrow[\substack{\text{polar}\\\text{aprotic}\\\text{solvents (eg DMSO)}}]{-\text{KCl}} \left[\langle\bigcirc\rangle-\text{SO}_2-\langle\bigcirc\rangle-\text{O}\right]_n$$

polyether sulphone
(PES)

A problem with step-growth polymerisations is that the viscosity of the reaction media increases making the removal of the last traces of the condensation product difficult; in the salt-formation reactions these drawbacks do not arise.

Another reason why condensation polymerisations were reclassified as step-growth was because some reactions did not involve condensation products, e.g.

$$\sim NCO + HO \sim \longrightarrow \sim NHCOO \sim$$

Isocyanate · · · · · · · · · · · · · · · · Urethane

$$\sim CH{-}CH_2 + H_2N \sim \longrightarrow \sim \underset{\underset{OH}{|}}{CH}{-}CH_2NH \sim$$

Epoxide or epoxy (oxirane) · · · · · · · · · · Epoxy resin

The principle behind step-growth polymerisation is that the condensation reactions do not alter in chemical reactivity throughout the course of the polymerisation. However, the rate at which the reagents (monomers) react will decrease for physical reasons, e.g. the reaction is second-order; the concentrations of the reactive groups are decreasing during polymerisation; and the viscosity of the system increases.

Thus the initial stages of the polymerisation will be:

Monomer + monomer $\longrightarrow$ dimer
Dimer   + monomer $\longrightarrow$ trimer
Dimer   + dimer $\longrightarrow$ tetramer
Dimer   + trimer $\longrightarrow$ pentamer (etc.)

In general:

n-mer   + m-mer $\longrightarrow$ $(n + m)$mer

Polymers of high MW only occur towards the end of the polymerisation (see chain growth, Figure 4.3).

Summarising the main features of step-growth polymerisation, we can see that:

1. The monomers must have a functionality (reactive groups) of two (or more if cross-linking is desired).
2. Any two molecular species can react together (i.e. monomer, dimer, trimer, etc.) and when they do so, they all react at the same rate and via the same mechanism.

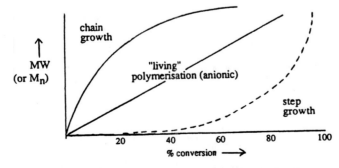

Figure 4.3 Step-growth and chain-growth polymerisation.

3. Reactions are usually slow at ambient temperatures but can be accelerated at higher temperatures without affecting the final MW of the product, other than by decreasing the time it takes to obtain it. Care must be taken to avoid non-polymerisation reactions that may occur at higher temperatures, e.g.

$$\sim CH_2-CH_2OH \xrightarrow{-H_2O} \sim CH=CH_2$$

4. The condensation reactions will have activation energies similar to normal organic reactions and the heats of reaction (polymerisation) will not be too excessive (cf. chain growth), i.e. heat removal will not be a problem in commercial production.

5. The MW of the product will increase steadily throughout the polymerisation (which can take several hours), increasing sharply towards the end of the reaction. It can be shown that the average degree of polymerisation (DP_n) is related to the fractional extent of the reaction (p) by the Carothers equation:

$$\overline{DP_n} = 1/(1-p)$$

and Table 4.1 can be constructed. It can be seen that even at 50% conversion $(p = 0.5)$ the average DP_n is only 2, i.e. dimers. The Carothers equation means that:

6. If a DP_n of say 500 is required (i.e. 99.8% conversion), then the reagents must be at least 99.8% pure.

7. The two reagent monomers must be in the correct molar stoichiometric proportions—at worst 100:99.8; this aspect can be accommodated if a monomer/monomer adduct can be made which is 1:1 (e.g. nylon 6,6 salt, see above, or a cyclic compound can be made which can be readily converted to difunctional monomer (e.g. caprolactam, which on hydrolysis gives $H_2N(CH_2)_5COOH$).

8. Up to now it has been assumed that the products of condensation (e.g. H_2O) are removed during the step-growth polymerisation, i.e. they are

Table 4.1 Relation between p and DP_n by the Carothers equation

% Conversion	p	DP_n	MW* of nylon 6,6
50	0.500	2	226
75	0.750	4	452
90	0.900	10	1 130
99	0.990	100	11 300
99.8	0.998	500	56 500
99.9	0.999	1000	113 000

MW* = MW of repeat unit multiplied by DP_n, and the repeat unit is $\text{+HN(CH}_2)_6\text{NHOC(CH}_2)_4\text{CO+}$ divided by 2, i.e. MW* = 113.

open systems. Thus, for the polyester equilibrium:

$$\sim COOH + HO \sim \overset{K}{\rightleftharpoons} \sim COO \sim + H_2O$$

if K is the equilibrium constant (it will have roughly the same value as for mono-acids and mono-alcohols), then it can be shown that:

$$\overline{DP_n} = K^{\frac{1}{2}} + 1$$

Therefore in order to get a reasonable DP_n, K must be large, e.g. $K = 1000$, and therefore $DP_n = 1000^{\frac{1}{2}} + 1 = 33$. However, the values for K for mono-ester formations are ~ 1–100, which will give too small values of DP_n; even the value of 33 above will not give a polyester with good mechanical properties. These systems are, of course, closed systems in which the condensation product is not removed, whereas in practice these are removed (often by vacuum) to push the equilibrium over to the right-hand side. This limitation can also be overcome by making the step-growth polymer via salt-formation (see above), in which the values of K are large and the salt need not be removed during polymerisation. It must be removed at the end of the polymerisation as it may interfere with the polymer properties (e.g. make it less stable, prone to oxidation, hydrolysis, etc.).

9. Most polymers are fabricated via melt processing and from Figure 4.2 it can be seen that the melt viscosity increases with MW. Although a worry with step-growth polymers is to get a high enough MW, a reproducible MW is essential for constant properties. Rather than just rely on purity, time, temperature, viscosity, etc., monofunctional reagents are often added for MW control (e.g. with nylon 6,6, 1 mole % acetic acid is added; in theory this is equal to limiting the polymerisation to 99% conversion).

4.3.2   Chain-growth polymerisation

Chain-growth polymerisation (unlike step-growth) is not usually spontaneous—it requires an initiator or catalyst, which may be adventitiously present (e.g. traces of oxygen, acids, etc.). The initiator can react with a monomer to form an activated monomer, which can further react with another monomer to form an activated dimer, etc.

There are thus three basic steps in chain growth polymerisations

1. Initiation or activation of the monomer
2. Propagation via activated growing chains
3. Termination.

Most alkenic or olefinic polymerisations (containing vinyl groups, $CH_2=CH-X$) proceed through this mechanism, giving a polymer with a $-C-C-C-$ backbone, although other polymers containing hetero-atoms can be formed via chain growth (especially from cyclic monomers). The activated intermediates are normally free radicals, anions, cations or co-ordination complexes. The structure of chain-growth polymers can be represented in various ways:

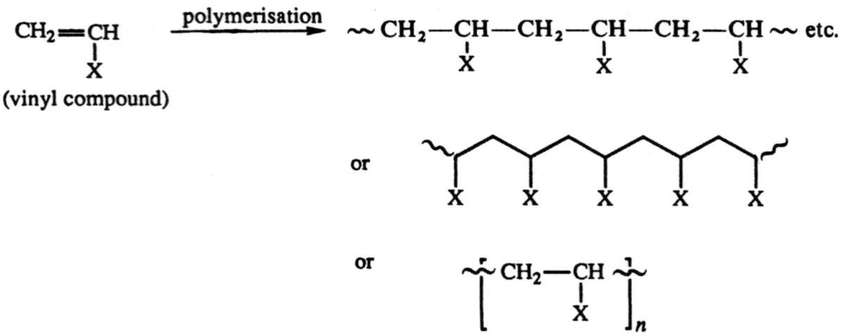

The nature of X determines which of these intermediates are most likely to promote polymerisation. X also governs the electron density of the vinyl group and therefore dictates whether the vinyl group wants to receive electrons (and become anionic) or lose electrons (and become cationic). Thus electron-withdrawing groups X, i.e.  , will tend to promote polymerisation via an anionic mechanism, the anionic intermediate being the most preferred, and electron-donating groups 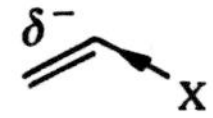 will promote polymerisation via a cationic intermediate, because a cationic intermediate is the most favoured.

If the X group is not strongly electron-withdrawing or -donating, then another pathway is available, free-radical polymerisation. This can occur if:

(a) The intermediate free radical is stable enough to act in the propagation step, through resonance structures, e.g. styrene

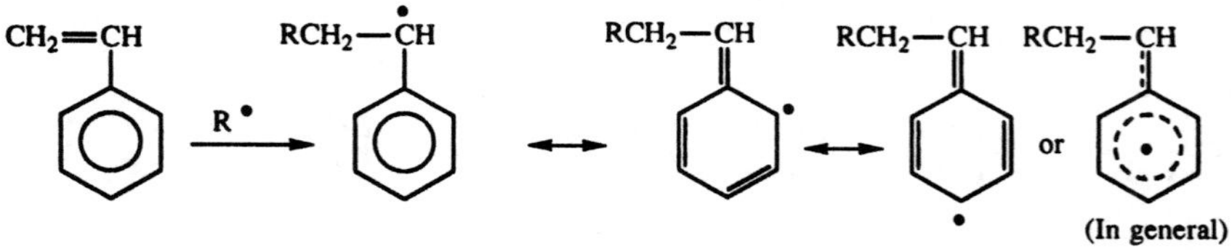

If the intermediate free radical is inherently unstable, the propagation step can be promoted by having the monomer in very high concentration (e.g. $CH_2{=}CH_2$ polymerises via a free-radical mechanism if the concentration of $CH_2{=}CH_2$ is high enough, i.e. at very high pressure $> 2\,000$ atm).

(b) If the ionic route is not possible because the X-group reacts with the anionic or cationic initiators (e.g. vinyl chloride ($CH_2{=}CHCl$) reacts with anionic initiators e.g. Li^+Bu^- to form LiCl).

If the vinyl compound contains no hetero atom, then co-ordination polymerisation is possible. It is called this because the initial step is the coordination of the olefinic bond to a (usually transition) metal centre via a π-donor bond from the monomer

$$M^{n+} + \overset{}{\underset{\substack{\text{(Transition} \\ \text{metal ion)}}}{}} \diagup\!\!\!\diagdown X \longrightarrow \overset{M^{n+}}{\underset{(\pi\text{ - bond)}}{}} \diagup\!\!\!\diagdown X$$

This route is quite limited, since if the X group contains a hetero atom (e.g. O, N, Cl) this will also co-ordinate to the metal centre and displace the vinyl group co-ordination. This means that only unsubstituted vinyl compounds can be polymerised by this method. This is good, because these monomers are often difficult to polymerise via free-radical or ionic routes.

Table 4.2 shows which monomers polymerise by which route. Free-radical polymerisation is the most widespread pathway for vinyl linear growth polymerisation; roughly half of the total output of polymers is made in this way. It is also the cheapest to operate on a commercial scale. Notice that styrene can be polymerised by all routes. This is because the phenyl ring can sufficiently stabilise the growing chain intermediate (free-radical, ionic or co-ordination) for propagation to occur with an incoming monomer.

Remember that it is the charge on the propagating species which governs whether polymerisation is by an anionic or a cationic mechanism.

4.3.2.1   Free-radical polymerisation.

The initial step is the generation of a free radical, which can interact with the monomer to generate another free radical. The most commonly used initiators are organic peroxides, azo compounds and aqueous redox systems (eg M^{n+}/H_2O_2). The thermal decomposition of peroxides and azocompounds has been thoroughly studied.

$$R-O-O-R \longrightarrow 2RO\cdot$$

$$R-N=N-R \longrightarrow 2R\cdot + N_2\uparrow$$

The decompositions are all first-order and therefore the rate of generation and hence the concentration of these free radicals can be calculated. This is important because the monomer/free radical ratio determines the MW and the extent of branching in the polymer.

The initiators in Table 4.3 are oil- (and monomer-) soluble; the following are water-soluble

$$HO-OH \longrightarrow 2HO\cdot$$

$$[O_3S-O-O-SO_3]^{2-} \longrightarrow 2SO_4^{\cdot-}$$
$$\text{\small Persulphate}$$

These systems are often used in conjunction either with transition-metal catalysts, e.g. Fenton's reagent (called a redox system):

$$H_2O_2 + Fe^{2+} \longrightarrow Fe^{3+} + HO^- + HO\cdot$$

Table 4.2   Initiation modes of various monomers

Monomer	Initiator			
	Free radical	Anionic	Cationic	Co-ordination
Ethylene 	✓			✓
Propylene (and other α-olefins) 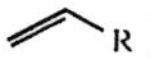				✓
Isobutylene 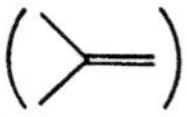			✓	
Styrene 	✓	✓	✓	✓
Butadiene and isoprene 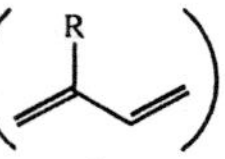	✓	✓		✓
Acrylates and methacrylates 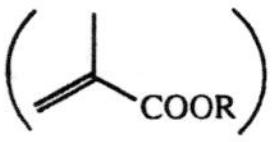	✓	✓		
Acrylonitrile 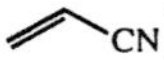	✓	✓		
Vinyl ethers 			✓	
Vinyl halides 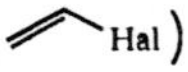	✓			
Fluorocarbons (e.g. TFE, $CF_2{=}CF_2$)	✓			
Vinyl esters (e.g. acetate) 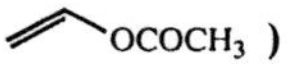	✓			
Formaldehyde ($CH_2{=}O$)			✓	
Formaldehyde trimer (trioxan) 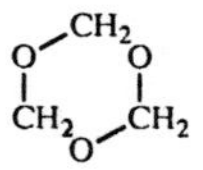		✓		
Ethylene oxide 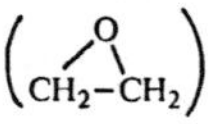		✓		
Cyclic ethers (e.g. THF) 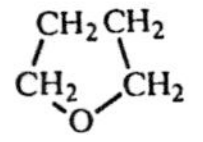			✓	
Cyclic lactams and lactones 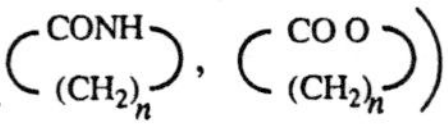		✓		
Cyclic siloxanes ($R_2SiO_{3\text{ or }4}$)		✓		

or with reducing agents, e.g. sulphites or mercaptans (thiols):

$$[S_2O_8]^{2-} + 2RSH \longrightarrow 2[HSO_4]^- + 2RS\cdot$$

Table 4.3 also includes the temperature required to obtain a 10-hour half-life. Although the actual polymerisations do not take this length of time (normally 1–2 h residence time) it does give an indication of the temperature at which to carry out the polymerisation. The choice of the initiator depends on which monomer is used and also which polymerisation method is used. Thus for bulk

Table 4.3 Examples of common free-radical initiators

Initiator	Temperature required for a 10-hour half life (guide to polymerisation temperature)	Polymerisation temperature range
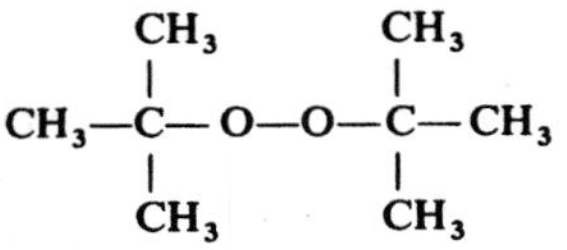 $$CH_3\text{—}\underset{\underset{CH_3}{\mid}}{\overset{\overset{CH_3}{\mid}}{C}}\text{—O—O—}\underset{\underset{CH_3}{\mid}}{\overset{\overset{CH_3}{\mid}}{C}}\text{—}CH_3$$ Di-*tert* butyl peroxide	126 °C	80–150 °C
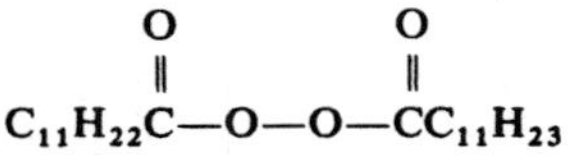 $$C_{11}H_{22}\overset{\overset{O}{\parallel}}{C}\text{—O—O—}\overset{\overset{O}{\parallel}}{C}C_{11}H_{23}$$ Dilauryl peroxide	62 °C	50–80 °C
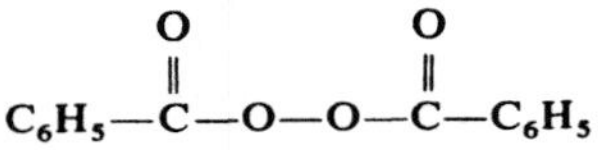 $$C_6H_5\text{—}\overset{\overset{O}{\parallel}}{C}\text{—O—O—}\overset{\overset{O}{\parallel}}{C}\text{—}C_6H_5$$ Dibenzoyl peroxide	73 °C	40–100 °C
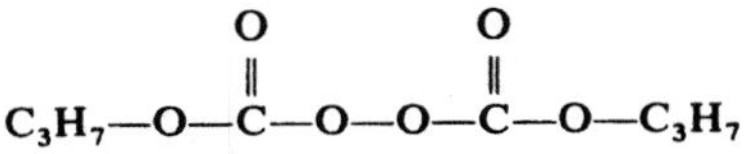 $$C_3H_7\text{—O—}\overset{\overset{O}{\parallel}}{C}\text{—O—O—}\overset{\overset{O}{\parallel}}{C}\text{—O—}C_3H_7$$ Dipropyl peroxydicarbonate	50 °C	Room temperature–70 °C
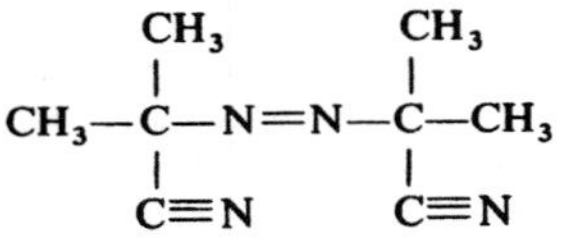 $$CH_3\text{—}\underset{\underset{C\equiv N}{\mid}}{\overset{\overset{CH_3}{\mid}}{C}}\text{—N=N—}\underset{\underset{C\equiv N}{\mid}}{\overset{\overset{CH_3}{\mid}}{C}}\text{—}CH_3$$ Azobisisobutyronitrile (AIBN)	64 °C	40–90 °C

or suspension polymerisation (see section 4.5.1) an oil- or monomer-soluble initiator is used, whereas with emulsion polymerisations water-soluble peracids or redox systems are most suitable.

Free-radical polymerisation proceeds as follows.

(1) *Initiation*
   (a) Initiator formation $(I \rightarrow R\cdot)$

   e.g. $(PhCOO)_2 \longrightarrow 2PhCOO\cdot \longrightarrow Ph\cdot + CO_2$

   (b) Chain initiation

   $$R\cdot + CH_2{=}CHX \longrightarrow RCH_2\dot{C}HX$$

(2) *Propagation*
   (a) Chain propagation

   $$RCH_2\dot{C}HX \xrightarrow{CH_2=CHX} RCH_2CHXCH_2\dot{C}HX$$

   $$\xrightarrow{n(CH_2=CHX)} R(CH_2CHX)_nCH_2\dot{C}HX$$

(b) Chain transfer (sometimes linked with chain termination)

$$\sim CH_2CHX \sim + R(CH_2CHX)_nCH_2\dot{C}HX \longrightarrow$$

Dead polymer molecule   Growing chain

$$\sim CH_2\dot{C}X \sim + R(CH_2CHX)_nCH_2CH_2X$$

Dead polymer

$$\Big| \quad CH_2{=}CHX$$

$$\downarrow$$

$$\sim CH_2CX \sim$$

$$\Big|$$

$$CH_2\dot{C}HX \quad \longrightarrow \text{ etc.}$$

Growing branched polymer chain

(3) *Termination*

$$R(CH_2CHX)_nCH_2\dot{C}HX$$

Growing chain

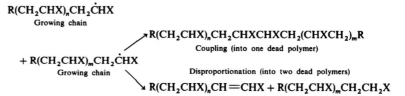

$R(CH_2CHX)_nCH_2CHXCHXCH_2(CHXCH_2)_mR$

Coupling (into one dead polymer)

$$+ R(CH_2CHX)_mCH_2\dot{C}HX$$

Growing chain

Disproportionation (into two dead polymers)

$R(CH_2CHX)_nCH{=}CHX + R(CH_2CHX)_mCH_2CH_2X$

The free radical is generated and then attacks a monomer giving another free radical, which then attacks a further monomer, etc. (propagation). The initiating radical is thus part of the polymer (R in the above scheme). Its concentration is so small that it rarely affects the mechanical properties; however, its concentration should be kept to a minimum since it may affect the polymer's stability to heat and UV light.

In the propagation step, normally a head-to-tail sequence is followed, i.e.   ⟋X ⟋X ⟋X ⟋X ;   rarely   does   head-to-head

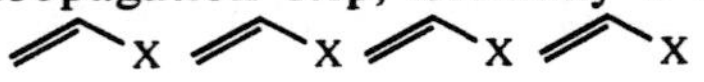

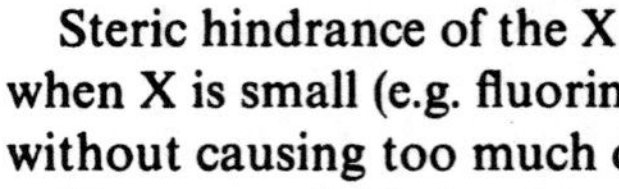

⟋X ⟋X ⟋X X   or   tail-to-tail   X ⟋X ⟋X ⟋X take place. This is because the free radical on the growing chains is $\sim CH_2{-}\dot{C}H$ (and not $\dot{C}H_2{-}CH \sim$, which would give a tail-to-tail polymer)

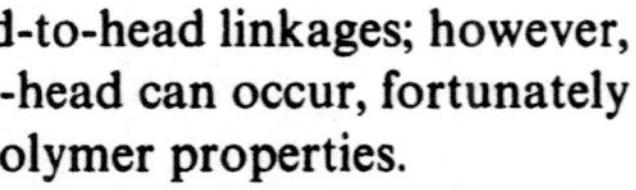

$$\underset{X}{\Big|} \qquad\qquad \underset{X}{\Big|}$$

since it is the more stable secondary radical.

Steric hindrance of the X groups prevents head-to-head linkages; however, when X is small (e.g. fluorine) up to 10% head-to-head can occur, fortunately without causing too much deterioration in the polymer properties.

However, single head-to-head linkages do occur as a result of termination coupling (see above). These defects should be kept small, since they can alter the stability of the polymer (weak links) or they may stop the polymer from crystallising (see section 4.6.3).

The propagation steps are rapid, with many thousands occurring in a few seconds. Termination is a bimolecular process involving coupling or disproportionation, depending on the nature of the radicals (e.g. styrene polymerisation tends to terminate by coupling whereas methyl methacrylate polymerisation terminates by disproportionation) and the reaction temperature (the higher the temperature the more disproportionation).

4.3.2.2 Anionic polymerisation.

4.3.2.2 Anionic polymerisation. Here the propagating species is an anion (see Table 4.2 for examples). This is less used in industry than free-radical polymerisation, principally because many monomers do not undergo anionic polymerisation, but also because rigorously pure monomers have to be used and also strict temperature control is necessary. However, when applicable, anionic polymerisation does give good control of the MW of the polymer with little branching.

The initial step is the generation of an anion with a strong base B^-

$$A^+B^- + CH_2=\underset{\underset{X}{|}}{C} \longrightarrow A^+[B-CH_2-\underset{\underset{X}{|}}{CH}]^-$$

In order to transfer an electron to the monomer, the basic strength of B^- must be high. Thus for vinyl compounds, ordinary bases like OH^- are too weak, and alkali metals or alkali-metals alkyls must be used, e.g. $Na^{\cdot}$ or Li^+Bu^-

$$Na^{\cdot} + CH_2=\underset{\underset{X}{|}}{C} \longrightarrow Na^+[CH_2-\underset{\underset{X}{|}}{CH}]^-$$

$$Li^+Bu^- + CH_2=\underset{\underset{X}{|}}{CH} \longrightarrow Li^+[Bu-CH_2-\underset{\underset{X}{|}}{CH}]^-$$

For cyclic monomers containing a heteroatom, the weaker alkali-metal hydroxides (or alkoxides) are suitable

$$RO^-Na^+ + \overset{\overset{\displaystyle O}{\diagup\diagdown}}{CH_2-CH_2} \longrightarrow ROCH_2CH_2O^-Na^+$$

Further examples of ring-opening polymerisation by anionic intermediates include:

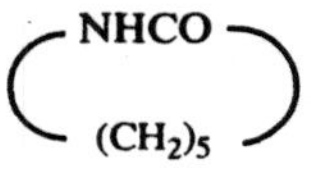

Caprolactam

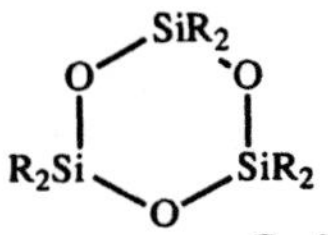

Cyclic siloxane
(using NaOR catalyst)

Interestingly, if caprolactam is hydrolysed to give $H_2N(CH_2)_5COOH$, then the polymerisation changes from chain-growth to step-growth polymerisation.

In the above examples, the initial anionic monomer is chemically similar to the original initiator (Li alkyls and Na alkoxides in the two examples above) and hence will react with another monomer; therefore propagation takes place, e.g.

$$\text{Li}^+[\text{BuCH}_2\text{—CH}]^- \xrightarrow{\quad X \quad} \text{Li}^+[\text{BuCH}_2\text{—CHCH}_2\text{CH}]^- \longrightarrow \text{etc.}$$
$$\underset{X}{|} \qquad\qquad\qquad \underset{X}{|} \;\; \underset{X}{|}$$

The unique nature of anionic polymerisation is that when performed in a non-reacting solvent and in the absence of impurities, there is no termination step (although this has also been observed for some cationic polymerisations). This type of polymerisation is therefore called *living polymerisation*. This phenomenon has several consequences:

(a) The DP_n can be determined from the stoichiometric ratio of the monomer versus the anionic initiator, e.g. if the mole ratio for monomer/activator is 200/1, then the DP_n will be 200

(b) Because all the chains grow at the same rate, the MW distribution (MWD) can be made very narrow by careful addition of a soluble initiator to the monomer with efficient mixing

(c) The MW of the polymer is directly proportional to conversion (cf. non-living step- and chain-growth polymerisation) (see Figure 4.2)

(d) This can be further chemically exploited by using the knowledge that the living polymer is a metal alkyl compound (i.e. similar to a Grignard reagent), e.g. $-\text{CH}_2^-\text{Li}^+$. Thus the following derivatising reactions are possible on the living polymer when polymerisation is complete

$$\sim\text{CH}_2^-\text{Li}^+ + \text{H}^+ \longrightarrow \;\sim\text{CH}_3$$

$$\sim\text{CH}_2^-\text{Li}^+ + \text{O}_2 \longrightarrow \;\sim\text{CH}_2\text{OH (via hydrolysis of} \sim\text{OLi)}$$

$$\sim\text{CH}_2^-\text{Li}^+ + \text{CO}_2 \longrightarrow$$

$$\sim\text{CH}_2\text{COOH (via hydrolysis of} \sim\text{CH}_2\text{COOLi)}$$

$$\sim\text{CH}_2^-\text{Li}^+ + \overset{\displaystyle O}{\overset{\displaystyle \diagup\;\diagdown}{\text{CH}_2\text{—CH}_2}} \longrightarrow$$

$$\sim\text{CH}_2\text{CH}_2\text{CH}_2\text{OH (via hydrolysis of} \sim\text{CH}_2\text{CH}_2\text{CH}_2\text{OLi)}$$

These end-capped polymers can then be used as a monomer for completely different polymers, and thus one can obtain copolymers (either diblock or graft copolymers; see Section 4.4.2.3).

All the above anionic polymerisations have been generated by one growing chain. However, there are techniques to generate two growing chains from the same initiator and these can be used to polymerise styrene or butadiene.

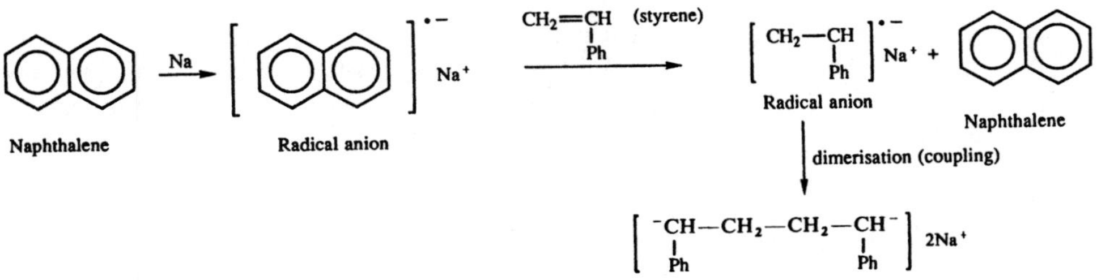

The main application of this method is to produce a polymer with two reactive ends, X———X (called a *telechelic* polymer), which can be chemically reacted further (as above) to produce a tri-block copolymer (see section 4.8.3).

If chain termination in anionic polymerisation is required to control MW, e.g. to control a low MW without having too much initiator, chain-transfer agents can be added, e.g. toluene:

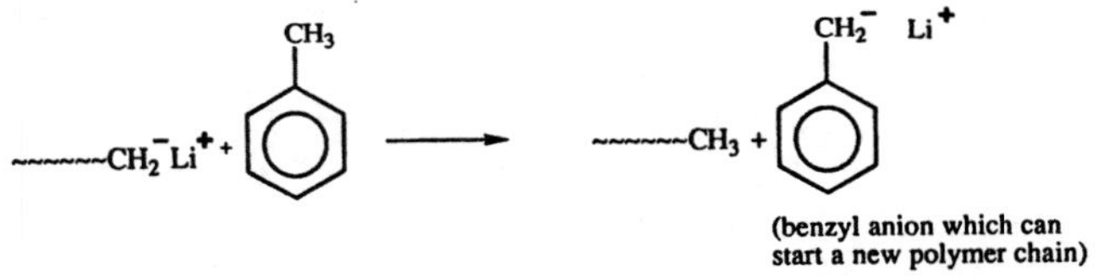

(benzyl anion which can start a new polymer chain)

Obviously traces of water and oxygen will terminate the polymerisation (and therefore control the MW), but they will also kill the catalyst. Therefore these reagents are not used for controlling the MW of the polymer.

4.3.2.3 Cationic polymerisation. Here the growing chain is a cation (see Table 4.2 for examples). The initiators are generated from Lewis acids together with traces of a cocatalyst:

$$BF_3 + H_2O \rightleftharpoons H^+[BF_3OH]^-$$

If too much water is added it will act as a catalyst poison.

$$AlCl_3 + RCl \rightleftharpoons R^+[AlCl_4]^-$$

These acidic catalysts initiate the polymerisation by electrophilic attack on the monomer, e.g. with isobutylene:

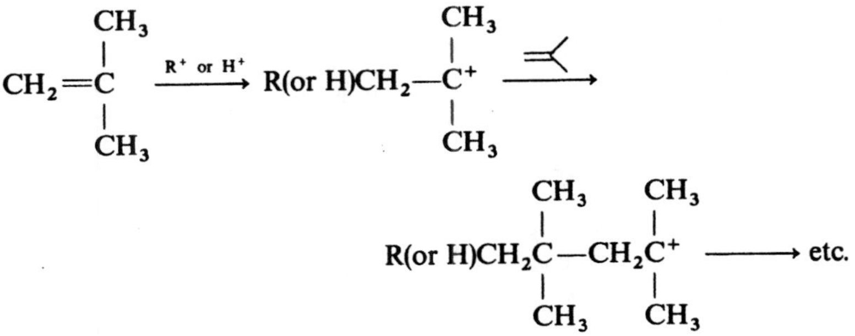

The growing cationic species —C^+ is rather unstable, which limits the variety of monomers that can be polymerised. Often it produces only low-MW oligomers, e.g. propylene only produces oils which are mixtures of dimers, trimers and tetramers. A notable exception is isobutylene, but even this polymerisation has to be carried out at a very low temperature ($-80\,°C$).

Most saturated cyclic compounds containing hetero atoms, e.g. O, S, NR, will

polymerise cationically if the cyclic strain is sufficient e.g. THF (⬠o), ethylene oxide (△ O) and cyclic imines (△ NR). For example, with THF:

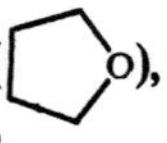

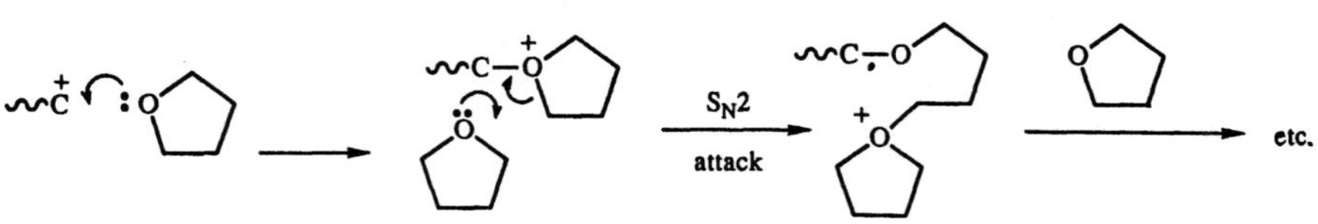

It can be seen that growth occurs by S_N2 attack of the incoming monomer on the propagating cation. It is therefore important that the associated anion (of the cation) should not have any nucleophilic tendency, which would prevent any attack by the monomer. This is satisfied by anions like $[AlCl_4]^-$, $[BF_3OR]^-$, $[SO_3F]^-$ and $[PF_6]^-$, all of which have the added advantage of being the conjugate base of a strong acid (i.e. strong base but weak nucleophile).

Termination cannot occur by coupling as occurs in free-radical polymerisation. It can occur by internal attack by the associated anion (see above).

$$\sim CH_2{-}\overset{\displaystyle R}{\underset{\displaystyle R}{\overset{|}{\underset{|}{C}}}}{}^+X^- \longrightarrow \sim CH{=}\overset{\displaystyle R}{\underset{\displaystyle R}{\overset{|}{\underset{|}{C}}}} + HX$$

or by chain transfer with an incoming monomer, thus starting a new growth centre

$$\sim CH_2{-}\overset{\displaystyle R}{\underset{\displaystyle R}{\overset{|}{\underset{|}{C}}}}{}^+ + CH_2{=}\overset{\displaystyle CH_3}{\underset{\displaystyle CH_3}{\overset{|}{\underset{|}{C}}}} \longrightarrow \sim CH{=}\overset{\displaystyle R}{\underset{\displaystyle R}{\overset{|}{\underset{|}{C}}}} + CH_3{-}\overset{\displaystyle CH_3}{\underset{\displaystyle CH_3}{\overset{|}{\underset{|}{C}}}}{}^+$$

Although not as easy to perform as with anionic polymerisation, living polymerisation via cationic growth is possible. This then allows further reactions to take place, e.g. with isobutylene, the following equilibria can occur:

$$\sim \overset{\displaystyle CH_3}{\underset{\displaystyle CH_3}{\overset{|}{\underset{|}{C}}}}{}^+X^- \rightleftharpoons \sim \overset{\displaystyle CH_2}{\underset{\displaystyle CH_3}{\overset{\|}{\underset{|}{C}}}} + HX$$

thus allowing the following reactions to be performed:

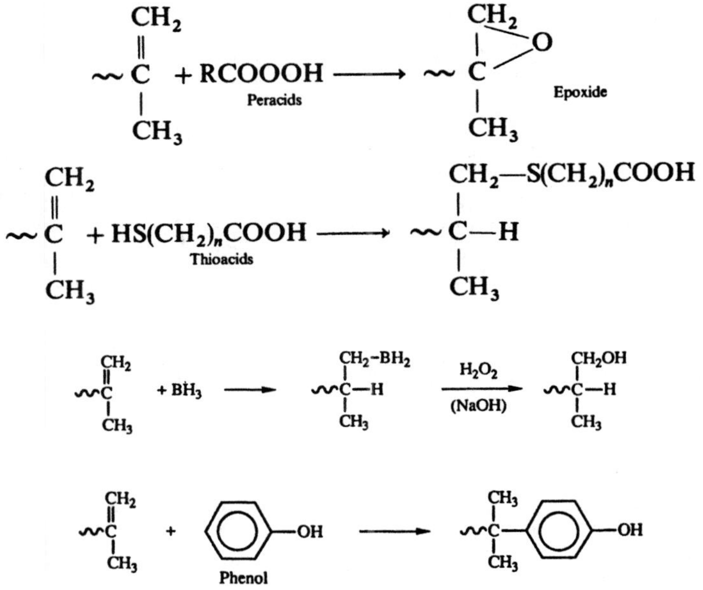

and these end-capped materials can be used as 'monomers' to prepare further block copolymers.

Two growing polyisobutylene (PIB) chains can be formed by starting with a bifunctional initiator, e.g.

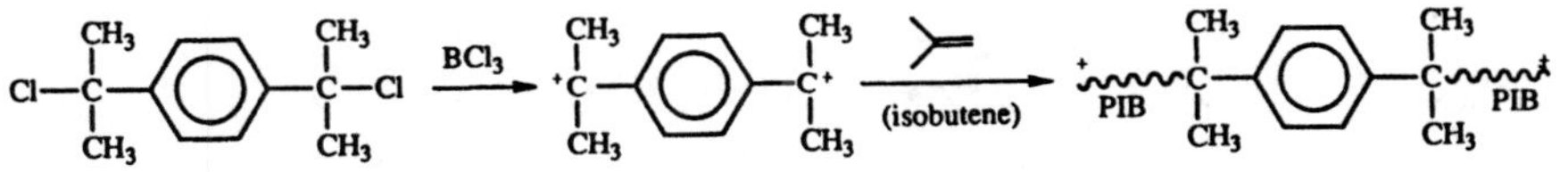

4.3.2.4 Co-ordination polymerisation.

This is most recently discovered class of polymerisation, but in actual tonnes of polymer produced, it outweighs all the other methods, principally because of the commercial polymerisation of ethylene and propylene.

This polymerisation involves a transition metal (TM) centre (groups 4, 5 and 6) together with a co-catalyst (main group metal alkyl), the latter operating as an activator or alkylating agent and the TM as the growth centre. It can therefore be thought of as a special case of anionic polymerisation with TM^+-[alkyl or polymer chain]$^-$. However, in co-ordination polymerisation, the TM can have a variable oxidation state and differing co-ordinating ligands, and these factors can alter the nature of the polymer produced.

Because of the sensitivity of these systems to trace impurities (water, oxygen, etc.) and the reactivity of the TM catalyst centre to ligands containing heteroatoms (e.g. Cl, O, N), only purely hydrocarbon olefins e.g. ethylene, propylene, 1-butene, etc. and dienes, e.g. 1,3-butadiene and, of course, styrene, can be polymerised. However, with special precautions (e.g. low temperatures) substituted olefins (e.g. methyl methacrylate) can be polymerised, although this latter process is not used commercially.

Although both homogeneous and heterogeneous co-ordination catalysts have been developed, it is mainly the heterogeneous systems that are used industrially. Because of this, the actual mechanism of polymerisation is difficult to understand, though with the insoluble heterogeneous system $TiCl_3 + AlEt_3$ the following route is believed to occur:

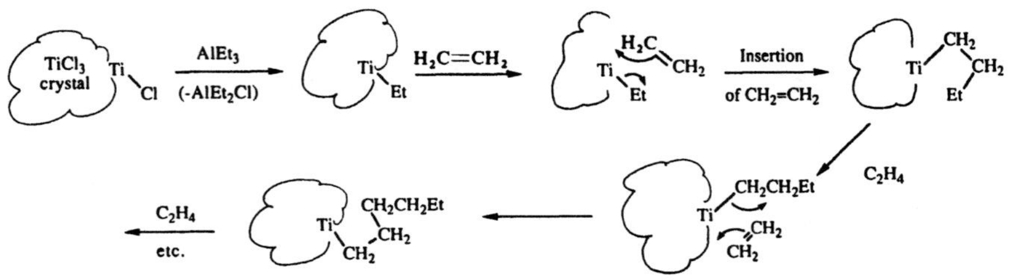

The termination step can occur by two routes

(a) Adding a chain transfer agent—this is usually hydrogen

$$\text{'Ti'}-CH_2 \sim \; + H_2 \xrightarrow{\text{Hydrogenolysis}} \text{'Ti'}-H + CH_3 \sim$$
$$\text{Dead polymer}$$

and the 'Ti'-H can act as a new centre for polymerisation

(b) Decomposition of the growing polymer chain by β-hydrogen elimination

$$\text{'Ti'}-\overset{\alpha}{C}H_2-\overset{\beta}{C}H_2 \sim \; \longrightarrow \; \text{'Ti'}-H + CH_2{=}CH \sim$$
$$\text{Dead polymer}$$

This mode of chain termination occurs at higher temperatures.

These coordination polymerisations are often referred to as *Ziegler–Natta polymerisations*, from the discoverers of polyethylene and polypropylene respectively. Although the catalyst system is very sensitive to impurities, excess aluminium alkyls can be used as scavengers, e.g. using a Ti/Al ratio of 1/10.

The growing chain has a very short lifetime (a few seconds) before it becomes a dead polymer. Therefore with these systems it is impossible to use the living-polymer approach to produce block copolymers.

$TiCl_3$ is an insoluble material and only the surface of the crystal is used. This means that more catalyst is used than is really necessary, resulting in more catalyst residues (Ti, Cl, Al) in the finished polymer. These can degrade the polymer if left in it, and also cause corrosion problems with polymer fabrication equipment. Although these catalyst residues can be extracted from the polymer (an expensive operation), a better method is to use supported catalysts, i.e. the core of the catalyst is an inert material and only the surface contains the catalyst centres, e.g. with silica (SiO_2):

This approach is now used commercially to produce polyethylene and poly-propylene. However, $MgCl_2$ core is used instead of SiO_2 because the catalyst is very much more active and produces much more polymer, so that the catalyst residues can be left in.

4.4.1   Homopolymers

These are polymers obtained by polymerising a single monomer by a chain-growth mechanism, e.g. ethylene, vinyl chloride or styrene to yield PE, PVC or PS:

$$nA \longrightarrow \sim AAAAAAAAA \sim$$

(a linear polymer; it is assumed that only head-to-tail linkages occur)
or a pair of monomers undergoing step-growth polymerisation:

$$nA + nB \longrightarrow \sim ABABABABAB \sim$$

(this type of polymers would be classed as alternating copolymers if a chain-growth mechanism occurred).

If A is ethylene (CH_2=CHR, where R = H), then the conformation of PE (i.e. the way in which the linear —C—C—C— backbone organises itself) is a planar zig-zag arrangement (Figure 4.4). The —C—C—C— backbone is in the same plane (this is an all *trans* arrangement). The linear arrangement for PE has the lowest energy and large polymer segments (e.g. ~ 100 carbon atoms long) will organise themselves in this zig-zag conformation (and crystallise, see section 4.6.3).

However, when R $\neq$ H, i.e. CH_2=CHR, then various head-to-tail combinations can occur depending on adjacent units. Thus, steric hindrance between

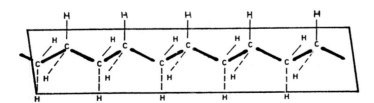

Figure 4.4   Structure of linear polyethylene.

the R groups governs the stereochemical manner in which the incoming monomer will add to the growing chain. This phenomenon is called *tacticity*. Three arrangements are possible, as shown in Figure 4.5. These conformations can be determined by ^{13}C NMR.

Actually, in the case of $R = CH_3$ i.e. polypropylene, the steric hindrance between the CH_3 groups causes the polymer chain to form a helical structure with CH_3 groups on the outside of the helix (Figure 4.6) (this is a *trans-gauche* conformation). For syndiotactic polypropylene the most favourable conformation for the C—C—C backbone is the all *trans* form (i.e. a planar zig-zag).

If the chain polymerisation mechanism is via free-radical intermediates usually the non-stereoregular conformation (i.e. atactic) results. However, the stereoregular isotactic (and syndiotactic) conformation of polypropylene and polystyrene can be obtained by using co-ordination catalysts. With many of these polymers the tacticity is not important and the mechanical properties

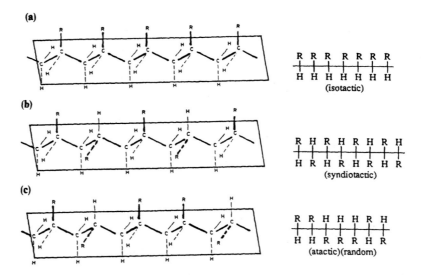

Figure 4.5   The three possible conformations of polypropylene: (a) isotactic; (b) syndiotactic; (c) atactic.

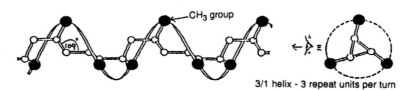

Figure 4.6   Conformation of isotactic polypropylene.

and melting points (often called softening points) are similar for all three conformations. However, with polypropylene (PP) the isotactic conformation—and the syndiotactic conformation, but this is difficult to make—can crystallise to give a stiff polymer, whereas the atactic PP is a viscous gum.

Once a polymer in a particular conformation (iso-, syndio-, a-tactic, or a mixture of tacticities) has been prepared, this conformation is fixed and cannot be altered by physical (e.g. heat or UV light) or chemical means.

4.4.1.1 Non-linear homopolymers.

These exist in both branched and cross-linked forms.

Branched

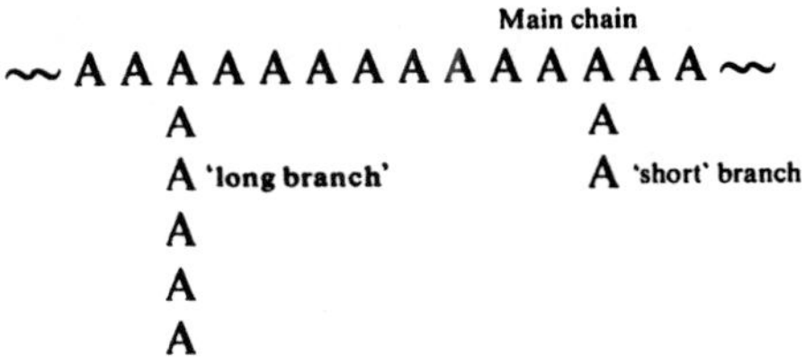

Main chain

$\sim$ A A A A A A A A A A A A A $\sim$
 A   A
 A 'long branch'   A 'short' branch
 A
 A
 A

This often occurs in free-radical polymerisation when either the initiator or the growing free-radical chain inadvertently attacks a dead polymer by an inter- or intra-molecular reaction (see section 4.3.2.1).

Cross-linked

$\sim$ A A A A A A A A A A A $\sim$
        | Cross link
$\sim$ A A A A A A A A A A A $\sim$

This type of cross-link is formed when the pre-formed polymer is reacted with free radicals to form polymer free radicals which couple together. This is often done when the polymer does not have any reactive groups (either pendant or main-chain) in the polymer chain, e.g.

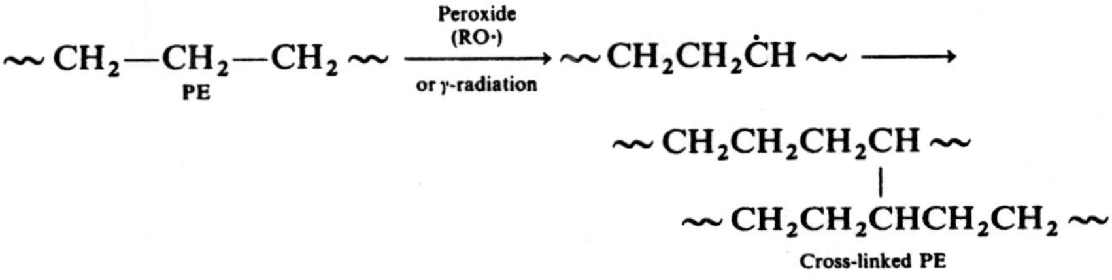

$$\sim CH_2{-}CH_2{-}CH_2 \sim \xrightarrow[\text{or } \gamma\text{-radiation}]{\substack{\text{Peroxide} \\ (RO\cdot)}} \sim CH_2CH_2\dot{C}H \sim \longrightarrow$$

$$\sim CH_2CH_2CH_2CH \sim$$
$$|$$
$$\sim CH_2CH_2CHCH_2CH_2 \sim$$
Cross-linked PE

Alkyd paints are a mixture of diacids (e.g. phthalic acid, benzene-1,2-di-carboxylic acid) a polyhydric alcohol (e.g. glycerol) and small amounts of long-chain unsaturated fatty acids (e.g. linseed oil, soybean oil). These latter compounds are cross-linked by reaction with oxygen (air) to form initially a peroxide which is decomposed by traces of metal ions (cobalt, manganese, lead):

$$\sim CH_2-CH=CH \sim \xrightarrow{\;O_2\;} \sim \underset{\underset{OOH}{|}}{CH}-CH=CH \sim \xrightarrow{\;M'\;}$$

$$\sim \underset{|}{\overset{}{C}H}-CH=CH \sim$$

which cross-links as above, causing the paint to dry or harden.

Cross-linking is also called curing or vulcanisation, and often reagents are added to couple or cross-link two (or more) polymer chains together, e.g.

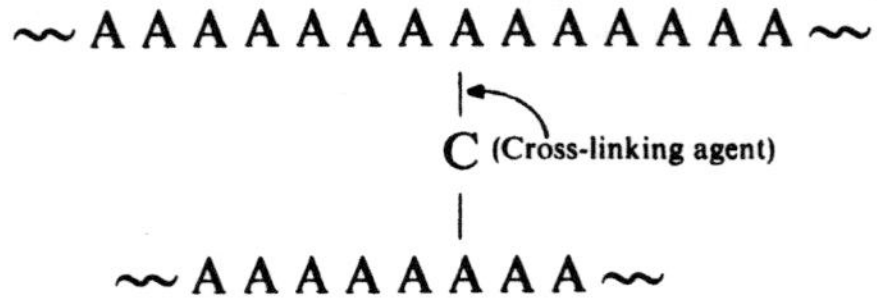

$$\sim A A A A A A A A A A A A A \sim$$
$$|$$
$$C \text{ (Cross-linking agent)}$$
$$|$$
$$\sim A A A A A A A A \sim$$

where C may be present in very small amounts (e.g. in the vulcanisation of natural rubber with sulphur, often less than 1% by weight is present).

The properties of these non-linear branched or cross-linked polymers compared with linear homopolymers can be dramatically different, especially when the non-linearity affects the ability to crystallise, and hence alters the mechanical properties. Other factors which will be affected include melt viscosity, barrier properties, etc.

4.4.2   Copolymers

Although many homopolymers are made industrially, e.g. polyethylene terephthalate (PET), polymethyl methacrylate (PMMA), PE and PVC, there is now a trend to produce copolymers using two or more monomers. The reason for this is that it is possible to alter the properties of the original homopolymer controllably, e.g. to change the stiffness (modulus), strength such as tensile strength, elasticity, fabrication methods, dye-ability of fibres, etc., without having to resort to new homopolymers. Thus one can effectively broaden the range of applications of the homopolymer.

A typical example is polystyrene (PS). The homopolymer is a brittle plastic with poor impact strength and poor solvent resistance. Copolymerisation and terpolymerisation (three monomers) using free radical initiators improves these properties. For example, copolymerisation with acrylonitrile (to give SAN) improves impact and solvent resistance, whilst copolymerisation with butadiene yields an elastomer (styrene–butadiene rubber, SBR) and terpolymerisation with acrylonitrile and butadiene yields ABS, a tough solvent-resistant plastic used for casings of, for example, telephones.

There are several types of copolymers.

4.4.2.1 Random copolymers. Random copolymers can be obtained when either of the two monomers has no affinity for the previous monomer insertion. If the reactivities of the two monomers differ greatly, the more reactive monomer is added gradually to the system, e.g. styrene–acrylonitrile, using controlled addition of the more reactive styrene.

The general structure of a random copolymer is:

$$\sim AABAAABBBABAAAB\sim$$

with various proportions of A and B.

Reasonably uniform random copolymers will display a single phase morphology, and because they will be amorphous and will not crystallise (see below) they will remain or become (if the original homopolymer was crystalline) transparent.

As described above, these copolymers often have superior properties over the parent homopolymer. It is impossible in the space available to list all of these, but a few examples will suffice.

(a) Introduction of 'defects' along the chain by copolymerisation inhibits the crystallisation of the polymer. This results in:

   (i) Reducing the modulus (i.e. stiffness), so that some copolymers have elastomeric properties. Table 4.4 gives the effect of the addition of ethylene to propylene polymerisation; notice the reduction of T_g to give an elastomer (see section 4.8).

   (ii) Inserting 5–10% of an α-olefin (e.g. 1-butene) into polyethylene yields a PE polymer that is intermediate between high-density linear PE (made by Ziegler catalysts) and low-density branched PE (made by high-pressure free-radical polymerisation). It is called linear low-density PE (LLDPE) (Figure 4.7).

   (iii) PTFE cannot be melt-processed because of the very high MW (10^6–10^7) resulting in an impossibly high melt viscosity. Addition of 1% of the perfluoro olefin CF_2=CF—O—CF_3 to form a random copolymer does not reduce the MW by much (to 0.5×10^6), but it reduces the melt viscosity from 10^{11} to 10^5 poise (at 370 °C), thus allowing melt-processing.

   (iv) PVC is difficult to melt-process because it decomposes at its melt-

Table 4.4 Effects of copolymerisation on properties

Monomer feed ratio ethylene/propylene	Crystallinity (%)		T_m (°C)	T_g (°C)
	iPP*	PE		
4/96	42		143	0 to 6
14/86	15		125	−10
22/78	4		119	−20
32/68	3		118	−33
44/56		2	123	−36
75/25		27	125	−50

* Isotactic polypropylene.

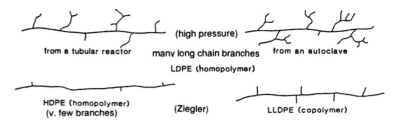

Figure 4.7   Schematic representation of types of polyethylene.

ing point (heat stabilisers can be used. It is also a brittle material. This can be overcome with plasticisers (see section 4.6.2.1), but another method is to incorporate randomly a small amount ($\sim 5\%$) of vinyl acetate (VA) which lowers the softening point and increases toughness (e.g. when the PVC is to be used for guttering, vinyl records). With larger amounts of VA ($\sim 30\%$), more flexible PVC products are obtained, such as floor tiles, shower curtains, etc.

(v) The polymer can become more transparent and flexible by destroying the crystallinity, e.g. with nylon 6,6 fibres, the introduction of small amounts of $H_2N(CH_2)_{10}COOH$ gives a transparent fibre useful for fishing lines.

(b) The second monomer, once incorporated into the chain, can be used for further chemical reactions, e.g.

(i) Introduction of dye-receptor groups; e.g. with polyethelene terephthalate (PET), small amounts of 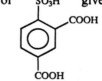 give a polyester

backbone with a few $-SO_3^-Na^+$ groups which interact with cationic dyes. With acrylic fibres (polyacrylonitrile, PAN) traces of

SO_3H or SO_3H also enable cationic dyes to be used.

(ii) with cationically polymerised polyisobutylene, vulcanisation (cross-linking, curing) is difficult because there is no unsaturation (cf. natural rubber, polybutadiene, etc.). Therefore a small amount (1%) of butadiene is added to give polyisobutylene (PIB) with butadiene incorporated into PIB chain, thus allowing vulcanisation with sulphur (similar to natural rubber).

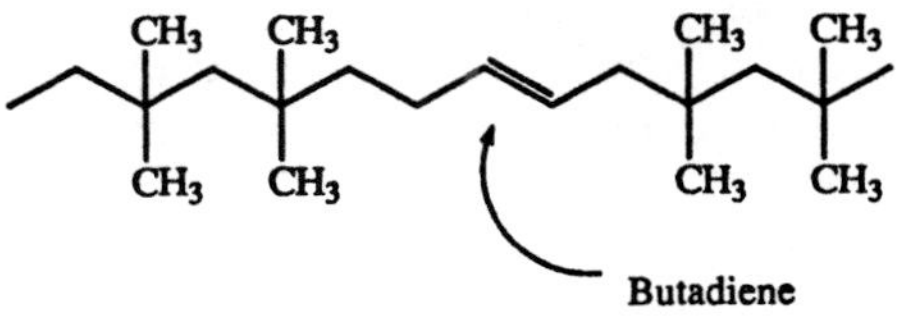

Butadiene

(iii) Many polymers are used with fillers. Often in order to get interaction of the polymer with the surface of the filler, a random copolymer with a small amount of monomer containing a reactive group (—COOH) is used.

There are many more industrial examples of random copolymerisation and the trend is growing. Although melt blending of two or more pre-formed homopolymers can be used to obtain an effect, this is an expensive operation and may cause decomposition of the individual polymers and also the blend may not be homogeneous (most polymers are non-miscible or non-compatible). Many of these properties of blends can be obtained by copolymerisation.

4.4.2.2 Alternating copolymers. These are defined as:

$$\sim A\,B\,A\,B\,A\,B\,A\,B\,A\,B\,A\,B\,A\,B\,A\,B\,A\,B\,A\,B \sim$$

and include step-growth polymers made from two different monomers A and B. In the case of nylon 6,6 the polymer is slightly different from nylon 6, which can be considered a homopolymer (T_m 265 °C versus 215 °C).

An alternating copolymer is a homogeneous material, like a perfectly random copolymer, and will possess a single glass transition temperature T_g (see section 4.6.2).

They can be made by free-radical chain-growth copolymerisation where the reactivity ratios of the monomers are such that the growing chain only reacts with the alternative monomer and the ratio of the monomer feed is controlled.

The reason for the formation of this type of copolymer is that a charge transfer complex between A and B is formed, AB, and it is this species which polymerises. Many alternating copolymers contain maleic anhydride, which is a strong electron acceptor and therefore will accept an electron from another monomer, often an olefin, e.g.

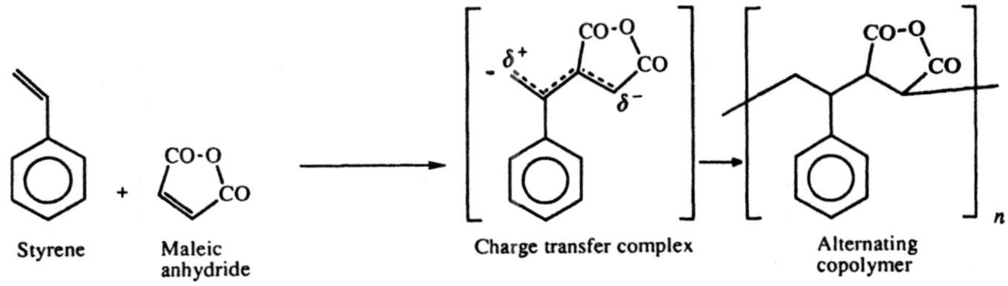

Styrene   Maleic anhydride   Charge transfer complex   Alternating copolymer

This type of charge-transfer free-radical polymerisation can be promoted by complexing the electron-acceptor monomer with, for example, aluminium alkyls, making it an even better electron acceptor), e.g. styrene + acrylonitrile with AlR_3, $(CH_2{=}CHCN \longrightarrow AlR_3)$.

These copolymers containing maleic anhydride can be used as adhesives since the anhydride group has a strong affinity for surfaces.

4.4.2.3 Block copolymers. These can take several different configurations (Figure 4.8).

Because homopolymers are usually non-miscible with each other, most block copolymers are two-phase systems. They are made by various methods.

(1) Using the living polymer method—usually by anionic polymerisation, e.g.

$$\text{Styrene} + \text{Li}^+\text{Bu}^- \longrightarrow \underset{\text{Polystyrene}}{\sim\sim\sim \overset{-}{\text{Li}}{}^+} \xrightarrow{\text{Butadiene}} \underset{\substack{\text{Polystyrene} \\ \text{segment}}}{\sim\sim\sim} \cdot \underset{\substack{\text{Polybutadiene} \\ \text{segment}}}{\sim\sim\sim}$$

This material is called polystyrene-block-polybutadiene or polystyrene-b-polybutadiene.

These polyoxide materials are surfactants (called Pluronics).

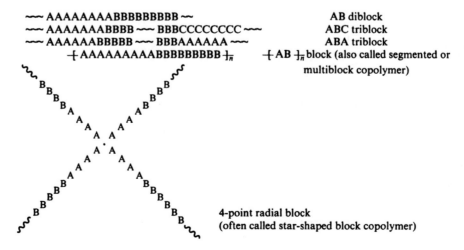

~~ AAAAAAAABBBBBBBBB ~~   AB diblock
~~ AAAAAAABBBB ~~ BBBCCCCCCCC ~~   ABC triblock
~~ AAAAAABBBBB ~~ BBBAAAAAA ~~   ABA triblock
─(AAAAAAAAAABBBBBBBBBB)$_n$ ─(AB)$_n$ block (also called segmented or
                                            multiblock copolymer)

4-point radial block
(often called star-shaped block copolymer)

Figure 4.8   Possible configurations of block copolymers.

Table 4.5   Architectural forms of polymers available by living polymerisation techniques

Polymer		Application
Functional ended		Dispersing agents Synthesis of macromonomers
$\alpha\omega$-Difunctional		Elastomers synthesis Chain extension Cross-linking agents
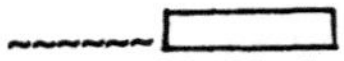 AB Block		Dispersing agents Compatibilisers for polymer blending
 ABA Block		Thermoplastic elastomers
	Graft	Elastomers Adhesives
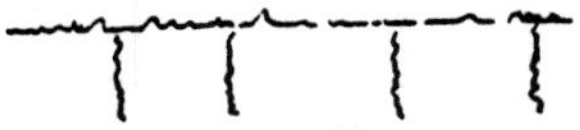	Comb	Elastomers Adhesives
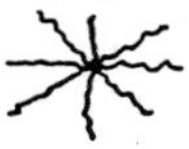	Star	Rheology control Strengthening agents
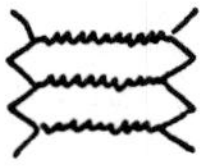	Ladder	High temperature plastics Membranes Elastomers
	Cyclic	Rheology control
	Amphiphilic network	Biocompatible polymer

Table 4.6   Initiators and monomers reactivity in anionic polymerisation

Monomer	pK_a*	Initiators
Styrenes ----------------	41–42	NH_2, RM**, Ar^-, M
Dienes	43	
Methacrylic esters	17–28	RMgX, fluorenyl$^-$
Acrylonitrile	32	$Ar_2C^{\dot{-}}$, $Ar_2CO^{\dot{-}}$
Vinyl ketones	19	
Oxiranes -------------	16–18	RO^-
Thiiranes	12–13	
Siloxanes	10–14	
Cyanoacrylates	11–13	HCO_3^-, H_2O
Nitroalkenes	10–14	
Lactones	4–5	RCO_2^-

* Negative logarithm of the acid dissociation constant for the conjugate acid of the anion involved in propagation of each monomer.

** M = Li, Na, K, Cs, Rb, etc.

Although the living anionic polymer approach appears to be very versatile, there is the limitation that the anionic chain end will only initiate the second monomer if the second monomer has the same (or stronger) acid strength as the initial monomer (i.e. the second monomer must have a lower pK_a value). Thus for the diblock copolymer polystyrene-b-polyethylene oxide (oxirane), styrene must be polymerised first, followed by ethylene oxide (oxirane), and not the other way round (see Table 4.6).

(2) Using a *telechelic* macromer or polymer (see section 4.3.2.2), e.g.

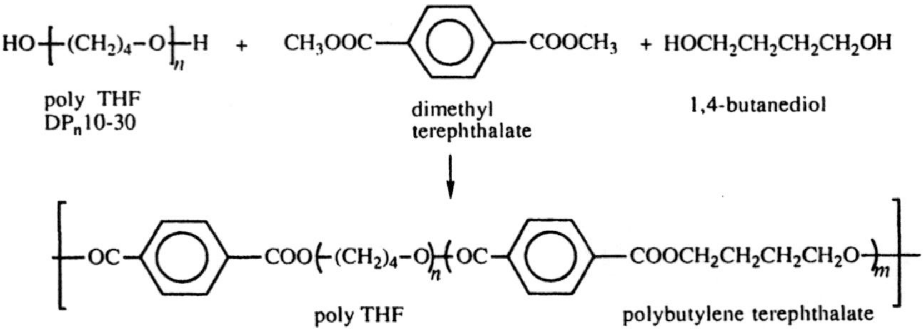

poly THF
DP_n 10-30

dimethyl terephthalate

1,4-butanediol

poly THF

polybutylene terephthalate

This is a thermoplastic elastomer (see section 4.8.3) called Hytrel (DuPont). Similarly:

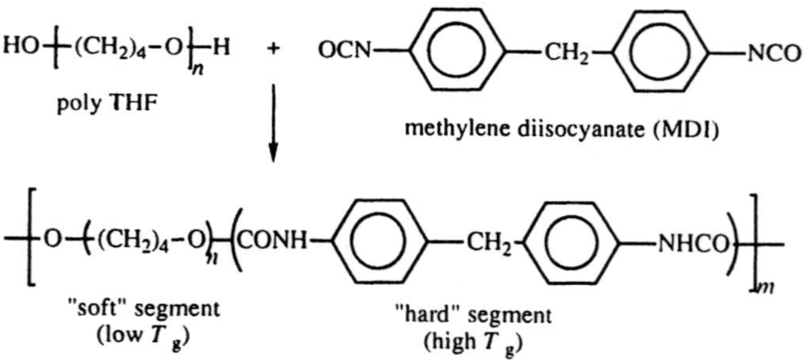

poly THF

methylene diisocyanate (MDI)

"soft" segment
(low T_g)

"hard" segment
(high T_g)

though actually the structure is more complicated than this. It is an example of a thermoplastic elastomeric fibre (Spandex) called Lycra (DuPont).

(3) Radical or star block copolymers can be made by treating the living anionic polymer with a multifunctional reagent e.g.

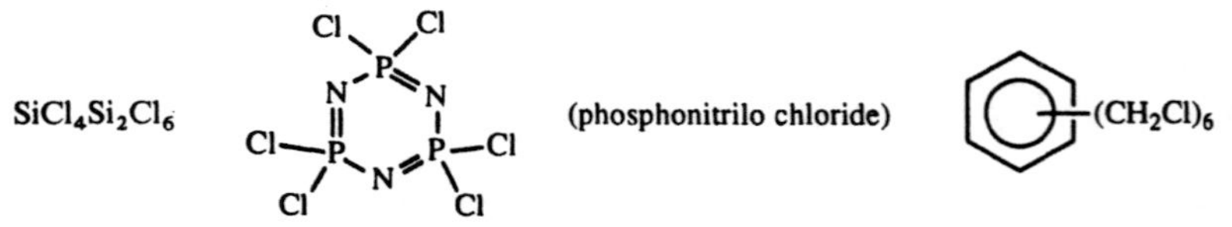

$SiCl_4 Si_2Cl_6$

(phosphonitrilo chloride)

$4 \ \sim\sim\sim\sim\sim^- Li^+ + SiCl_4 \longrightarrow (\sim\sim\sim\sim\sim\sim)_4 Si + 4LiCl$

Obviously by using a di-functional reagent Cl ~~~~ Cl with an anionic living polymer, an ABA block copolymer can be formed.

(4) A simple (and cheap) method of making AB block copolymers is to start off with a step-growth homopolymer, e.g. polyamide or polyester, and mechanically blend (with heat at ~300 °C) the polymer with another different polyamide or polyester. Because of the lability of the amide or ester bond at high temperature, interchange between the polymers takes place, with the removal of any volatiles, and a block copolymer with segments of the original polymer and the added polymer is produced, because of the non-compatibility of the segments. This approach has been used, employing mechanochemical main-chain breakage of chain-growth polymers, e.g. natural rubber (which is *cis*-1,4-polyisoprene,

) in the presence of methyl methacrylate, to produce a block copolymer. Chain breakage of the natural rubber yields free radicals which polymerise the methyl methacrylate.

(5) There are basically three methods for preparing triblock copolymers.
   (a) Three-step sequential polymerisation using a monofunctional initiator (section 4.4.2.3)—the only technique for producing ABC copolymers.
   (b) Two-step sequential polymerisation using a difunctional initiator (see section 4.3.2.2).
   (c) Two-step polymerisation using a monofunctional initiator followed by coupling of the chain ends (see star copolymers, Figure 4.8).

4.4.2.4 *Graft copolymers.* As the name implies, these are made by grafting a second polymer on to the main chain of the original, e.g.

```
~ AAAAAAAAA ~      or     ~ AAAAAAAAAAAAAAA ~
      B                         B          B
      B                         B          B
      B                         B          B
      B                         B          B
      B                         B          B
```

Their preparation involves initiation by a growth centre in the main chain and then reaction with a second monomer, e.g.

(1) Generate a free radical by:
   (a) Peroxide treatment

$$\sim CH_2CH{=}CHCH_2 \sim \xrightarrow[(-ROH)]{RO^{\bullet}}$$

1,4-Polybutadiene

$$\sim \dot{C}H{-}CH{=}CH{-}CH_2 \sim \xrightarrow{\text{Second monomer (B)}}$$

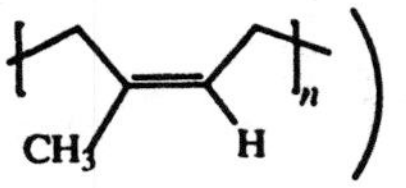

(b) Radiation with X-ray or γ-rays in the presence of a second monomer. The problem with these two methods is that polymerisation of the second monomer can occur independently, thus producing a blend (actually this results in an interstitial polymer blend).

(2) Create an anionic centre by treating with LiBu, e.g.

$$\sim CH_2CH=CHCH_2 \sim + Li^-Bu^+ \longrightarrow \sim \overset{Li^+}{\overset{|}{CH}}-CH=CH\cdot CH_2 \sim \xrightarrow{styrene}$$

In these examples, often a chelating agent (e.g. $Me_2NCH_2CH_2NMe_2$) is added to the LiBu to increase the basicity of the $Bu^-\left(Li^+ \overset{N}{\underset{N}{\diagdown}} \quad Bu^-\right)$.

(3) Construct a polymer containing reactive centres e.g.

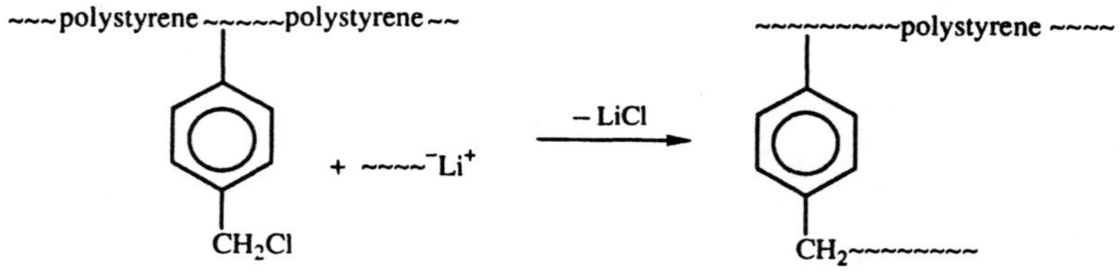

4.4.2.5 Properties of block and graft copolymers. The properties of block and graft copolymers are similar. However, graft copolymers are difficult to character- ise because it is often difficult to estimate the number of grafts per molecule, if any, and also the length of the graft. The polydispersity (MWD) of both the main chain and the graft chain also complicates the exact nature of the polymer. Nevertheless, very useful systems can be obtained, for example prepared polybutadiene (elas- tomer) on to which has been grafted either styrene or a combination of acrylonit- rile and styrene, by using free-radical techniques, to yield grafted polymers. This is the basis of high-impact polystyrene (HIPS) and ABS polymers respectively.

The main features of these copolymers are:

(1) They are two-phase systems (heterogeneous), and the polymer on cooling from the melt will separate into domains. These domains, however, are very small—less than 0.1 μm—and can be crystalline, glassy or rubbery depending on the nature of A and B and the temperature. Compared with random or alternating copolymers, the stiffness of block and graft co- polymers is controlled by the glass transition temperature (see sec- tion 4.6.2) of each block (Figure 4.9) thus giving a constant stiffness over a wide temperature range (cf. random or alternating copolymers).

(2) If only the A or B block is soluble in a solvent, then the solution of the polymer will have a 'milky' appearance, consisting of micelles. These materials have been used as viscosity improvers over a wide temperature range for multigrade lubricating oils.

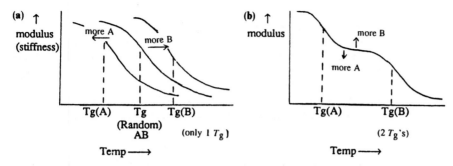

Figure 4.9 T_g values and stiffness of (a) random and (b) block or graft copolymers.

(3) Although most polymers are non-compatible, block and graft copolymers will be miscible with their constituent homopolymers

Thus it is possible to use an AB block or graft copolymer to 'homogenise' a blend of A and B homopolymers. In these applications they are called compatibilisers.

As Figure 4.10 shows, the tensile strength of polystyrene (PS) is reduced less by the addition of the graft copolymer of LDPE-PS. Electron microscopy of the material shows a reduction in the domain size (or phase size) and improved homogeneity.

(4) It is possible to produce AB block and graft copolymers where A and B have completely different characteristics. This allows a wider range of applications than with homopolymers, e.g.

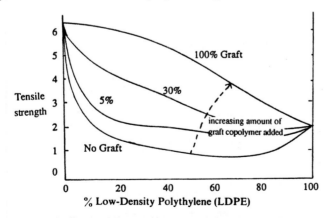

Figure 4.10 Tensile strength of LDPE/PS blends containing various concentrations of added LDPE/PS graft copolymer.

(a) A styrene–butadiene random latex polymer can be prepared by using a water-soluble free-radical initiator (emulsion polymerisation). When a block copolymer of butadiene–ethylene oxide is added to this latex, the polybutadiene block interacts with the latex—there are sufficient free radicals around to give grafting—to produce styrene–butadiene latex with a hydrophilic surface as a result of the polyethylene oxide structure. This material has been used to make wettable bandages and also disposable diapers (nappies).

(b) A similar material is obtained by treating starch with Ce^{4+} to generate a free radical which can act as a growth centre for grafted polyacrylonitrile. Because starch is very soluble in water, these materials have a high affinity for water, absorbing ~ 20 times their weight, yet still remaining insoluble. They have been called 'super slurpers' and are used in diapers and growth media for house plants, etc.

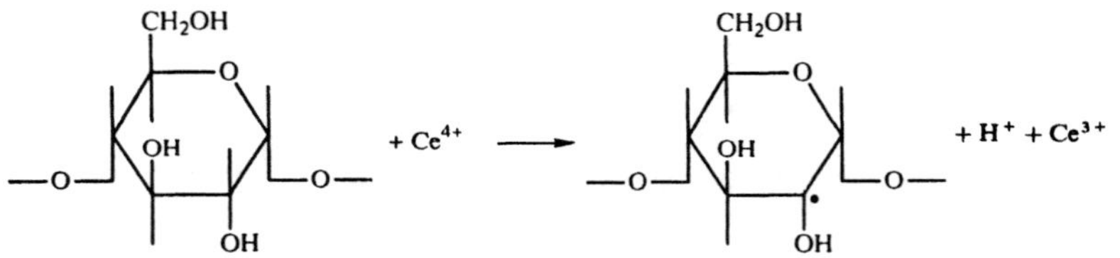

Those block copolymers with domains that are hydrophobic and hydrophilic are called amphiphilic and have excellent biocompatibility (see Table 4.5).

Thus, summarising, the main applications for commercial block and graft copolymers are:

(1) Thermoplastic elastomers (see section 4.8.3)
(2) Toughened plastics
(3) Surfactants
(4) Compatibilisers for polymer blends.

4.5
Manufacture of polymers

Approximately 100 million tonnes of polymers are made annually, in plants ranging from 240,000 tonnes/year continuous single-stream polypropylene plants to a single batch preparation of a few kilograms of advanced performance composites. The highest tonnage polymers are, as listed in section 4.1.1, LDPE, HDPE, LLDPE, PP, PVC and PS.

The most important parameters in making polymers are quality control and reproducibility. They are different from simple organic compounds such as acetone, where often a simple distillation gives the desired purity. There are many different grades of the 'same' polymer, depending on the final application, e.g. different MW, MWD, extent of branching, cross-linking, etc., and these variations are multiplied when copolymers (random, alternating and block) are considered. Many of the properties are fixed during polymerisation and cannot be altered by post-treatment. Blending is sometimes carried out to obtain desired properties or just to up-grade production polymer that may be slightly off-specification.

A polymerisation process consists of three stages: (a) preparation of monomer; (b) polymerisation; and (c) separation of the polymer.

(a) Monomer preparation.   This is not discussed here, other than to emphasise that the purity of the monomer is paramount. Thus monofunctional impurities in step-growth, and radical scavengers, chain-transfer impurities and catalyst poisons (e.g. water in ionic propagation) in chain polymerisations are all significant. Manufacture of many monomers is described in chapter 11 of Volume 1.

(b) Polymerisation.   As stated above, uniformity of polymer properties is absolutely necessary; this not only includes MW, etc., but other factors such as colour, shape of polymer particle (if the polymer is not pelletised or granulated), catalyst residues, odour (especially when used for food application), etc.

The polymerisation operation has to cope with the following parameters.

(1) Homogeneous or heterogeneous reactions (see below).
(2) In homogeneous systems, large increases in viscosity, affecting kinetics of polymerisation, heat transfer and efficiency of mixing.
(3) Most chain-growth polymerisations are exothermic, giving out several hundred kilowatts per tonne of polymer produced; this heat has to be removed, since most polymerisations are performed at constant temperature (isothermal). Heat removal is achieved by heat transfer through reactor walls, reactor design (long tubular reactors with high surface area/volume ratio), added cold reagents (monomer and/or solvent) or by using the latent heat of evaporation of monomer or solvents.
(4) Control of MW and MWD, branching and cross-linking in the polymer. The polymerisation process affects these—whether batch, semi-batch (polymerising reaction mixture is fed through to further reactors) or continuous. The residence time of polymerisation, whether narrow or broad, also determines MW and MWD.

As a general rule, once a process has been perfected it is not altered, unless economics dictate a change—for example, gas-phase polymerisation performed in the absence of a solvent, thus eliminating solvent purification, recovery, fire hazard, etc.

(c) Polymer recovery.   Unless the polymerisation takes place in bulk, separation from the solvent has to be carried out. The conventional methods for

low-MW compounds, e.g. crystallisation, distillation, precipitation, etc., cannot normally be used because polymers possess properties such as high viscosity and low solubility in solvents, and are sticky and non-volatile. Nevertheless, precipitation by using a non-solvent followed by centrifuging, or by coagulation of an emulsion or latex (see section 4.5.5) and removal of the solvent by steam-stripping (to keep the temperature down and prevent decomposition) can be used. Devolatilisation of the solvent or unused monomer from the polymer can be performed during pellitisation in an extruder.

4.5.1 Polymerisation techniques

Most polymerisations are performed in the liquid phase using either a batch or a continuous process. The continuous method is preferred because it lends itself to smoother operation leading to a more uniform product, because of modern on-line analysis techniques. It also has lower operating costs.

However, continuous processes have difficulties. The residence time of the polymer in the reactor will be variable, unless plug-flow is adopted using tube reactors. This may result in:

(1) Catalyst residues in the polymer, e.g. peroxides, which may degrade the polymer during granulation or processing.
(2) The polymer may have a broad MWD with some decomposition, e.g. cross-linking (for long residence times).
(3) The polymer may adhere to the reactor walls, requiring shut-down for cleaning.
(4) Repeated changes in polymer grade may be required; during change-over the polymer will be a mixture making the polymer unsuitable for use.

There are five general methods of polymerisation:

(1) Bulk (or mass)
(2) Solution
(3) Slurry (or precipitation)
(4) Suspension (or dispersion)
(5) Emulsion

Further lesser-used methods include:

(6) Interfacial
(7) Reaction injection moulding (RIM)
(8) Reactive processing of molten polymers.

4.5.1.1 Bulk polymerisation. No solvent or diluent is used, just the monomer (which can be a liquid or a gas) and catalyst. The polymer (and catalyst) can be soluble in the liquid monomer. The main problem is the increase in viscosity during the course of polymerisation, resulting in poor mixing and heat-transfer. Auto-acceleration (also known as the Trommsdorf effect) occurs because the initiation and propagation steps proceed as normal, but the termination step is inhibited because of the high viscosity. A large increase in the rate of polymerisation is frequently observed.

Two reactors may be used—one from the start to, say, 20% conversion and finishing with a second reactor. Very rarely is polymerisation carried out to 100% in chain-growth polymerisations; the monomer is removed and recycled. In step-growth (where 100% is necessary) it starts in solution and ends in bulk, e.g. nylon salt (see section 4.3.1), which is soluble in water; the water is evaporated off (including the condensation water):

$$[H_3N(CH_2)_6NH_3]^{2+}[OOC(CH_2)_4COO]^{2-} \xrightarrow{-H_2O} \text{nylon 6,6}$$

The polymer can be insoluble in monomer. In this case a slurry is formed. Again, two reactors may be used where the second reactor may have a different shaped stirrer to stop large chunks of the polymer being formed.

A recent technique is bulk gas-phase polymerisation, and this is now used extensively for the polymerisation of ethylene and propylene, and copolymers of these monomers. Several designs are possible, e.g. stirred powder polymer or starting off the polymerisation in liquid propylene and completing in the gas phase, but the most popular is the *fluidised-bed technique* (Unipol Process, Union Carbide Corporation, USA) (Figure 4.11). The catalyst is a slurry of

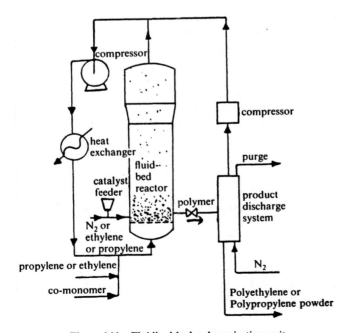

Figure 4.11 Fluidised-bed polymerisation unit.

TiCl$_3$ (Ziegler catalyst) which is pumped (together with the activator Al alkyls) into the fluidised bed, which contains the polymer powder kept in suspension by an ethylene or propylene gas stream. The TiCl$_3$ adheres to the polymer and starts to produce more polymer. Statistically, the larger the size of the polymer particles, the longer they have been in the reactor (and hence the higher the conversion—also called 'mileage', activity × time), and because of their weight they will settle towards the bottom of the reactor, where they will be removed. Abrasion will provide smaller polymer particles and the incoming TiCl$_3$ will adhere to these particles and the process is repeated.

The big advantage of this technique is that no solvent is used, with the attendant purification, recovery, etc. The exothermicity of the system is solved by cooling the ethylene; if propylene is used, liquid propylene is pumped into the reactor, utilising latent heat of evaporation.

Precautions have to be taken to prevent the polymer particle from becoming 'sticky' and to make certain that no 'hot spots' occur (e.g. temperature $\not> 100\,°C$) thus causing melting of the polymer and fusion of the polymer particles.

4.5.1.2 Solution polymerisation. This technique uses a solvent which is miscible with the monomer, and the resulting polymer is also soluble.

Just as in the bulk process, the main worry is the increase in viscosity as the polymerisation proceeds. Auto-acceleration can also take place, though it is not as prevalent as in the bulk method. Impurities in the solvent can affect chain transfer, poisoning of the catalyst, etc.

The solvents can be water- or organic-based. Monomers and their polymers which are soluble in water include polyacrylic acid ($\diagup\!\!\!\diagdown_{\text{COOH}}$), polymethacrylic acid($\overset{\text{CH}_3}{\diagup\!\!\!\diagdown_{\text{COOH}}}$), polyacrylamide ($\diagup\!\!\!\diagdown_{\text{CONH}_2}$), etc. The majority of other monomers/polymers are soluble in organic solvents, e.g. polystyrene, poly(methyl methacrylate), etc. The main advantage with water is that it is cheap, with no fire risk.

4.5.1.3 Slurry (or precipitation) polymerisation. These polymerisations consist of two phases—the diluent phase and the solid polymer. The diluent is a solvent for the monomer and the catalyst or initiator, and a non-solvent for the polymer. The polymer particles are not stabilised (cf. suspension and emulsion polymerisation, see below) and tend to agglomerate together to form a polymer paste or slurry. The absence of a stabiliser or suspending agent means that the polymer is quite pure. Some bulk polymerisations when the polymer is insoluble in the monomer behave similarly to slurry polymerisations, e.g. polyacrylonitrile ($\diagup\!\!\!\diagdown_{\text{CN}}$), PVC and polypropylene.

As soon as polymerisation starts, the polymer particles start to precipitate from the solution. Although one can calculate which is the best solvent to use (e.g. from the solubility of the polymer in various solvents), often the solvent is chosen empirically. The greater the solubility of the polymer, the smaller the polymer particles, i.e. 'fines' may be produced; these are undesirable because they are difficult to handle and can cause explosion hazards.

Fortunately, the viscosity of the slurry is not usually a problem and the heat of the reaction can be counterbalanced by the refluxing solvent.

4.5.1.4 Suspension (or dispersion) polymerisation. This is sometimes called bead or pearl polymerisation because of the smooth spherical shape of the polymer particles produced.

In suspension polymerisation, the monomer and polymer are insoluble in the solvent (which is often water). The catalyst (or initiator) is soluble in the monomer (see Table 4.3), which is suspended as small droplets in the solvent (water) phase by vigorous agitation aided by adding suspending agents, which prevent the monomer (and hence the polymer) from coagulating together. These suspending agents (sometimes called stabilising agents) are absorbed on the droplet and polymer particle surface and can be a variety of either water-soluble compounds, e.g. water-soluble natural polymers such as starch, or synthetic (polyvinyl alcohol) or inorganic (e.g. clays, aluminium hydroxides) insoluble powders, or alternatively surfactants.

The polymerisation can be thought of as a whole series of micro-bulk reactions and the resulting polymer particles are ~ 1 mm in diameter. These settle out as soon as the agitation stops. The heat of the reaction is taken care of by adding cold solvent (water) or refluxing. The viscosity of the mixture is low.

The filtered or centrifuged polymer has to be dried to free it from the diluent, which is expensive. Although a high packing-density polymer with a close polymer size range is often desired, in some cases, especially PVC, a porous particle is necessary for the subsequent absorption of plasticisers (see section 4.6.2.1).

4.5.1.5 Emulsion polymerisation. This technique makes latex polymers consisting of colloidal polymer particles. Again, as with suspension, the system contains two liquid phases consisting of the liquid monomer, or a solution of the monomer in a diluent, and (usually) water. The water is the continuous phase, i.e. it is in excess, and the polymer is insoluble in water.

Similarly to suspension polymerisation, the mixture is vigorously agitated and also a surfactant is used, this time called an emulsifier, which is often an anionic (e.g. Na salts of long-chain carboxylic or sulphonic acids, $RCOO^-Na^+$ or $RSO_3^-Na^+$ or non-ionic (e.g. $R-(CH_2CH_2O-)_n H$, where $n \approx 10$) surfactant.

The main difference from suspension polymerisation is that a water soluble initiator is used, e.g. persulphate, H_2O_2/M^{n+} (see section 4.3.2.1)—remember in

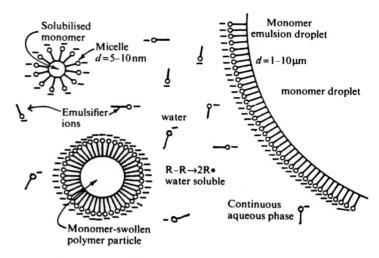

Figure 4.12 Likely structure of monomer–solvent droplets.

suspension polymerisation an oil (or more correctly, monomer) soluble initiator is used. As a consequence of this the final polymer particles have
diameter $\sim 0.1\ \mu m$ (cf. ~ 1 mm for suspension polymerisation).

From observations such as the number of polymer particles are 1 000 times
that of the monomer–solvent droplet, the way in which the polymerisation
proceeds is thought to be as shown in Figure 4.12. The water phase contains
the initiator and small amounts of dissolved monomer (depending on solubility), chain-transfer agents to limit MW, and the emulsifier. Most of the
monomer is present in the large droplet, but some of the monomer is present as
micelles, solubilised and stabilised by emulsifier.

The water-soluble free radicals (e.g. SO_4^- from persulphate) react with the
monomer within the micelles, since they have a much greater surface area
compared with the large droplets, and polymerisation commences and de-
pletes the monomer within the micelles. Replenishment of the monomer within
these micelles, to form more polymer, comes from the large droplets with the
aid of the emulsifier until the size of these monomer-swollen polymer particles
is such that their surface areas become negligible compared with the new
monomer–emulsifier micelles, which are continually being formed. The size of
these polymer particles is $\sim 0.1\ \mu m$.

Eventually the monomer in the large droplet is exhausted and the polymerisation ceases. The role of the emulsifier is to assist transfer of the monomer
from the large droplet through the water phase to the micelle, thus allowing
attack by the water-soluble free radical. It also stabilises the growing monomer-swollen polymer particle so that a stable emulsion is produced.

Many of these latexes are used as such for surface coatings and paints and
also water-borne adhesives, etc., but when bulk polymer is required coagu-

Table 4.7 Advantages and disadvantages of various methods of polymerisation

	Technique	Advantages	Disadvantages
Bulk (no solvent)	Batch	(1) Low impurity level (2) Casting possible	(1) Thermal control difficult (2) Autoacceleration possible
	Continuous	(1) Improved thermal control (2) No solvent purification, recovery, etc.	(1) Isolation difficult (requires monomer devolatilisation) (2) Requires agitation and monomer recycling (3) 'Inerts' build-up, therefore flaring (purging)
Solution (polymer soluble in solvent		(1) Improved thermal control (2) Can use polymer solution directly (3) Can be used with high-m.p. polymers (4) Can use ionic solvents (e.g. water) to promote step-growth (e.g. nylon)	(1) Difficult to remove solvent (2) Solvent recovery costly (3) Chain transfer from solvent may limit MW (4) Viscosity increases (autoacceleration?) (5) Often isolation of polymer produces irreversible change and cannot be redissolved
Slurry (polymer insoluble in solvent)		(1) little viscosity increase	(1) particle size and size distribution difficult (2) solvent removal: organic (fire risk) aqueous (expensive)
Suspension (solvent and monomer non-miscible)		(1) low viscosity (2) simple polymerisation (3) easy thermal control (4) may be of direct usable particle size	(1) highly sensitive to agitation rate (2) particle size difficult (3) possible contamination from suspending agent (4) washing, drying and compaction necessary
Emulsion (solvent and monomer non-miscible)		(1) low viscosity (2) good thermal control (3) latex may be directly usable (4) 100% conversion may be achieved (5) high MW possible at high rates (6) Small particle size product can be obtained (7) Soft or tacky polymers can be made	(1) emulsifiers, surfactant and coagulants must be removed (2) high impurity levels may degrade certain polymer properties (3) high cost (4) washing, drying compaction may be necessary

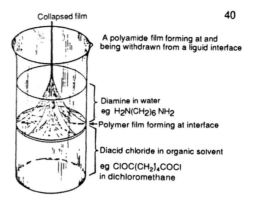

40

Figure 4.13 The 'nylon rope trick'.

lation can be achieved by adding electrolytes, thus increasing the ionic strength and destroying the emulsifying action.

Many polymers are made by this technique, e.g. PVC, polyvinyl acetate (PVA), polystyrene and its copolymers (ABS, SAN and SBR), and poly(methyl methacrylates).

4.5.1.6 Advantages and disadvantages of various methods of polymerisation. Table 4.7 gives the advantages and disadvantages of the above processes. These are the main methods for making polymers. Other, more specialised techniques include the following.

4.5.1.7 Interfacial polymerisation. This is very similar to suspension and emulsion polymerisation in that a two-phase system consisting of monomer (and solvent) and a non-miscible second solvent (again, often water). The 'nylon rope trick', Figure 4.13, illustrates this, but is not used commercially.

The reactants in interfacial polymerisation are normally agitated, and the main use is for the preparation of polycarbonates from the gas phosgene ($COCl_2$) and a bisphenol.

Sodium salt of
bis-phenol A

Polycarbonate of bis-phenol A (this
is the most popular polycarbonate)

The system consists of small droplets of dichloromethane (CH_2Cl_2) containing $COCl_2$ suspended in a water solution containing alkaline phenolate. In order for reaction to take place, it is necessary to have a phase transfer agent that will facilitate transport of the phenolate into the organic solvent, thus promoting polymerisation (see Figure 4.14).

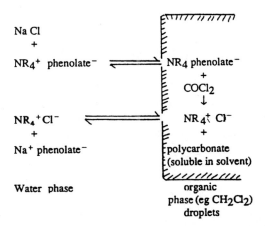

Figure 4.14 Interfacial polymerisation system for the preparation of polycarbonates from phosgene.

Na^+ phenolates$^-$ are very insoluble in CH_2Cl_2, but the tetra-alkyl ammonium salts (e.g. $NR_4^+OPh^-$) are soluble and therefore on addition of, say $NBu_4^+Cl^-$ the following equilibrium is established:

$$NBu_4^+Cl^- + Na^+OPh^- \rightleftharpoons NBu_4^+OPh^- + NaCl$$
(Phase transfer agent)   (Soluble in CH_2Cl_2)

and the passage of the phenolate into the organic phase is promoted.

The resulting polycarbonate polymer is soluble in this particular solvent (CH_2Cl_2). If control of the MW of the polycarbonate is required, a monofunctional phenol (C_6H_5OH as C_6H_5ONa) is added. The polymer solution–water is often spray-dried to give the polymer.

The main advantage of this technique is that rapid polymerisation can take place at controlled low temperatures. The disadvantage is that the inorganic salt (NaCl) has to be washed out.

Phase transfer agents are used nowadays to promote reactions in which the two reagents, or solutions of these reagents, are non-miscible with each other, e.g. oleophilic and hydrophilic reagents or solutions. They do this by making the water-soluble reactant (often salts, the hydrophilic reagent) soluble in the organic phase (which contains the oleophilic reagent). Typical phase transfer agents include tetra-alkyl ammonium salts and crown ethers. This technique can replace the use of expensive polar non-protic solvents in which both reagents are soluble at high temperatures, e.g. DMF, DMSO.

Just as phase transfer agents allow reactions to occur in liquid–liquid non-miscible systems, they also promote reaction between insoluble solid–liquid systems by solubilising the surface layers of the solid. Thus the following polymer can be obtained by using solid K_2CO_3 and a phase transfer agent:

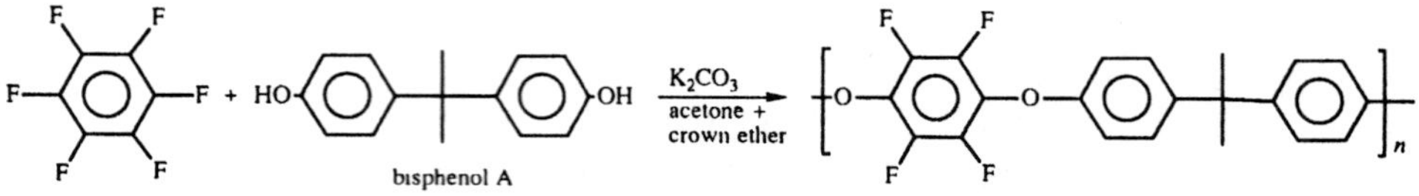

bisphenol A

4.5.1.8 Reaction injection moulding (RIM).

4.5.1.8   Reaction injection moulding (RIM).   This is a recent development which has been made possible by advances in catalysis, polymer engineering (e.g. metering, valves), polymer physics (rheology—the study of the flow properties of monomers and polymers), and kinetics, together with computer control of the overall process parameters (e.g. temperature, mould pressure). A further advance is to perform RIM in the presence of short fibres; this is called re-inforced RIM or RRIM.

The process involves the rapid mixing of two or more reactive monomers, or macromers or oligomers, and the injection of this reactive mixture into a mould. Other components may be added to the monomers, e.g. foaming or blowing agents, polymers which are soluble in the monomers, dyes, etc. The two reagents may be catalysts (or catalyst components) and monomers in separate reservoirs.

Adiabatic polymerisation takes place rapidly—i.e. the mixture heats up—and the article after ejection from the mould must have sufficient 'green strength' to be self-supporting, allowing any post-reactor polymerisation to occur if necessary.

The advantage of this method is that the polymerisation in the mould is very rapid (often <1 min), the temperature rise contributing to the polymerisation, but not so high as to cause degradation of the polymer. Because of the low viscosity of the monomers, large complicated articles can be moulded, which would be impossible by standard injection moulding from the heated polymer melt. Also, using a heated mould would incur long fabrication times because the mould must be cooled after the injection of the polymer. Another benefit from the RIM technique is that, because high pressures are not used (cf. standard injection moulding), cheaper moulds can be used, e.g. aluminium compared with stainless steel.

The process is used for:

(1) Making polyurethanes via the reaction

$$\sim NCO + HO \sim \longrightarrow \sim NHCOO \sim$$

using di-isocyanates and di- (or tri-) ols as reagents. Traces of water can be added with the diols to generate CO_2 (i.e. foamed polyurethanes)

$$\sim NCO + H_2O \longrightarrow \sim NH_2 + CO_2\uparrow$$

thus giving good mould replication. Depending on the reagents large, rigid or elastomeric products can be obtained.

(2) Nylon 6 $\binom{NHCO}{(CH_2)_5} \xrightarrow[\text{catalyst}]{\text{Anionic}} \text{--}[(CH_2)_5CONH]_n$

(3) Epoxy compounds and unsaturated polyesters.
(4) Unsaturated polyesters.

4.5.1.9 Reactive processing of molten polymers. Blending of polymers is often performed in the heated barrel of an injection-moulding or extrusion machine. In these processes, heat and residence time (<5 min) are kept to a minimum to avoid degradation of the polymers. Reactive processing of polymer mixtures, on the other hand, is performed at high temperatures (up to 500 °C) and for long times (up to 45 min), sometimes with the addition of catalysts in order to promote controlled modification of the polymers.

Originally, simple single-screw extruders were used, but now custom-built apparatus with intermeshing twin screws for efficient mixing is employed. If necessary, if gaseous products are formed, devolatisation with use of a vacuum can be performed, or the converse, the addition of gaseous or liquid reagents through ports in the extruder.

A big advantage of this method is that no solvents are used, eliminating recovery, fire hazards, environmental problems, etc., and it can handle highly viscous mixtures. In principle, the moulded articles can be made directly, but usually polymer manufacturers just pelletise or granulate the product for sale to the fabricators.

Examples include:

(1) Obtaining very high-MW polymers from step-growth polymers, e.g. polyesters.
(2) Grafting, for example, maleic anhydride or acrylic acid on to polypropylene

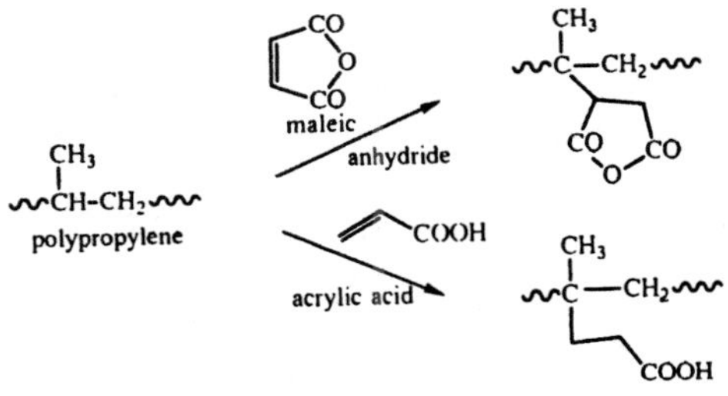

These grafted PP materials form superior blends with nylon 6.6 (cf. ungrafted PP). The main application is wheel hubs for cars.

(3) To increase the heat softening points of acrylates

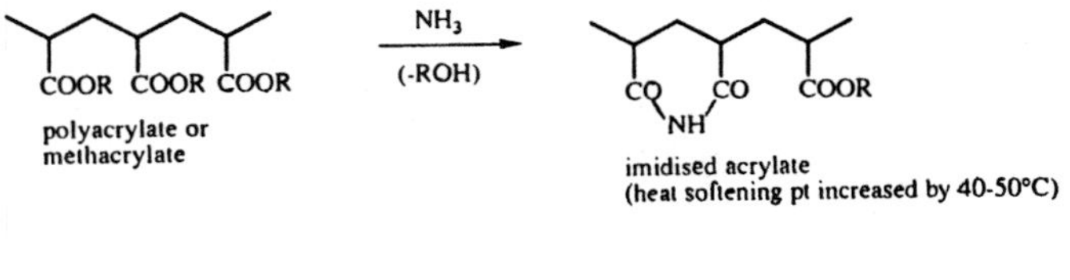

Table 4.8 Some properties correlatable to molecular weight distribution (MWD)

Tensile strength	Impact strength
Modulus of elasticity	Tear strength
Relaxation times	Low temperature toughness
Melt viscosity	Resistance to environmental attack
Hardness	Stress cracking
Flex life	Drawability
Biological activity	Coefficient of friction
Softening temperature	Spinnability
Elongation at tensile break	Permeability

4.6
Properties of polymers

In Figure 4.2, the influence of MWD on some of the properties of a polymer was shown. This list is extended in Table 4.8.

Not all these properties can be considered here. An important parameter is the effect of temperature on a polymer, e.g. can the polymer be melt-processed; what is the maximum temperature the polymer can stand without losing strength, etc.?

4.6.1   Effect of temperature on polymers

It has been found that the viscosity of any substance (e.g. organic solids, metals, water, etc., including polymers) at its melting temperature, T_m, is $\sim 10^{13}$ poise. However, when non-polymeric materials are heated to 20–50 °C above T_m this viscosity decreases to 1–100 poise, whereas polymers (with reasonably high MWs) only fall to 10^5–10^{11} poise. Moreover, in some polymers the value for T_m is very ill-defined, whereas in others it is quite sharp. The reason for this is because some polymers are crystalline (sharp T_m) whilst some are non-crystalline i.e. amorphous or glassy (diffuse T_m). This means that the polymer molecules, or segments of the polymer molecules, arrange themselves into 'ordered' crystalline regions or domains, whereas in some other polymers little or no order is observed.

The extreme is when a polymer does not have any T_m at all. Thus when viscosity versus temperature is plotted, two different types of behaviour can be observed (Figure 4.15). With amorphous polymers, the rate of decrease in viscosity is not linear; there are temperature regions where the reduction in viscosity is greater (though not as sharp as crystalline materials).

Viscosity is the resistance to flow; mathematically it is equal to stress/rate of strain. Modulus is very similar: resistance to deformation on applying an applied force or stress. Its value is stress/strain, and at a given stress, modulus = stress/$(\Delta L/L_0)$ (see section 4.8.1), where L_0 = initial, unstretched sample dimension or length, and ΔL = the deformation (or elongation if a rod or bar is under test). Modulus is a measure of the stiffness or rigidity of a polymer, and it is temperature- and time-dependent.

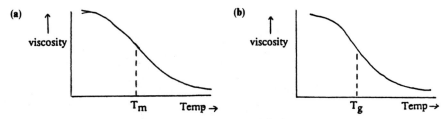

Figure 4.15 Melting behaviour of (a) crystalline and (b) amorphous polymers.

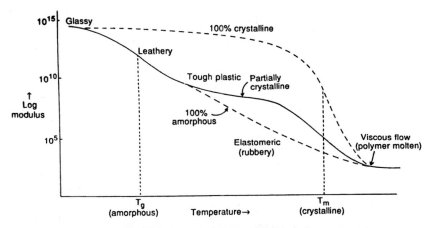

Figure 4.16 Plot of 10 second tensile modulus against temperature.

When a stress is applied to a polymer, the time for the polymer to readjust should be stated e.g. for a 10 second modulus, the measurement is taken 10 s after the applied stress. With amorphous and crystalline polymers, when a 10 second tensile (i.e. stretching) modulus of a polymer is plotted against temperature the plot shown in Figure 4.16 is obtained.

Two transitions can be identified: T_g (called the glass transition temperature) and T_m (melting or softening temperature) which occur because of the amorphous and crystalline regions of the polymer respectively. Most polymers are either 100% amorphous or partially (say 5–80%) crystalline. Very rarely can 100% crystallinity in polymers be achieved, whereas with low-MW solids and metals it is common.

The ability of a polymer to crystallise implies that segments of the polymer chain organise themselves into a regular repeating pattern (see Figure 4.17). This is only possible if the temperature is high enough (to allow thermal motion of these segments) and there are enough holes or free volume. Free volume is the amount of space not occupied by the polymer molecules themselves.

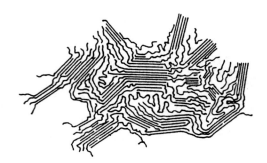

Figure 4.17   Two-dimensional representation of a polymer with ordered crystalline (micelle) and tangled amorphous (fringed) regions.

4.6.2   Glass transition temperature, T_g

This is perhaps the most important parameter to consider when deciding on an application for an amorphous polymer; it is less important with semi-crystalline polymers. Thus a high-MW amorphous polymer with a T_g of $-70\,°C$ would have elastomeric (rubbery) properties at room temperature, whereas if $T_g = 100\,°C$, then a hard plastic would result. For high-temperature applications (e.g. 'under the hood' automobile applications), an amorphous polymer of T_g 200 °C would be required.

The T_g is the temperature above which the molecules start to move— sufficiently, that is, for them to knock other polymer segments out of the way, thus creating new holes. Thus, a polymer can only crystallise when the temperature is above the T_g for that polymer. Obviously it cannot crystallise at temperatures above T_m, the melting temperature of the crystallites (Figure 4.18).

The T_g of a homopolymer is a unique value for the polymer, and it is often quoted. Unfortunately it is time (of measurement) dependent and the absolute value of T_g is perhaps impossible to measure because it depends on the rate of

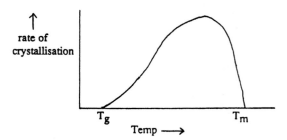

Figure 4.18   Crystallisation of a polymer.

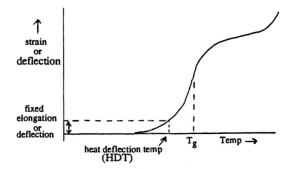

Figure 4.19   Plot of strain or deflection against temperature.

heating or cooling. With rapid heating, at a given temperature, segments may not have time to move into free-volume spaces (holes) before the spaces close, thus preventing movement. Therefore a higher temperature is necessary to cause movement. Thus experimentally determined T_g values increase as the rate of heating increases or T_g values decrease as the rate of cooling decreases.

There are many other techniques for measuring T_g; indeed, just about any property which is dependent on the movement of polymer segments will show a change at T_g. Thus, when plotted against temperature the following will give a value for T_g: modulus and specific volume, heat capacity, refractive index, dielectric loss, X- and β-ray adsorption, gas permeability, ^1H and ^{13}C NMR. However, as stated above, if the rate of measurement is rapid, e.g. ^{13}C NMR, the value of T_g could be 20–50 °C higher than for 10 second modulus measurements.

For mechanical applications, a useful plot is the deflection (or deformation or strain) for a given stress (or applied force) versus temperature (see Figure 4.19). Also included in this plot is the heat deflection temperature (HDT). This is measured by a fixed weight on a specified measured bar of polymer. The temperature at which a fixed deflection occurs is the HDT. It is a guide to the maximum working temperature of the polymer. Experimentally it is 10–20 °C below T_g.

As mentioned above, because polymers are never 100% crystalline, all polymers show a T_g, the importance of which depends on the amount of crystallinity. If this is high, e.g. nylon 6,6, PTFE, or PE, it is not too important, but for 100% amorphous polymers (e.g. PS, PMMA and several advanced engineering polymers), because they do not have a T_m, the value of T_g will dictate their final end use, particularly with temperature.

4.6.2.1   Parameters affecting T_g.   These can be sub-divided into chemical effects (i.e. how various chemical or molecular groups within the polymer alter T_g) and also, given that a polymer with a particular structure has been made,

Table 4.9   Thermal transitions of polymers

Polymer	Repeating unit	$T_g(°C)$	$T_m(°C)$
Polydimethylsiloxane	—OSi(CH$_3$)$_2$—	−127	−40
Polyethylene	—CH$_2$CH$_2$—	−125	137
Polyoxymethylene	—CH$_2$O—	−82	181
Natural rubber (poly *cis*-isoprene)	—CH$_2$C(CH$_3$)=CHCH$_2$—	−73	28
Polyisobutylene	—CH$_2$C(CH$_3$)$_2$—	−73	44
Poly(ethylene oxide)	—CH$_2$CH$_2$O—	−41	66
Polypropylene	—CH$_2$CH(CH$_3$)—	−13	176
Poly(vinyl fluoride)	—CH$_2$CHF—	−41	200
Poly(vinylidene chloride)	—CH$_2$CCl$_2$	−18	200
Poly(vinyl acetate)	—CH$_2$CH(OCOCH$_3$)—	32	
Polychlorotrifluoroethylene	—CF$_2$CFCl—	48	220
Poly−(ε−caprolactam)	—(CH$_2$)$_5$CONH—	52	223
Poly(hexamethylene adipamide)	—NH(CH$_2$)$_6$NHCO(CH$_2$)$_4$CO—	50	265
Poly(ethylene terephthalate)	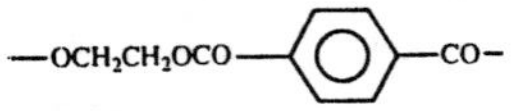 —OCH$_2$CH$_2$OCO—⟨O⟩—CO−	69	270
Poly(vinyl chloride)	—CH$_2$CHCl—	81	273
Polystyrene	—CH$_2$CPh—	100	240
Poly(methyl methacrylate)	—CH$_2$C(CH$_3$)(CO$_2$CH$_3$)—	105	200
Cellulose triacetate		105	306
Polytetrafluoroethylene	—CF$_2$CF$_2$—	−150 (some controversy)	327
Polyether ether ketone (PEEK)	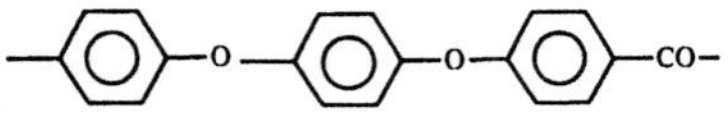 ⟨O⟩—O—⟨O⟩—O—⟨O⟩—CO—	143	334
Polycarbonate (of bisphenol A)	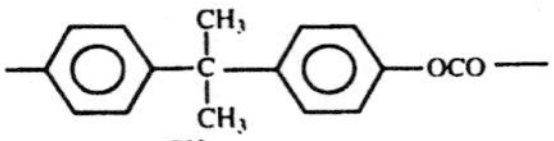 ⟨O⟩—C(CH$_3$)$_2$—⟨O⟩—OCO—	149	267
Polyphenylene oxide (or polyphenylene ether) (PPO)	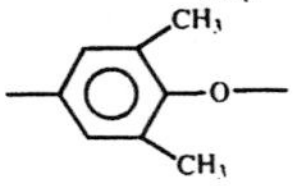 ⟨O⟩(CH$_3$)$_2$—O—	211	268
Polyphenyl sulphide (PPS)	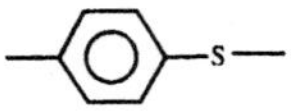 ⟨O⟩—S—	200	285
Polyetherimide (PEI—Ultem)	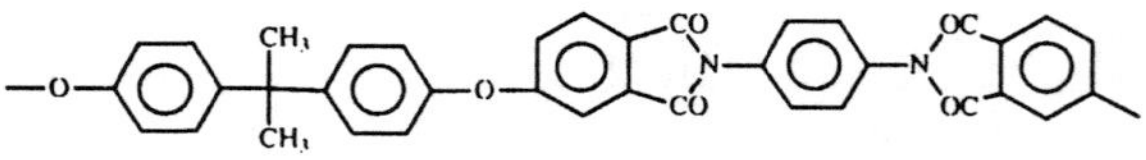 —O—⟨O⟩—C(CH$_3$)$_2$—⟨O⟩—O—⟨imide⟩—N—⟨O⟩—N—⟨imide⟩—	217	
Polyether sulphone (PES)	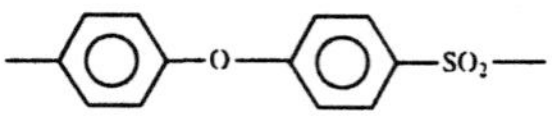 ⟨O⟩—O—⟨O⟩—SO$_2$—	225	
Aromatic polyamide (*meta*) (Nomex)	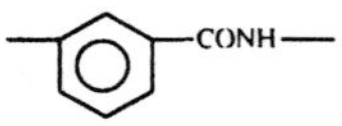 ⟨O⟩—CONH—	270	
Polyamide imide (PAI)	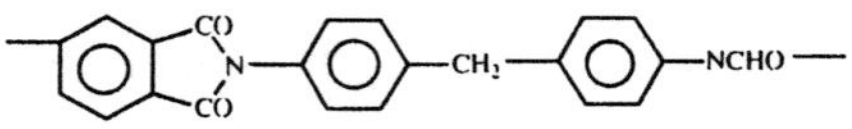 ⟨imide⟩—N—⟨O⟩—CH$_2$—⟨O⟩—NCHO—	280	
Aromatic polyamide (*para*) (Kevlar)	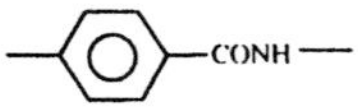 ⟨O⟩—CONH—	Decomposes >400	

Left margin groupings: Commodity polymers; Engineering polymers; Advanced polymers.

how can the T_g (which is fixed for a particular homopolymer) be altered. Thermal transitions of many polymers are given in Table 4.9.

Molecular or chemical structure of the polymer chain

(1) *Chain flexibility*

When chain rotation can occur easily (no steric hindrance) then a low T_g is observed.

$\sim CH_2$⟨CH$_2 \sim$ Polyethylene (PE) $T_g \approx -125\,°C$

$\sim Si$⟨O$\sim$ Polydimethylsiloxanes (SI) $T_g = -123\,°C$

(with CH$_3$ groups above and below the Si)

$\sim CH-CH_2-O\sim$ Polyethylene oxide (PEO) $T_g = -47\,°C$

Para-substituted aromatic groups increase T_g.

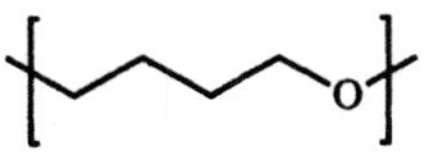

poly THF, $T_g = -40°C$ 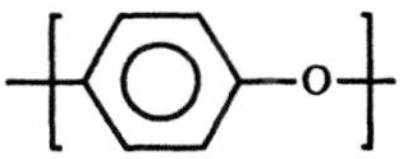   polyphenyl oxide (PPO) $T_g = 105°C$

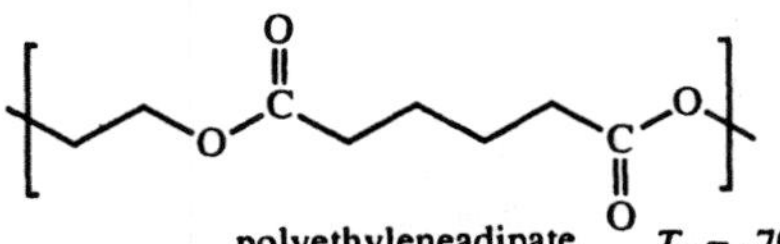

polyethyleneadipate (plasticiser for cling wrap film) $T_g = -70°C$

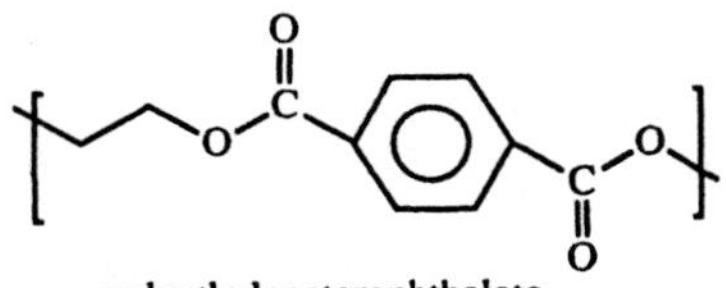

polyethyleneterephthalate (PET) $T_g = 80°C$

Steric restriction in polymers of the type $-CH_2-\overset{\text{X}}{\underset{|}{CH}}-$ increases with increasing size of the group X, and hence raises T_g.

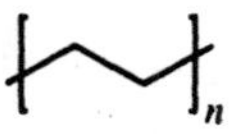

polyethylene $T_g = -125°C$

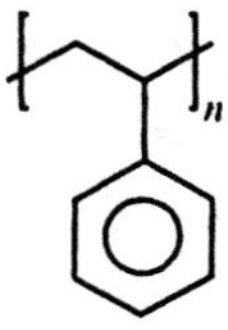

polystyrene $T_g = 100°C$

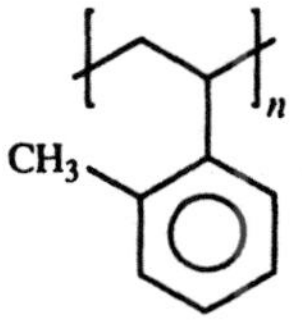

$T_g = 115°C$

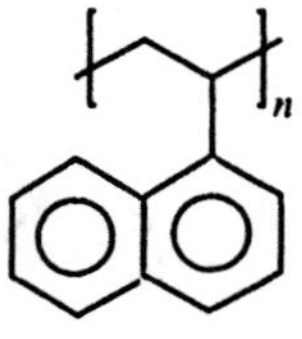

$T_g = 135°C$

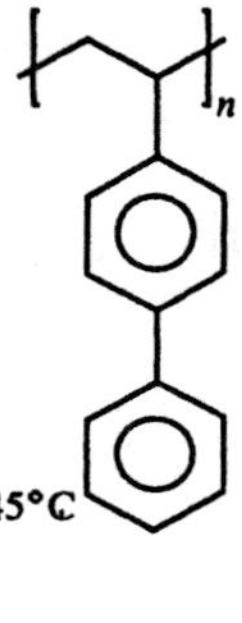

$T_g = 145°C$

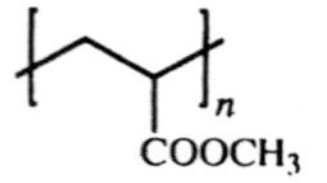

polymethyl acrylate
(PMA)$T_g = 0°C$

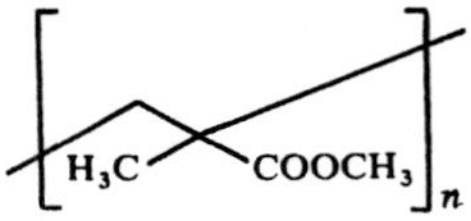

polymethyl methacrylate
(PMMA)$T_g = 105°C$

Not only the size of the R group is important, but also its shape, e.g. in the following polybutyl methacrylates:

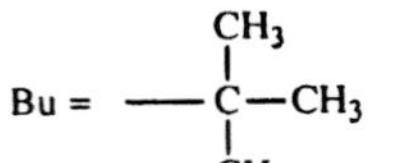

$$Bu = \begin{array}{c} CH_3 \\ | \\ —C—CH_3 \\ | \\ CH_3 \end{array}$$

tertiary butyl
$T_g = 43°C$

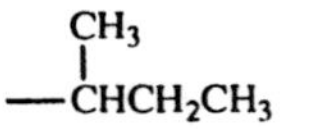

$$\begin{array}{c} CH_3 \\ | \\ —CHCH_2CH_3 \end{array}$$

secondary butyl
$T_g = -22°C$

$$—CH_2CH_2CH_2CH_3$$

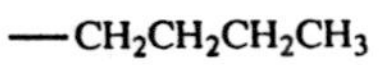

primary butyl
$T_g = -56°C$

(ie rubbery at room temp)

(2) *Interchain forces (see p. 108)*
These can be either H-bonding or dipole–dipole attraction. These forces increase T_g.

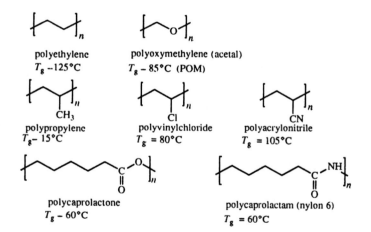

polyethylene
$T_g -125°C$

polyoxymethylene (acetal)
$T_g - 85°C$ (POM)

polypropylene
$T_g - 15°C$

polyvinylchloride
$T_g = 80°C$

polyacrylonitrile
$T_g = 105°C$

polycaprolactone
$T_g - 60°C$

polycaprolactam (nylon 6)
$T_g = 60°C$

(3) *Symmetry of the monomer unit*
If the monomer unit is asymmetric (with only one substituent) it will tend to have a higher T_g than the symmetrical disubstituted unit:

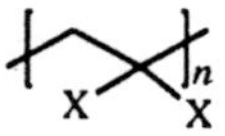

(symmetric)

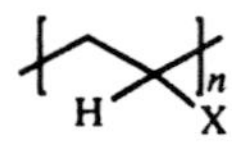

(asymmetric)

$X = CH_3 \; -70 °C$ (Polyisobutylene) (PIB) $-15 °C$ (Polypropylene)
Cl $-18 °C$ (PVDC) $80 °C$ (PVC)
F $-40 °C$ (PVDF) $110 °C$ (PVF)

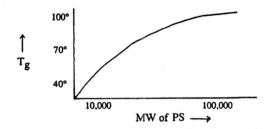

Figure 4.20   Plot of T_g against molecular weight for polystyrene (PS).

Alteration of T_g for a given structure

(1) *MW of the polymer*. The lower the MW, the lower the T_g (Figure 4.20). An explanation is that as the number of chain ends increases (lower MW), the amount of free volume space increases (chain ends can sweep out much more space than internal units).

(2) *Branching*. Although the effect of branching (which occurs frequently in free-radical vinyl chain-growth polymerisations) decreases the ability of the main chain to move (by steric restriction—see above), the dominant effect is the presence of more chain ends. Thus branching decreases T_g.

(3) *Cross-linking*. This can be thought of as increasing the MW of the polymer, hence increasing T_g, but a more correct explanation is that the cross-links bring the polymer chains closer together, thereby increasing the T_g (Figure 4.21). The classical example of curing or cross-linking is with natural rubber using sulphur. With about 0.3% curing (i.e.–S–S– cross links every 300 repeat units) the rubbery nature is preserved. However, on increasing the cross-link density the hard, brittle, thermoset ebonite is obtained. Thermoset materials (see section 4.7) do not have T_g or T_m values; they tend to decompose before softening, because the required temperature is so high.

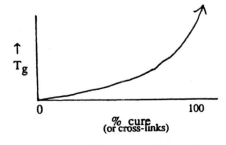

Figure 4.21   Effect of cross-links on T_g.

(4) *Copolymers and blends.* These are sometimes called polymer alloys. Amorphous random $(AB)_n$ copolymers (see section 4.4.2.1) exhibit a single T_g value. A rough estimate of T_g can be obtained from measurements on the component homopolymers $(A)_n$ and $(B)_n$. The T_g of the random copolymer is always intermediate between the T_g values of the homopolymers, and its value is closer to the homopolymer with the highest proportion of monomer units (see Figure 4.9). A single T_g also occurs with miscible (compatible) blends of two polymers though this is quite rare. This phenomenon can be very useful, e.g. the two amorphous

polymers 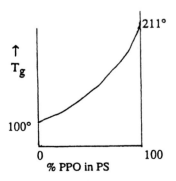 polyphenylene oxide (PPO) (T_g 211 °C) and

polystyrene (PS) (T_g 100 °C) are (surprisingly) miscible with each other. In order to melt-process it is necessary to heat PPO (and indeed any amorphous polymer) to ~150 °C above its T_g to get a reasonable melt viscosity. However, at these temperatures PPO decomposes (the $-CH_3$ groups tend to oxidise, etc.) The addition of PS lowers the T_g and also the viscosity of the melt, and makes processing possible without sacrificing the mechanical properties of PPO too much (see Figure 4.22). Another bonus is that PS is cheap compared with PPO.

For semi-compatible (or miscible) systems, one of the components tends to be distributed as droplets in a continuous matrix of the other component. The droplet size and degree of separation depend on % composition, MW, method of sample preparation, etc. Two T_g values, or more if there are more components, will be observed for these systems.

With block or graft copolymers, or blends, where the component parts are non-miscible, two distinct T_g values will be obtained. The addition of compatibilisers to homogenise blends (see Figure 4.9) does allow some broadening of these individual T_g values.

Figure 4.22 T_g values of blends of polystyrene (PS) and polyphenylene oxide (PPO).

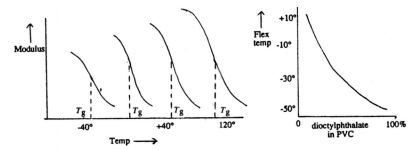

Figure 4.23 Interaction between a polymer and a plasticiser.

(5) *Plasticisation.* This is the addition of low-MW materials which, when added to rigid and brittle polymers, reduce the T_g thus allowing the parent polymer to be processed at a reasonable temperature (e.g. PVC), and also converting the polymer into a flexible solid.

For a solvent to be an effective plasticiser, there must be some interaction between the polymer chain and the solvent (see Figure 4.23); e.g. with PVC ↔ carbonyl group in ester plasticiser:

A simple explanation is that the solvent (the T_g of which is usually very low, typically -100 to $-200\,°C$) reduces the T_g of the miscible mixture (i.e. solvent/polymer) to a low single T_g. Using a free-volume approach, the free volume of the solvent (plasticiser) is large allowing more movement and flexibility of the polymer chain.

Water is a common plasticiser in nature, allowing cellulose polymers (e.g. wood) to bend. Water can also act as a plasticiser with hygroscopic synthetic polymers, e.g. nylon 6,6 and the T_g can be reduced from $\sim 50\,°C$ to room temperature in going from a dry atmosphere to a very humid one. In water, the polymer chains become more flexible, and nylon shirts and blouses lose their creases when hung up to dry (drip-dry).

4.6.3 Crystallisation of polymers

In order for polymer chains to organise themselves into ordered structures and crystallise, the chains must be able to:

(1) Be held together by intermolecular forces, so as to maintain a lattice structure (see also section 4.9.2).

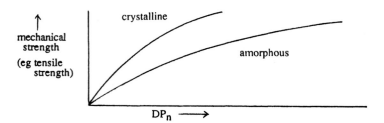

Figure 4.24 Mechanical strength of crystalline and amorphous polymer.

(2) Closely pack together to form this regular lattice structure.
(3) Have sufficient flexibility and mobility to create this lattice structure.

As expected, many of these conditions are opposite to those promoting glassy or amorphous polymers.

As stated earlier (section 4.6.2), when a polymer can crystallise the importance of the amorphous phase and the way it affects various properties (e.g. modulus) is diminished. This is because even when a small amount of crystallinity occurs (say 5%), the presence of these high-modulus regions together with the associated 'tie' molecules (polymer chains which participate in both the crystallites and the amorphous regions) enhances the mechanical strength of the polymer matrix (Figure 4.24).

With amorphous polymers, chain entanglement generates strength, but this requires a higher DP_n than for semi-crystalline polymers.

4.6.3.1 Intermolecular forces between polymer chains (see section 4.9.2). Three types of intermolecular forces may be encountered.

(1) *Dispersion forces* (also called van der Waals or London attraction forces). These are present in every polymer system, but are the only type present in non-polar hydrocarbon polymers, e.g. PE, PS, and polybutadiene. They are the weakest forces and their strength is critically dependent on the chains achieving the correct packing distance in the crystal lattice.

Table 4.10 Bond lengths and bond dissociation energies of intermolecular forces

Intermolecular force		Bond length $\mathring{A}$ ($=0.1$ mm)	Bond dissociation energy (kJ/mole)
	Dispersion	~5	~0.4
	Dipole–dipole	~5	~16
H-bond	$\begin{cases} -O-H\cdots N- \\ -O-H\cdots O- \\ -N-H\cdots O- \end{cases}$	2.8 / 2.7 / 2.9	12–25 / 12–25 / ~16
	compared with		
	C – C covalent bond	1.54	350

(2) *Dipole–dipole forces.* These are due to differences in the electronegativity of the covalently bonded atoms.

(3) *Hydrogen bonds.* These are a special type of dipole–dipole, but stronger because the H atom is an electropositive element.

In polymer systems all these forces are additive and accumulative, giving polymers their properties and crystalline regions (where these forces are maximised) their strength.

4.6.3.2   Crystallinity and melting point, T_m.   On heating a crystalline polymer, it fuses or melts at T_m. This is not a sharp melting point but occurs over several degrees. This process is described by the free energy of melting (ΔG_m) and can be written as:

$$\Delta G_m = \Delta H_m - T_m \, \Delta S_m$$

But at T_m, $\Delta G_m = 0$ and by rearranging, it can be seen that $T_m = \Delta H_m / \Delta S_m$. Thus the value of T_m will be increased by increasing ΔH_m or decreasing ΔS_m. ΔH_m is a measure of the intermolecular forces which, if strong, will raise T_m (e.g. H-bonding in water raises T_m to a higher value, 0 °C, than that expected from its MW). The ΔS_m term depends on the degree of polymer chain movement and any restriction imposed upon it will decrease ΔS_m and hence increase T_m.

Thus, comparing the requirements for crystallinity (see section 4.6.3), the intermolecular forces increase crystallinity and raise T_m, but the chain rigidity, although it raises T_m, can inhibit crystallinity.

4.6.3.3   Factors affecting the ability of a polymer to crystallise and the effect on T_m.   This section should be compared with that on amorphous polymers (section 4.6.2.1)

(1) *Chemical composition of the polymer chain*
   The symmetry of the chain influences both T_m and the ability to form crystallites
   (a) Symmetric —C—C— backbone homopolymers, e.g. HDPE, PTFE and other polymer chains containing —O—, —C—O—, or —C—NH— crystallise, often in extended zig-zag conformations.
      If the backbone contains angled bonds, eg *cis*-double bonds, *ortho*- or *meta*-phenylene groups, or *cis*-orientated puckered cyclohexane rings, then close packing is difficult and these polymers do not crystallise (Figure 4.25).
      In a similar manner, irregularities obtained by branching or random grafting of small groups inhibit crystallisation (see Figure 4.7), e.g. LDPE, ~ 30 branches/1000 carbon main-chain atoms; $\sim 40\%$ crystalline; specific gravity 0.92; HDPE, ~ 3 branches/1000 carbon main-chain atoms; $\sim 90\%$ crystalline; specific gravity 0.97.
      Similarly, chlorination and chlorosulphonation of PE (i.e. intro-

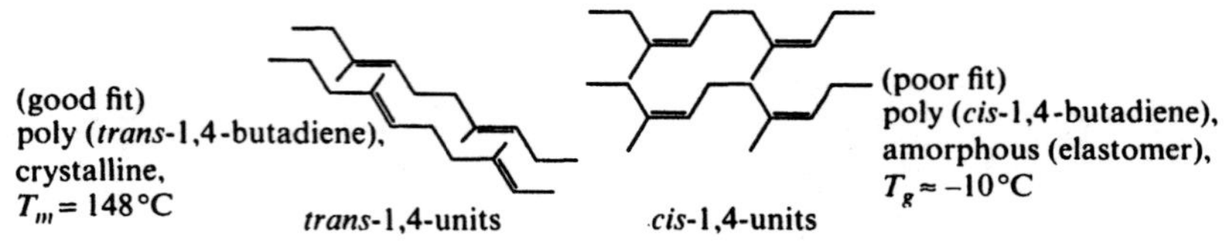

(good fit)
poly (*trans*-1,4-butadiene),
crystalline,
$T_m = 148\,°C$

trans-1,4-units   *cis*-1,4-units

(poor fit)
poly (*cis*-1,4-butadiene),
amorphous (elastomer),
$T_g \approx -10\,°C$

Figure 4.25   Skeletal representations of units of 1,4-dienes derived from 1,4-butadiene, chloro-
prene, isoprene and other 1,4-dienes.

duction of —Cl and —SO$_2$Cl groups on to the backbone of PE)
converts crystalline PE into a rubber when about 20% of the
backbone carbon atoms are substituted.

Likewise, polyvinyl alcohol (PVA), which is made by hydrolysis of
polyvinyl acetate and is highly crystalline, becomes amorphous
when treated with aldehydes:

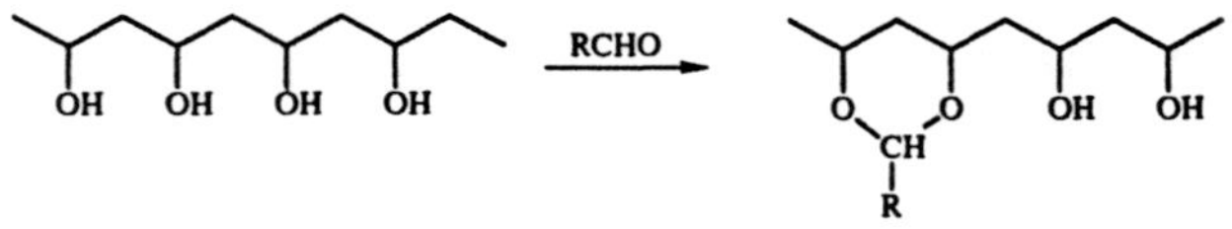

(when R = n-Bu, this polymer is used to laminate sheets of thin glass
together in automobile safety windscreens).

Random copolymers do not crystallise; for example, the crystallinity of PE can be reduced by copolymerising with propylene or
vinyl acetate, to give ethylene/propylene (EP) and ethylene/vinyl
acetate (EVA) rubbers respectively. Block copolymers, on the other
hand, do crystallise if the blocks are long enough (e.g. thermoplastic
elastomers—this type of material will be discussed further in section
4.8.3).

(b) Stereoregularity of the main chain. Although branching decreases
crystallinity (see above), if a stereoregular arrangement of these
small branches occurs then the main chain can take up a helical
conformation, allowing a reasonably close-packed system, and
crystallinity may occur (see Figures 4.5 and 4.6 for PP and PS).

(c) Cross-links and large branch points either do not fit easily into a
crystalline lattice or reduce the flexibility of the polymer chain.
Therefore, their occurrence leads to amorphous polymers, or at least
to polymers containing small or imperfect crystals.

(d) Intermolecular forces. PE, with flexible chains, crystallises even
though only dispersion forces (see section 4.6.3.1) are present. With
heteroatoms in the main chain, dipole–dipole or H-bonding interactions can occur, which will increase the intermolecular forces.
Step-growth polymers are more likely to be crystalline than chain-
growth polymers because of their regular structure, lesser degree of
branching and stronger intermolecular forces. Thus nylons which

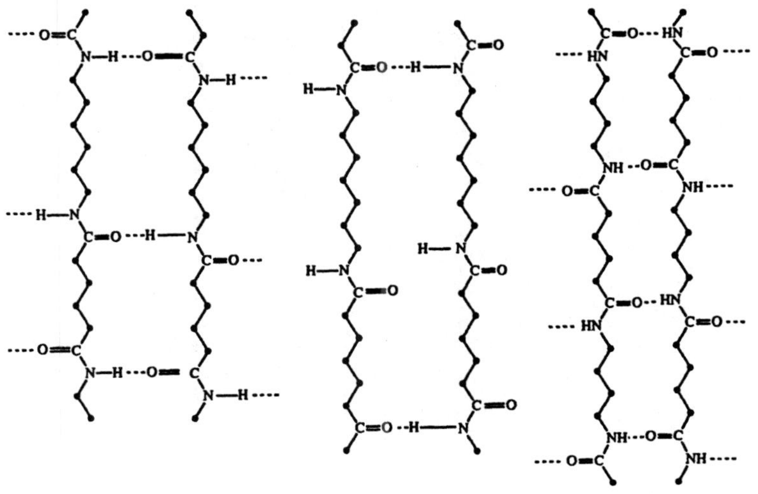

Figure 4.26 Hydrogen bonding in various nylons.

Nylon 6,6
$T_m = 265\,°C$

Nylon 7,7
$T_m = 137\,°C$

Nylon 4,6 (good fit)
$T_m = 295\,°C$

have a planar zig-zag conformation like PE, are highly crystalline, and the T_m is raised by:

(i) The number of amide groups per —CH_2— group. Thus:

Nylon:	Nylon 4	Nylon 6	Nylon 7	Nylon 11	Nylon 12
T_m:	262 °C	215 °C	235 °C	194 °C	179 °C

(cf. T_m for linear PE = 137 °C).

(ii) If there is a good fit for hydrogen bonding (Figure 4.26).

Nylon 4,6 has a perfect fit and has been recently introduced as a new engineering nylon by Dutch State Mines (DSM).

(e) The stiffer the main central chain, the less likely the polymer is to crystallise by twisting and folding into the necessary conformation. Table 4.9 shows that many of the aromatic-backbone polymers are amorphous. Fortunately, many of these materials have T_g values above 200 °C, giving high heat distortion temperatures (HDTs) (see Figure 4.19).

It could be argued that a high T_g and T_m are good because the HDT will be higher. However, it must be remembered that the polymer must be fabricated, and a rough temperature guide for a suitable melt viscosity for processing is $T_g + 150\,°C$ or $T_m + 50\,°C$. If T_g and T_m are too high, then the polymer may decompose during fabrication.

(2) *External factors influencing the ability of a polymer of crystallise*

The above are all 'chemical effects' and depend upon the structure of the polymer. The following parameters influence whether crystallisation takes place or not.

(a) Nucleation. The initiation of crystallites must occur by nucleation, homogeneous or heterogeneous. The former is controlled by the chemical nature of the polymer and determines the size of the crystallites (called spherulites, see section 4.6.4.1), and this size affects the mechanical properties of the polymers. Heterogeneous nucleation can occur by impurities (dirt, etc.), or by controlled additives; for example, PET and isotactic PP are very slow to crystallise, but addition of small amounts of (in)organic materials (e.g. sodium benzoate and dibenzylidene sorbitol respectively) creates small crystallites. These enhance not only the transparency of the polymer (see section 4.6.4) but also, in some cases, the flexural modulus. The exact reason for nucleation with these particular additives is unknown.

(b) The more rapid the cooling from the polymer melt the less crystallinity occurs; conversely, holding the polymer between T_g and T_m for a long time (annealing) promotes crystallinity (see Figure 4.18). The other extreme is spat cooling, in which molten metals, which crystallise very readily, are cooled rapidly on a large rotating copper drum containing liquid nitrogen (cooling rate 10^6 °C/s), giving amorphous metals.

Some polymers take a long time to crystallise. Symmetrical polymers with small protuberances crystallise rapidly, whereas bulky substituents tend to lower the rate of crystallisation. Thus, it is difficult not to obtain crystalline PE and nylon 6,6, whereas PVC, polycarbonate and syndiotactic PS are difficult to crystallise. This may have important consequences because the dimensions of the polymer article—remember that the density of the crystalline phase is higher than amorphous, hence less volume—and the mechanical strength of the polymer may change with time because of slow crystallisation. Fortunately, the higher the final crystallinity, the less time it requires to achieve this: e.g. PE, 90% crystallinity, less than 1 s; PVC, ~ 10% crystallinity, > 1 year.

(c) Orientation (i.e. stretching) of the polymer fibre or film aligns the polymer molecules and reduces their movement, thus promoting crystallinity.

If this is done from the polymer melt, i.e. above T_m, then the polymer molecules have sufficient mobility to rearrange to an amorphous, isotropic, non-directional arrangement. However, if rapid quenching is used immediately after orientation crystallisation is promoted. This retained orientation can also be obtained by stretching between T_g and T_m (see section 4.6.2); e.g. bi-axial (two-directional) orientation of films (PET, PP) or bottles (by blow moulding, for example PET bottles) or the uni- (or mono-)axial orientation of fibres (PP, PET or nylon) (see section 4.9.2).

Cold drawing or forging. Drawing below the T_g of semicrystalline fibres reorganises the isotropic spherulite structure into a fibrular structure along the axis of the fibre, giving high strength materials (e.g. PE, or nylon for tyre cords).

(d) Solvents can induce (promote) crystallisation. Thus, when a solvent is in contact with an amorphous but potentially crystallisable polymer, diffusion of the solvent into the amorphous regions lowers the T_g, allowing movement and reorganisation into crystalline lattices. The solvent itself is excluded from the lattice.

4.6.4   *Properties assoicated with amorphous and crystalline polymers*

(1)   Fracture.   As a general rule, amorphous polymers tend to be brittle below their T_g; e.g. most rubbers are brittle at liquid nitrogen temperatures. Fortunately, this does not apply to some of the amorphous engineering polymers possessing high T_g values (see Table 4.9), which remain tough 200 °C below their T_g. This could be due to their high molecular orientation due to the *para*-phenylene groups in the backbone.

The brittleness of amorphous polymers can be reduced by blending with rubbers, which have even lower T_g values. High-impact polystyrene (HIPS)—homopolymer PS is brittle—is made by polymerising PS on to preformed polybutadiene. If a crack is formed, when it meets a rubber particle, the energy is absorbed and the crack stops.

Composites are made by reinforcing polymers with high-strength fibres (e.g. glass, carbon or Kevlar fibres), and these composites are not brittle, assuming that good adhesion between the polymer and fibre exists.

(2)   Transparency.   Generally, amorphous polymers tend to be transparent whereas crystalline polymers are translucent or opaque. This is because the crystalline regions will have a different refractive index from the amorphous regions and hence will refract light. Thus amorphous PS, PMMA, and PC polymers are transparent whereas crystalline nylon, PE, and PP are translucent or opaque.

Although PET and PP are crystalline, films of these materials are transparent. This is not just because they are thin, e.g. 50 μm, it is because they have been bi-axially orientated—stretched in two directions—which orientates the polymer molecules in the plane of the film. When light passes through the film, because of this orientation no refraction occurs and hence the film appears transparent. Solvent-cast films of PET or PP, on the other hand, are opaque.

(3) Solubility of the polymer in solvents.   The solubility of the amorphous regions is usually higher than that of the crystalline regions. This is because in the amorphous regions, the intermolecular forces between the polymer chains tend to be low, whereas in the crystalline regions the solution has to overcome the heat of fusion of the crystallites. If there is no affinity of the solvent for the polymer, e.g. non-polar crystalline PE, PP and PTFE, then these polymers will not dissolve in any solvent below their T_m. Conversely, for polar polymers, solvents which have specific interactions with polar groups in the polymer (e.g. H-bonding) will promote solubilisation (see plasticisation, section 4.6.2).

(4) Gas barrier properties (in films).   As expected, amorphous films tend to be more permeable to gases (e.g. oxygen, carbon dioxide and water) than crystalline polymers. Thus orientated films have better barrier properties than solvent-cast undrawn films. This is because amorphous polymers have more free-volume space than crystalline polymers. For a specific polymer, the lower the temperature below the T_g of that polymer, the less will be the movement of the polymer chains. This will restrict available pathways through the film, giving a higher gas barrier. It therefore follows that polymers with low T_g values (e.g. elastomers) will have low barrier properties at room temperature.

Because even crystalline films contain large amorphous phases, many approaches to either reducing the movement of the polymer chains or filling up the spaces to reduce permeability have been tried. These include increasing interchain attraction using ionomers (i.e. polymers containing ionic groups, e.g. $-SO_3^-$, on the backbone of the polymer), blends—both miscible and non-miscible—and plate-like fillers (e.g. mica).

(5) *Heat deflection (or distortion) temperature (HDT).*   This is the temperature at which a measured deflection occurs when a controlled stress (force) is applied on a bar or plaque of polymer, i.e. it is a variation of a fixed modulus with temperature (see Figure 4.19).

Experimentally it is $10-20\,°C$ below T_g for amorphous polymers and slightly above T_g for crystalline polymers. Remember that in crystalline polymers amorphous regions are present. However, when they are reinforced by 20% by weight of glass fibres, the HDT for crystalline polymers is greatly increased, almost reaching the T_m, whereas with amorphous polymers, the HDT is only increased slightly (see Table 4.11). Although these are advanced engineering polymers, this effect also occurs with any semi-crystalline polymer, e.g. nylon and PP, assuming good adhesion between fibre and polymer. This can be explained by examining the modulus/temperature curves for amorphous and crystalline polymers in Figure 4.27 (see also Figure 4.16). The reinforcing action of the glass fibres increases the HDT for crystalline polymers, thus extending the temperature application range for these polymers.

Table 4.11   HDT values of glass-reinforced polymers

Polymer		T_g (°C)	HDT		T_m (°C)
			Neat (°C)	20% glass-filled (°C)	
$\begin{array}{c}\text{—⬡—O—⬡—SO}_2\text{—}\end{array}$ PES		225	203	208	Amorphous
			Compare		
—⬡—O—⬡—O—⬡—CO— PEEK		143	165	282	334
PAI(Polyamide-imide)		280	274	278	Amorphous
PEI(Polyetherimide, Ultem)		217	200	223	Amorphous
			Compare		
—⬡—S— PPS		200	135	241	285

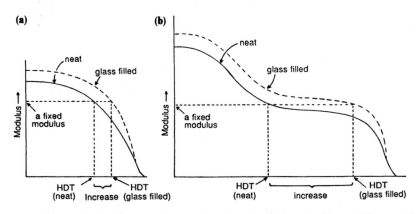

Figure 4.27   Modulus/temperature curves for (a) amorphous and (b) partially crystalline polymers.

(6) *Precision moulding and warpage.*   Amorphous polymers, which include thermosets, are excellent for precision moulding or casting, and show very little warpage when injection moulded. This is because little contraction (0.5%) occurs in amorphous polymers, whereas in crystalline polymers $\sim 2\%$ contraction occurs because they are denser than the molten polymer.

The coefficient of expansion of solid polymers is a problem when large-dimension fabricated articles (e.g. car body panels) are being used, and when they are in competition with metals. This is because the expansion of polymers can be 1000 times that of metals and allowances have to be made for this expansion. In injection moulding, depending on the mould design, some orientation of the polymer molecules also occurs. Therefore there is the added complication of different coefficients of thermal expansion or contraction in different directions. The resulting warpage can be reduced by either using high loadings of fillers or controlling the MWD of the polymer. With glass-fibre reinforced polymers this warpage is again acute, because of the possible alignment of the glass fibres and differential contractions between polymer and fibre. Correct mould design is therefore essential.

4.6.4.1 Crystallinity and morphology and their effect on the properties of polymers. Morphology of polymers is the shape, arrangement and role of polymer crystals which are embedded in the amorphous regions of the polymer. As mentioned previously, 100% crystallinity in polymers rarely occurs, though special cases include solid-state polymerisation from crystalline monomers (e.g. POM), liquid-crystal polymers, or cooling slowly from the polymer melt under high pressure (PE).

The fringe–micelle hypothesis of crystalline and amorphous regions (see Figure 4.17) is not the normal pattern, although it may occur in plasticised polymers or polymers with low crystallinity (< 5%). A more correct picture has come from the slow cooling of very dilute solutions (e.g. 0.01 wt%) of polymers in solvents. Single crystals are formed which can be 1–100 nm in diameter, but with a thickness of only $\sim$ 10 nm (100 Å). These single crystals are called lamellae and appear like a pack of cards, sometimes forming hollow pyramids. Their thickness increases with increasing crystallisation temperature and time. These crystals are too small for X-ray studies, but electron-diffraction patterns are interpreted as indicating that the polymer chains are aligned perpendicular to the surface of the crystal.

Three possible models have been suggested to represent lamellae crystal theories:

(1) Regular structure with adjacent re-entry folds
(2) Irregular, with adjacent re-entry folds of varying lengths
(3) Switchboard with non-adjacent fold-entry.

As the polymer chain lengths greatly exceed the crystal thickness (100 Å) (e.g. for a MW of $10^5 - 10^6$, the extended chain length should be $10 - 100\,\mu m$) the chains must adopt a chain-fold morphology.

Further studies, from measured and theoretical density measurements, have shown that these single crystals are $\sim 80\%$ crystalline with the folds in the polymer chain being the main source of amorphous content, together with chain-ends and chain irregularities.

Crystallisation of polymer melts in the absence of a solvent gives rise to a crystalline mass that can be explained by the chain-folded scheme. On cooling of the molten polymer, initially spherical bundles of fibres of diameter $1–100\,\mu m$ can be seen under an optical microscope. These are called spherulites. The spherical shape tends to be lost at the final stages of crystallisation, when the spherulites expand and impinge on each other (Figure 4.28).

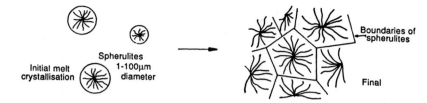

Figure 4.28 Crystallisation of spherulites.

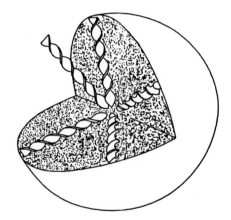

Figure 4.29 Growth of spherulites.

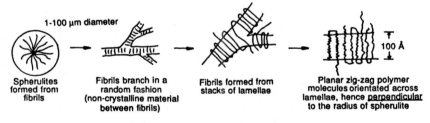

Figure 4.30 Development of lamellae.

The size of these spherulites can be controlled by seeding or nucleation, which can be deliberate or adventitious (see section 4.6.3.3), and also the temperature of crystallisation. Temperatures near T_m produce large spherulites ($>100\,\mu m$) whereas temperatures just above T_g produce small spherulites ($<1\,\mu m$). It is thought that the spherulites grow from a central point in a radial pattern in the form of helical fibrils (Figure 4.29). Furthermore, from electron-diffraction studies (after fracture and replication of the specimen) it can be seen that these fibrils are composed of stacked lamellae (Figure 4.30). The thickness of these is the same as with dilute solution

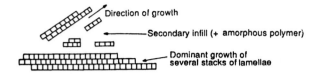

Figure 4.31   Form of many solidified polymers.

crystallisation. The distance between the lamellae can be obtained from small-angle X-ray scattering.

However, because under normal processing conditions polymer melts crystallise fairly rapidly, the polymer chains are not confined to a single lamella but can bridge between lamellae on different fibrils and even between different spherulites at the boundaries. These molecules are called tie molecules and contribute greatly to the strength of the polymer—the longer the polymer chains (i.e. the higher the MW) the more chance of bridging, and also the more rapid the cooling the greater is the possibility of the chain being entrapped in several lamellae and fibrils. Additionally, with slow cooling, defect polymers or low-MW polymers which do not partake in the crystallisation tend to congregate at the spherulite boundaries, together with impurities, e.g. residual solvent and additives (anti-oxidants, pigments, etc.).

The above tends to give an idealised story; the fibril-like structure appears only because of the stacking arrangement of the lamellae. Many solidified polymers have the form shown in Figure 4.31. Thus the larger the spherulites, the more perfect the crystalline structure and the higher the modulus (rigidity or stiffness) of the polymer; but it will have a lower extension to break (see sections 4.8.1 and 4.9.2) and a lower tensile strength and be more brittle than a rapidly cooled polymer, which will have a smaller spherulite structure, which will in turn make the polymer ductile and tough. Another aspect is that, in films and fibres, if the spherulite is of the same thickness or diameter respectively, with slow cooling the impact strength and tensile yield strength (extension to break) decrease. Fortunately this often does not happen, because these materials, as a consequence of their shape, cool rapidly from the melt.

Summarising, the consequences of crystallisation are:

(1) Enhanced stiffness
(2) Extended operating temperature range
(3) Reduced creep
(4) Reduced solvent uptake
(5) High orientation (in films and fibres) can be achieved
(6) Usually opaque (exceptions include orientated polymers and small crystallites).
(7) Fabricated using melt processing ($\sim 50\,°C$ above T_m)
(8) Kinetics of crystallisation controls solidification
(9) Greater mould shrinkage.

4.6.5 Liquid crystalline polymers (LCP)

These are sometimes called self-reinforcing polymers (SRP). Although amorphous and semicrystalline polymers are the principle microstructural types of polymers, a third morphological form is possible: liquid crystalline polymers. These polymers have crystalline characteristics (i.e. they have structural order) even in the molten state, i.e. above T_m. Molecular liquid crystals have been known for over 100 years; in 1888 it was noticed that cholesterol esters, on melting, form colourful opaque liquids which, when further heated, became transparent as expected from disordered, isotropic liquids. This phenomenon is now used in liquid crystal thermometers.

The feature of liquid crystals is that they contain mesogenic groups (i.e. liquid crystal groups) that can form mesogenic phases. These groups can be incorporated into a polymer either as part of the main chain or as side groups on a flexible backbone connected via flexible groups; the flexibility allows orientation of the mesogenic groups above the T_g of the parent polymer (see Figure 4.32).

Mesogenic groups are organic moieties which have the following features:

(1) Long shape relative to diameter, and can contain 'flat segments' such as aromatic rings
(2) Some rigidity along the long axis, i.e. containing double bonds or *para*-phenylene bonds
(3) Strong dipoles and/or easily polarised groups (see Table 4.10)
(4) Directional (anisotropic) intermolecular forces
(5) Weak polar groups at the extremities.

Examples of mesogenic groups are shown in Figure 4.33.

Polymers containing these groups are thermotropic, and two intermediate liquid crystilline phases can be identified at differing temperatures (e.g. by DSC)—see Figure 4.34.

However, molecular cholesteric compounds form a special type of liquid crystals different from thermotropic form, in which the crystals line up in sheets which are in a rotating orientation thus giving rise to coloured interference patterns, i.e. the colour is not due to chromophores or dyes (see Figure 4.35).

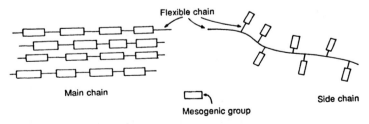

Figure 4.32 Structure of liquid crystalline polymers.

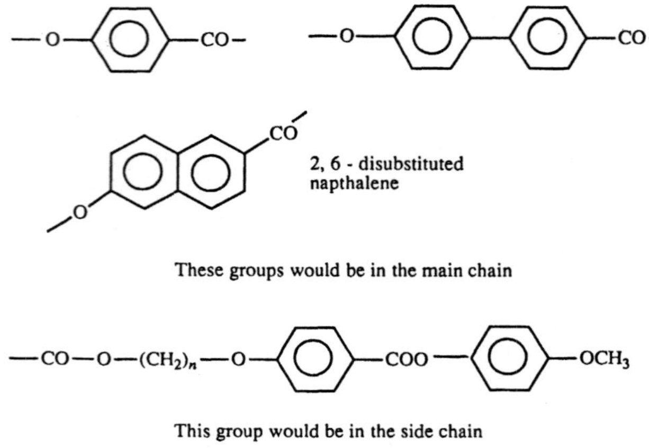

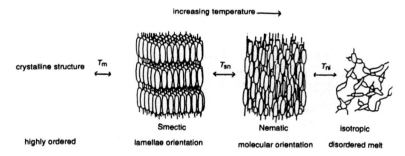

2, 6 - disubstituted
napthalene

These groups would be in the main chain

This group would be in the side chain

Figure 4.33   Examples of mesogenic groups.

increasing temperature ⟶

crystalline structure T_m T_{sn} T_{ni}

Smectic	Nematic	isotropic
highly ordered	lamellae orientation	molecular orientation	disordered melt

Figure 4.34   Intermediate phases in thermotropic polymers.

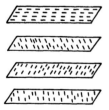

Figure 4.35   Structure of molecular cholesteric compounds which gives rise to interference patterns.

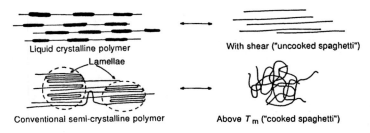

Figure 4.36 Liquid crystalline region above T_m.

When the polymer contains mesogenic groups, the liquid crystal temperature range can be quite large ($\sim 100\,°C$). Schematically the liquid crystalline region above T_m compared with a conventional polymer can be illustrated as shown in Figure 4.36.

When shear is applied to the polymer in the liquid crystalline state (e.g. by extrusion, spinning, or injection moulding) the mesogenic groups are orientated in the direction of flow and this orientation is retained on cooling.

On examination of the thermodynamics of melting of main-chain liquid crystal polymers, it was found that ΔS_m was very small, resulting in a high value for T_m (see section 4.6.3.2); this can be so high ($> 400\,°C$) that decomposition occurs before T_m is reached, hence prohibiting melt fabrication. This was overcome (i.e. T_m was lowered) by introducing small amounts of defects into the backbone, e.g. by:

(1) introducing flexible 'spacers (e.g. $-CH_2-$, $-OCH_2CH_2O-$) between some of the mesogenic groups
(2) replacing *para*-phenylene with some *meta*-phenylene groups

Some LCPs also retain their liquid crystallinity when dissolved in a solvent, i.e. they are lyotropic. Because of the high intermolecular forces, these compounds are often difficult to dissolve, but when it is possible, solution spinning into fibres can be achieved. Thus Kevlar can be dissolved in concentrated

H_2SO_4 and passed through a spinneret into water to obtain fibres. Normally Kevlar cannot be melted and it decomposes above $500\,°C$.

Figure 4.37 summarises the properties of LCPs.

Thermoplastics are materials that are usually processed by heating to the molten state and then converting to the finished article. Thermoset polymers, on the other hand, do not melt; the finished product is made in a mould by an

**4.7
Thermoset polymers and cross-linking**

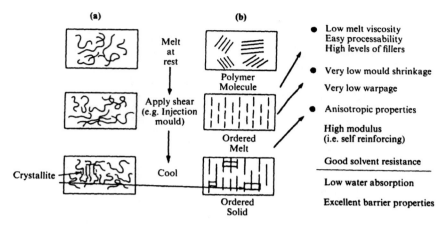

Figure 4.37   Differences in behaviour between LCPs and isotropic polymers and summary of key properties of LCPs. (a) Conventional polymer; (b) self-reinforcing polymer.

irreversible chemical process. This is because a proportion of the monomers have a functionality greater than two (see section 4.3.1), resulting in a three-dimensional infinite matrix. Both step-growth and chain-growth polymerisation can be used to make thermosets.

4.7.1 Examples of step-growth thermosets

(1) Phenol–formaldehyde (to give Bakelite).   The key reaction is the formation of a methylol intermediate formed by the reaction of formaldehyde with an active hydrogen group

$$\sim H + CH_2O \longrightarrow \sim CH_2OH$$

which, on heating, can further condense

$$\sim CH_2OH + H \sim \xrightarrow{-H_2O} \sim CH_2 \sim$$

Thus moelcules containing multiple active hydrogen groups form cross-linked structures:

⊛ = active hydrogen

phenol - formaldehyde (PF)

urea - formaldehyde (UF)

melamine - formaldehyde (MF)

These polymerisations can be stopped at an intermediate stage whilst still flexible and moulded to the finished article by heat and pressure. There is some shrinkage in the mould.

Applications of PF resins include pan handles, electrical equipment (e.g. switches—an important aspect is that decomposition to carbon, resulting in arcing, should not occur). UF is used for trays, knobs and toilet seats, and the main use for MF is in dinnerware where hardness, chemical resistance and finish are essential. However, the main application for these materials is for laminates (e.g. Formica) and wood composites (e.g. chipboard).

(2) Epoxy moulding and adhesives. Epoxides (oxiranes) react with active hydrogen groups:

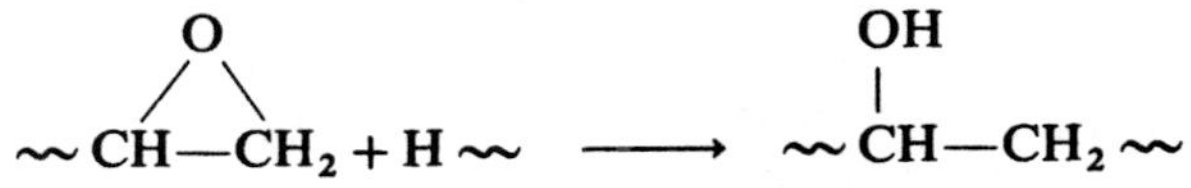

$$\sim CH\overset{O}{-}CH_2 + H \sim \longrightarrow \sim CH\overset{OH}{-}CH_2 \sim$$

Monomers containing two epoxide groups react with compounds containing two or more amine groups (aliphatic or aromatic) or dicarboxylic acids or carboxylic anhydrides, to effect cross-linking:

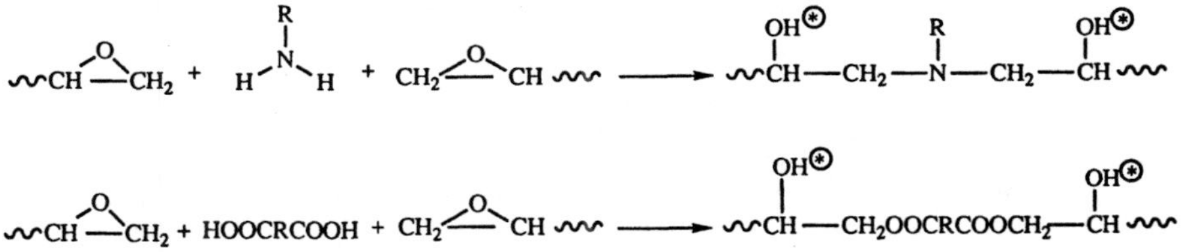

$$\sim CH\overset{O}{-}CH_2 + H\overset{R}{\underset{H}{N}} + CH_2\overset{O}{-}CH \sim \longrightarrow \sim CH\overset{OH^{\circledast}}{-}CH_2 - \overset{R}{N} - CH_2 - \overset{OH^{\circledast}}{CH} \sim$$

$$\sim CH\overset{O}{-}CH_2 + HOOCRCOOH + CH_2\overset{O}{-}CH \sim \longrightarrow \sim CH\overset{OH^{\circledast}}{-}CH_2OOCRCOOCH_2 - \overset{OH^{\circledast}}{CH} \sim$$

Hydrogen atoms marked ⊛ react further with the epoxide groups and the carboxylic anhydride to form a thermoset.

These reactions are catalysed by tertiary amines (R_3N) and are used for rapid-setting varieties. Epoxides have reasonable resistance to heat and humidity (cf. formaldehyde thermosets) and epoxy laminates are used for more demanding applications (aerospace, electrical circuit boards, etc.). Good adhesion of the epoxy polymer with glass and carbon fibres allows reinforcement, and this overcomes the inherent brittleness normally associated with thermoset polymers (composites).

(3) Polyurethanes. Isocyanate groups react with any groups that contain active hydrogens, e.g.

$$\underset{\text{Isocyanate}}{\sim NCO} + \underset{\text{Alcohols}}{HO \sim} \longrightarrow \underset{\text{Urethanes}}{\sim \overset{\circledast}{NH}-CO-O \sim}$$

$$\underset{\text{Amines}}{\sim NCO + H_2N \sim} \longrightarrow \underset{\text{Ureas}}{\sim \overset{\circledast}{NH}-CO-\overset{\circledast}{NH} \sim}$$

The NH$\circledast$ group further reacts with the isocyanate resulting in cross-linking. The common di-isocyanates are:

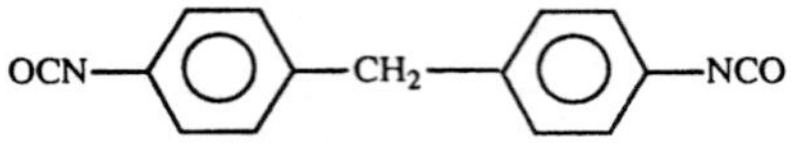

Diphenylmethane di-isocyanate (MDI)

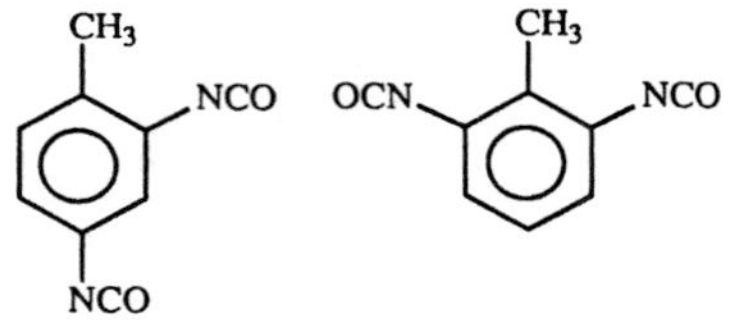

Toluene di-isocyanate (TDI)
(a mixture of 2,4- and 2,6-isomers)

MDI and TDI, being aromatic di-isocyanates, yield polyurethanes that tend to yellow on prolonged exposure to sunlight. For paint applications, aliphatic and saturated cyclic di-isocyanates are used:

$OCN-(CH_2)_6-NCO$
Hexamethylene di-isocyanate

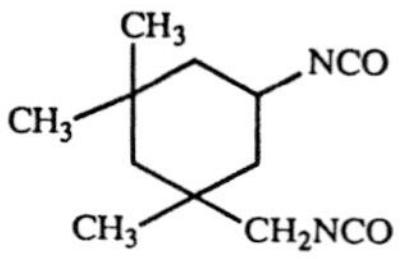

Iso-phorone di-isocyanate

Reaction of the di-isocyanates with polyols gives cross-linked structures. Glycerol gives a rigid polyurethane. For the preparation of soft flexible polyurethanes, telechelic macromers of MW 500–3 000 are used, e.g.:

$$HO-\!\!\left[CH_2CH_2CH_2CH_2-O\right]_n\!\!H \qquad \text{Poly-THF}$$

$$HO-\!\!\left[CH_2CH_2-O\right]_n\!\!H \qquad \text{Polyethylene oxide}$$

$$HO-\!\!\left[\underset{\underset{CH_3}{|}}{CH}-CH_2O\right]_n\!\!H \qquad \text{Polypropylene oxide}$$

$$HO-\!\!\left[R-OOC(CH_2)_xCOO-R\right]_n\!\!OH \qquad$$ Aliphatic polyesters (e.g. the polyester formed from adipic acid and excess ethylene glycol)

Foamed polyurethanes are obtained by adding traces of water, which liberates carbon dioxide with the isocyanate:

$$\sim\!\!\sim NCO + H_2O \longrightarrow \sim\!\!\sim NH_2 + CO_2\uparrow$$

or by injecting a volatile liquid or gas during the polymerisation (e.g. CCl_3F, b.p. 24 °C). Many rigid and flexible foamed articles are made by reaction injection moulding (RIM, see section 4.5.8). The foam produced counterbalances the shrinkage in the mould.

4.7.2   Examples of chain-growth thermosets

Thermosets can be obtained by chain-growth polymerisation by using divinyl compounds. Since a monovinyl compound can be considered as bifunctional, addition of a divinyl compound will cause cross-linking, examples include:

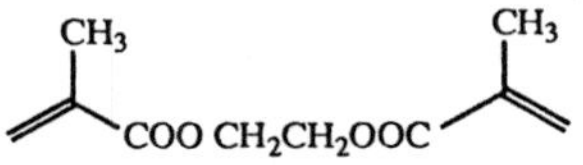

The ester made from methacrylic acid and ethylene glycol

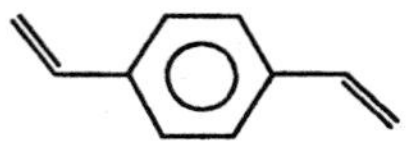

Divinyl benzene (DVB)

The bismethacrylate when added ($< 5\%$) to mono-(meth)acrylates introduces a small amount of cross-linking, which improves:

(a) The strength and rigidity, lowers the creep of PMMA and also enhances the solvent resistance.
(b) The elastomeric characteristics of butyl acrylate; CH_2=CH–$COOBu$ has a low T_g ($-56\,°C$) and is an elastomer at room temperature.

The free-radical polymerisation of styrene containing $\sim 5\%$ DVB produces a cross-linked polymer that can be sulphonated (with SO_3/H_2SO_4) to give —C_6H_4—SO_3H groups. These materials are used as cationic exchange resins:

$$\sim\!\sim\!\sim C_6H_4SO_3{}^- + M^+ \longrightarrow \sim\!\sim\!\sim C_6H_4SO_3{}^- M^+$$
Insoluble resin

When $\sim 25\%$ DVB is used, together with an inert solvent (e.g. toluene), a highly cross-linked polystyrene is produced. The removal of the toluene yields a rigid, porous, high-surface-area polystyrene. Sulphonation (see above) gives a heterogeneous sulphonic acid which is often used as a catalyst in organic solvent media, in which they are insoluble, but become swollen with solvent, similarly to ion-exchange resins in water. These materials are called macroreticular resins.

Although high-MW rubbers (i.e. compounds with low T_g) have elastomeric properties, in order to improve the dimensional stability under stress (i.e. when stretched), a small amount of cross-linking is nearly always introduced (see section 4.6.2.1).

Many paints and coatings cross-link by free-radical aerial oxidation (see section 4.4.1.1). Paints tend to be a mixture of monomers containing unsaturated esters, polyurethanes, phenolics, expoxies, etc., depending on the application.

High-performance fibre-reinforced composites are made by exploiting both thermoplastic and thermosetting properties. A continuous fibre is coated with the molten polymer (e.g. by a *pultrusion* technique, which involves drawing the fibre through a molten bath of the polymer) and is fabricated into a shape, i.e. a pre-peg is made. Further heating causes the polymer to cross-link or cure into a rigid reinforced structure. The polymers used are low-MW aromatic polymers (Table 4.9) containing terminal acetylenic ($—C{\equiv}CH$) or cyanate ($—OCN$) groups. On heating, these groups trimerise, forming stable aromatic structures.

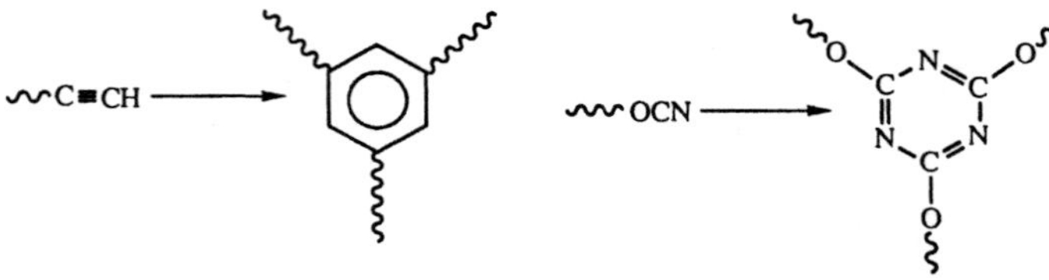

An important feature of these reactions is that no volatiles are given off during processing that may weaken the structure.

4.7.3   General properties of thermosets

Curing of a thermoset is a succession of chemical reactions giving an interconnected, three-dimensional, rigid, amorphous material of infinite MW. During curing and cross-linking, the T_g of the system increases until there is no softening phenomenon—if heated the polymer just chars and carbonises.

The differences between thermoplastics and thermosets are shown in Table 4.12.

4.8 Elastomers (rubbers)

Most materials (e.g. metals, glass or plastics), when stressed (i.e. stretched) to an extension of $<1\%$, return to their original dimensions when the stress is removed. This is due to small reversible interatomic forces. When this extension is exceeded, the deformation increases more than the applied stress and either it becomes non-reversible (i.e. plastic behaviour) or the material fractures (e.g. glass or metals).

Table 4.12 Advantages of thermoplastics and thermosets

Thermoplastic advantages	Thermoset advantages
Ease of mouldability	Non-fusible
Toughness	High operating temperatures
Good surface finish	Dimensional stability
Easy to pigment	High rigidity and low creep
Very thin sections possible (films)	Hardness
	Lower monomer costs (but fabrication slower)
	Low flammability
	Solvent resistance
	Arc/track resistance

Elastomeric polymeric materials behave differently—reversible extensions of up to tenfold are possible. This difference is not due to the above interatomic movement, but is due entirely to rearrangement by rotation of the covalent bonds within the main polymer backbone chain. The random conformations of a non-crystalline amorphous polymer can be calculated or computed (see Figure 4.38). This diagram has assumed tetrahedral bond angles (109.6°) and the more thermodynamically stable *trans* arrangement (see Figure 4.6), giving rise to the zig-zag structure.

When the chain is elongated, the entropy of the system will decrease. The retractive force is simply the tendency for the chain to return to the un-deformed state of maximum entropy.

Interestingly, raising the temperature of the system will increase the re-arrangement of these bonds and thus increase the tendency to a more random state. This means that the polymer will shrink at higher temperatures, i.e. decrease in length at constant stress or increase in stress at constant length.

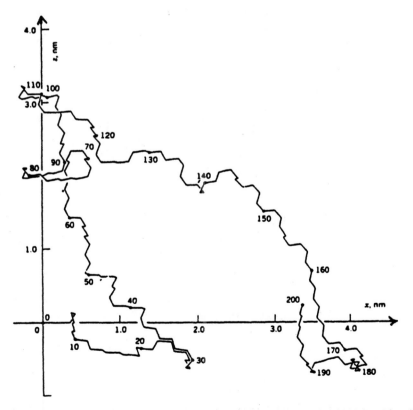

Figure 4.38   A computer-generated conformation of a polymethylene chain of 200 C—C bonds. The projection shown is in the plane of the first two bonds.

4.8.1   Requirements for polymeric elastic behaviour

Although many polymers do, or can be made to, exhibit elastomeric characteristics (e.g. by varying the temperature, see Figure 4.16), the requirements
considered here are for normal operating conditions, although it must be
emphasised that they are different for different rubbers, e.g. for tyres, where
heat build-up is an important consideration, versus O-rings, where wear and
chemical/solvent resistance are important.

(1) The polymer must have a low tensile modulus (sometimes called
modulus of rigidity, elastic modulus or Young's modulus). Typical
values of Young's modulus for various materials, in N/m^2, are:

Mild steel	2.2×10^5
Glass	6.0×10^4
Polystyrene	3.2×10^3
Natural rubber	2
Water	~ 0

Strain (ε) (also known as relative deformation, linear dilation or
extension) is equal to $\Delta L/L_0$, where $\Delta L =$ elongated length (L) $-$ original length (L_0), and λ is the elongation or extension ratio, $\lambda = L/L_0$.
Therefore:

$$\varepsilon = \frac{\Delta L}{L_0} = \frac{L - L_0}{L_0} = \lambda - 1$$

Figure 4.39 is a stress–strain curve of a non-Hookean solid, a typical
curve for an elastomer in uniaxial extension.

$$\text{Stress} = \frac{f}{A} = \frac{\text{Applied force}}{\text{Cross-sectional area}}$$

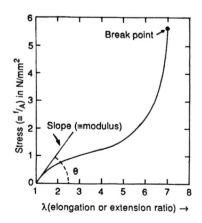

Figure 4.39   Relation of elongation to stress.

$$(N/m^2 = Pa = 145 \times 10^{-6} \text{ p.s.i.}).$$

The initial modulus (i.e. small deformation) is the slope of the stress–strain curve at the origin, i.e.

$$\text{Young's modulus } (E) = \frac{\text{Stress}}{\text{Strain}} = \frac{f}{A\varepsilon} = \tan\theta$$

The chemical requirement for this low modulus is that the polymer chains must be flexible at ambient temperature, thus enabling them to take up different arrangements to an imposed stress. All configurations should be available. This means that the bonds must have free rotation, not only of the main-chain bonds, but also of the small side chains, i.e. no stiffening of the polymer chain should occur, either chemical (e.g. *para*-phenylene groups) or associative (crystalline) and the polymer chains should have little intermolecular attraction (see section 4.6.3.1).

These conditions are satisfied by amorphous polymers (see section 4.6.2), and the higher the temperature above T_g the more flexible the chains become, i.e. the lower the T_g, the more elastomeric the polymer; in practice it should be at $T_g + 50\,°C$. This means that most amorphous polymers will be elastomeric at some elevated temperature range, but only room-temperature applications are considered here.

The rotational energy barrier can be determined for certain groups:

$$\underset{\text{kJ/mole} \quad 11.7}{H_3C \curvearrowleft CH_2} \sim \underset{8}{H_3C \curvearrowleft CH=CH} \sim \underset{4.6}{H_3C \curvearrowleft \overset{\overset{O}{\|}}{C}} \sim \underset{4.1}{H_3C \curvearrowleft O} \sim$$

Although the energy barrier for rotation for CH_3 adjacent to heteroatoms is low, their presence increases intermolecular attraction and promotes crystallinity. The low energy barrier of $—CH_3$ groups next to a double bond (e.g. in the polymer $—CH_2—CH=CH—$) suggests that these should give good elastomers, however, only in the *cis*-configuration; the *trans*-arrangement crystallises (similar to PE, see section 4.6.3.3). Natural rubber is poly(*cis*-1,4-isoprene).

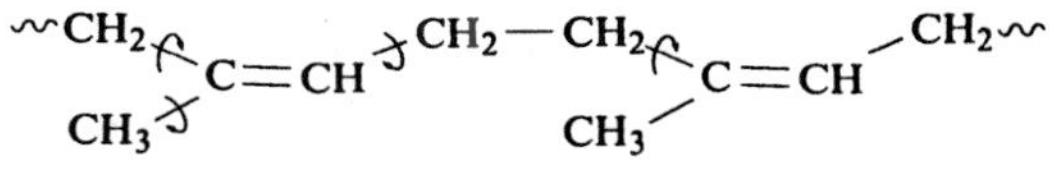

poly (*cis* - 1,4 - isoprene)—natural rubber

Although olefinic unsaturation may be good for promoting elasticity its presence stimulates atmospheric ozone attack (0.1 p.p.m. in air). When a rubber band (which is often a random copolymer of butadiene–styrene or natural rubber) is stretched, microcracks appear, exposing fresh surfaces, and ozonolysis causes chain scission. Although anti-oxidants are added to many elastomers, they are not normally used in rubber bands.

Another consequence of these low energy barriers is a more rapid contraction (or springiness) following extension. This is called resilience, and can be simply measured by the extent of rebound of a rubber ball. Poly(cis-1,4-butadiene) has a rebound resilience of $> 80\%$, whereas in polybutylene rubber with no unsaturation (see section 4.3.2.3) it is $< 10\%$.

(2) High elongation to break under stress. If stress is applied to a polymer over a prolonged time, then the polymer chains just flow past each other. The higher the MW, the more possibility of chain entanglement inhibiting this flow, but never quite eliminating it. This is overcome by controlled cross-linking (or vulcanisation or curing) to form a network structure. It has been found for C—C elastomers that the optimum MW is 100 000–150 000, and that the segment between the cross-links should have a DP_n of ~ 300, which, given that the MW of the repeat unit is ~ 100, means that each polymer chain has 4–5 cross-links occurring at, preferably, regular spacings.

For very high-MW polymers, the modulus is too high and often chain scission occurs during mechano-chemical processing. Conversely, amorphous polymers with low MW do not form good elastomers simply by increasing the cross-linked density—the T_g is raised too much and the material becomes brittle, and also the loose, unconnected chain ends do not contribute to the elasticity.

4.8.2 Chemical cross-linking to produce elastomers

Chemical cross-linkings can be achieved by several methods, depending on the chemical make-up of the polymer chain.

(1) Unsaturated rubbers, e.g. poly(cis-1,4-butadiene) or isoprene, or random SBR, are usually vulcanised with sulphur

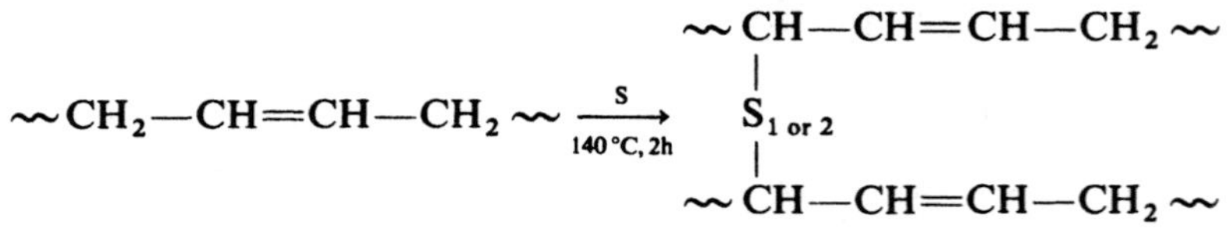

$$\sim CH_2—CH{=}CH—CH_2 \sim \xrightarrow[140\,°C,\,2h]{S} \begin{array}{c} \sim CH—CH{=}CH—CH_2 \sim \\ | \\ S_{1\ or\ 2} \\ | \\ \sim CH—CH{=}CH—CH_2 \sim \end{array}$$

This is used together with ZnO, acid retarders, accelerators e.g.

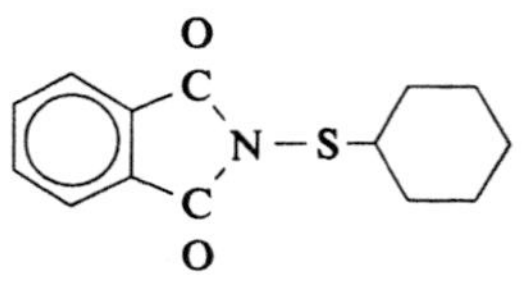

Also included in this class are random ethylene–propylene–diene monomer (EPDM) rubbers prepared by Ziegler co-ordination catalysts. Only the one double bond* of the diene monomer is incorporated into the EP chain; the other allows sulphur vulcanisation. Examples of diene monomers include:

1,4-hexadiene 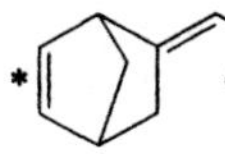   4-ethylidenenorborn-2-ene

(2) Saturated rubbers, e.g. random copolymerised EP rubbers (prepared by Ziegler catalysts), require a peroxide cure (using benzoyl peroxide). Poly siloxane polymers are cured in a similar manner.

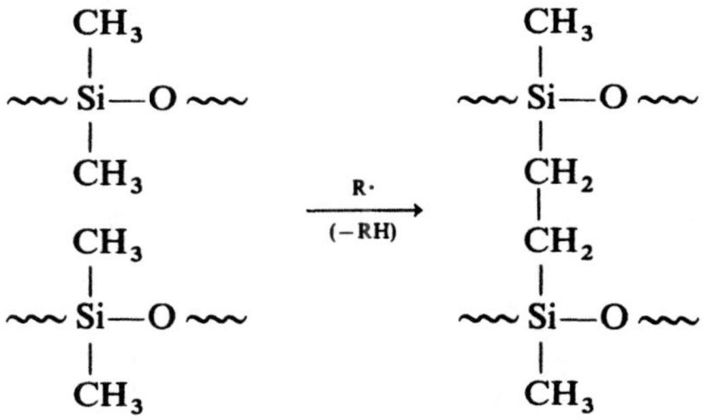

(3) Halogen-substituted main chains e.g.  include poly-chloroprene (neoprene), which is a mixture of 1,4-and 1,2-structures:

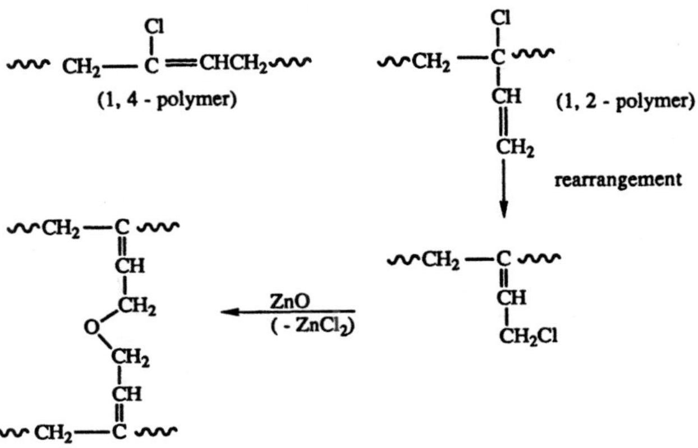

On heating with ZnO the 1,2-structure rearranges and reacts with ZnO to form an ether link.

'Vinylic' chlorides $CH_2{=}\overset{\underset{\displaystyle |}{Cl}}{C}H$ are very unreactive (similar to chlorobenzene), whereas 'allylic' chlorides $CH_2{=}CH{-}\overset{\underset{\displaystyle |}{Cl}}{C}H_2$ are very reactive. With fluorocarbons (e.g. Viton, a copolymer of vinylidene difluoride ($CH_2{=}CF_2$) and hexafluoropropylene ($CF_2{=}CF{-}CF_3$)), an amine cure is used.

$$\text{\small$\sim\!\!$CH}_2{-}CF_2{-}CF_2{-}\overset{\underset{\displaystyle |}{CF_3}}{CF}\text{\small$\sim\!\!$} \xrightarrow[(-HF)]{(heat)} \text{\small$\sim\!\!$}CH{=}CF{-}CF_2{-}\overset{\underset{\displaystyle |}{CF_3}}{CF}\text{\small$\sim\!\!$}$$

$$\downarrow \begin{matrix} H_2N \\ H_2N \end{matrix})$$

$$\text{\small$\sim\!\!$}CH_2{-}\underset{\underset{\displaystyle \begin{matrix}HN\\HN\end{matrix})}{\displaystyle |}}{CF}{-}CF_2{-}\overset{\underset{\displaystyle |}{CF_3}}{CF}\text{\small$\sim\!\!$}$$

$$\text{\small$\sim\!\!$}CH_2{-}\overset{}{CF}{-}CF_2{-}\overset{\underset{\displaystyle |}{CF_3}}{CF}\text{\small$\sim\!\!$}$$

(4) With elastomeric polyurethanes and polysiloxanes, a controlled amount of cross-linking can be achieved by adding small amounts of polyfunctional monomers during the preparations e.g. tri-isocyanates and $RSiCl_3$ respectively.

An important aspect is that crosslinking increases resistance to solvent attack (swelling); this is because many applications involve solvent contact (hoses, O-rings, etc.). How other properties are altered with increasing cross-link density is shown in Figure 4.40.

(5) Often rubbers are compounded with fillers, and carbon black is used with many unsaturated rubbers. It is believed that the carbon black contains surface carboxylic, phenolic, etc., groups which chemically interact with the rubber during mastication, possibly through mechanical breakdown of the polymer chains giving free radicals. The carbon black particles act as giant cross-links.

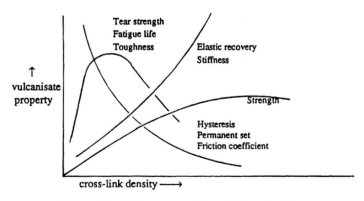

Figure 4.40 Consequences of chemical cross-linking.

(6) From Figure 4.39 it can be seen that the force (stress) required for a given strain (elongation) increases at large elongations. This is because on stretching the polymer chains align together, thus reducing the confrontational entropy of the chains. The tendency for the polymer chain to crystallise is increased even above the T_m of the 'normal' isotropic state. This is because ΔS_m in the equation $T_m = \Delta H_m / \Delta S_m$ is decreased (see section 4.6.3.2), thus raising T_m and encouraging the polymer to crystallise. This is called strain-induced crystallisation. This is a desirable phenomenon because these crystallites improve the ultimate properties of the polymer by acting as physical cross-links or reactive fillers. The ultimate modulus and tensile strength are increased by this self-reinforcement.

This type of strain-induced crystallisation only occurs when good alignment of the polymer chains is possible. With bulky polymers, e.g. styrene–butadiene random copolymer rubber, the phenyl groups inhibit crystallite formation, and in these cases fillers are used to impart strength.

4.8.3 Thermoplastic elastomers (TPE)

Once an elastomer has been vulcanised or cured by cross-linking, fabrication is difficult and it behaves like a thermoset. Thermoplastic elastomers overcome these difficulties by allowing the polymer to be processed at high temperatures like a thermoplastic, but on cooling the polymer behaves like a conventional rubber. This is achieved by making the cross-links reversible, similarly to strain-induced crystallisation. The cross-links can be glassy or crystalline polymer domains with a high T_g or T_m joining elastomeric segments together. These domains are rigid at the application temperature, but are molten at the fabrication temperature.

The most important TPEs are butadiene–styrene and isoprene–styrene block copolymers (see section 4.4.2.3) made by the anionic polymerisation technique. Three methods are used.

(1) Initially generating living PS, followed by PB and finally PS:

$$ S \xrightarrow{\text{LiBu}} SSSSS^- \xrightarrow{\text{B}} SSSSBBBB^- \xrightarrow{\text{S}} SSSSBBBBSSSS $$

This is the preferred method.

(2) Coupling of the intermediate with a difunctional reagent:

$$ 2SSSSSBBBBB^- + Me_2SiCl_2 \xrightarrow{-2LiCl} SSSSBBBBB\underset{\underset{\textstyle Me}{|}}{\overset{\overset{\textstyle Me}{|}}{Si}}BBBBSSSSS $$

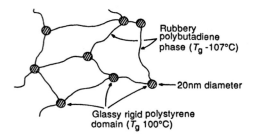

Figure 4.41   Typical structure of a butadiene–styrene block copolymer.

(3) Using a difunctional anionic initiator and starting with B and finishing with S (see section 4.3.2.2):

$$B \longrightarrow {}^-BBBBBBB^- \longrightarrow SSSSSBBBBBBSSSSS$$

A typical overall ratio of SBS would be 14/72/14 and the MW of the blocks would be 5000/50 000/5000. At temperatures $>100°C$ ($\sim 200°C$, fabricating temperature), PS domains melt allowing processing and at room temperature the PB phase is an elastomer (see Figure 4.41). Just like 'chemical' cross-linking, in which loose unconnected ends are not desirable, the alternative arrangement BBBBSSSSBBBB does not possess good TPE properties.

Similar TPE systems include:

(1) Hytrel: this consists of a $[AB]_n$ block copolymer (see Figure 4.8):

$$[Polybutylene\ terephthalate–PolyTHF]_n$$
Crystalline (hard)   Rubbery (soft)

(2) Spandex fibres, e.g. Lycra (see section 4.4.2.3):

$$[Urethane–PolyTHF–Urethane]_n$$

Strong H-bonding between the urethane (and urea) produces 'tie' points and an effective network system, whilst the polyTHF ($DP_n \sim 25$) remains rubbery. This product can be spun from aqueous DMF into fibres.

These types of TPEs are part of an expanding area, and new products are appearing annually, e.g.

Polydimethylsiloxanes/Aromatic polyesters or carbonates
(soft)   (hard)

Polyperfluoropropylene oxide/Aromatic polyimides

CF₃—CF—CF₂   (soft)   (hard)

Although natural plant fibres (e.g. cotton, flax, jute and hemp), animal fibres (e.g. wool and silk) and fibres based upon chemically modified natural polymers (e.g. viscose rayon—regenerated cellulose fibres—and cellulose acetate, e.g. Tricel) are extensively used, the production of synthetic fibres has recently overtaken that of natural fibres, principally cotton, so that ~ 20 million tonnes of synthetic fibres are now made per annum.

In this section, only apparel or textile fibres will be discussed, and these are dominated by polyester (PET or PETP), nylon (nylon 6,6 or 6), acrylic (polyacrylonitrile, PAN) and polypropylene (PP) fibres. Initially, trade names for these fibres were particularly powerful and highly advertised, despite the fact that other manufacturers produced very similar products, although this trend is diminishing. Examples include:

Polyester	Terylene (ICI), Trevira (Hoescht)
	Dacron (DuPont)
Nylon	Bri-Nylon, Tactel (ICI), Antron (DuPont)
	Celon (Courtaulds), Enkalon (Enka, Germany)
Acrylic	Dralon (Bayer), Acrilan (Monsanto)
	Courtelle (Courtaulds), Orlon (DuPont)
Polypropylene	Meraklon (Montedison)

Other organic fibres include elastomerics (Spandex, see section 4.8.4) and engineering fibres (e.g. Kevlar, see section 4.6.5) and the recently introduced high-strength PE fibres.

Almost any polymer can be converted into a fibre if:

(1) Its MW is above M_c, i.e. when chain entanglement takes place. The value of M_c is given by a rapid increase in melt viscosity (at a constant temperature) with increasing MW.
(2) It melts without decomposition, or $\left.\begin{array}{l} \\ \end{array}\right\}$ for spinning
(3) It is soluble in some solvents

However, the fact that only a few apparel fibres are used commercially depends on many other factors, e.g. modulus, tensile strength, durability, processibility, etc., and cost; for example, DuPont introduced the polyamide Quiana made from

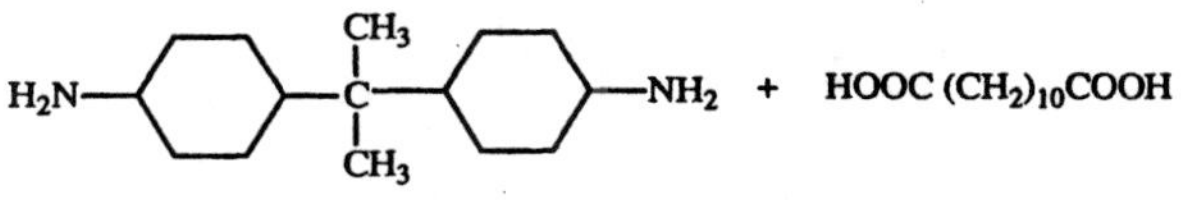

$$H_2N - \bigcirc - \underset{\underset{CH_3}{|}}{\overset{\overset{CH_3}{|}}{C}} - \bigcirc - NH_2 \quad + \quad HOOC\,(CH_2)_{10}COOH$$

which had silk-like properties, but was withdrawn in the 1980s for economic reasons.

4.9.1 Spinning processes

There are three principal methods for making continuous mono-filaments. One is melt spinning, from the molten polymer, and the other two are solution

Table 4.13 DP_n values of some polymers

Polymer	DP_n
PE	> 62 000
PP	> 45 000
Polyacrylonitrile (PAN)	1000–2000
Aliphatic polyamides (nylon)	> 500
PET	> 250

spinning—either dry-spinning, where the solvent is evaporated in a current of hot air, or wet-spinning into a bath of a non-solvent which coagulates or precipitates the fibre after the spinneret. In all these examples, hundreds of holes in the spinneret are used and there may be more than one spinneret.

4.9.2   Prerequisities for fibre formation

(1) MW and fibre formation.   The fibre forming tendency for a polymer increases with MW. The optimum MW depends on the structure—the weaker the intermolecular forces the higher the MW required (Table 4.13).

(2) Molecular structure and fibre properties.   The chemical components of a polymer determine not only fibre capabilities, but also properties like modulus, strength and thermochemical limits, and also dyeability and solubility in solvents. This is because the structure determines the ability of the polymer to crystallise. Also, generally the less flexible the polymer chain, the higher the tensile strength (tenacity) and modulus. Similarly, the less the intermolecular attraction the more soluble in solvents.

  (a) Linearity of polymer chain.   As expected, linear polymers give optimal properties, not only in molecular shape, but also in the absence of long branches and cross-links (unless thermo-reversible, e.g. Spandex). These are the same properties that promote crystallinity (see section 4.6.3).
  (b) Intermolecular forces (see Table 4.10).
      (i) Intermolecular dispersion forces, e.g.

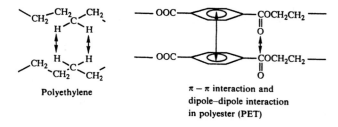

Polyethylene

$\pi - \pi$ interaction and dipole–dipole interaction in polyester (PET)

(ii) Dipole–dipole.

Polyacrylonitrile (*trans - trans* structure)
— this strong bonding limits solubility,
and only polar non-protic solvents, e.g. DMF,
can be used

(iii) H-bonding. This occurs in nylons (see Figure 4.26), also poly-vinylalcohol (PVAL) and cellulose (viscose rayon).

polyvinylalcohol
(PVAL)

Cellulose (~ 45 kJ/mole
attraction between chains)

(3) Orientation and crystallinity. If a fibre is spun from the melt through a spinneret it does not have time to crystallise completely (only possible between T_g and T_m, Figure 4.18), even though the polymer molecular chains are uncoiled and slightly orientated in the direction of flow, i.e. in the axis of the fibre. This also occurs if the fibre is spun from solution.

A higher orientation is achieved when the polymer fibres are drawn or stretched, i.e. extended by 2–7-fold. The cooling time is too quick for spherulite formation (see Figure 4.28) and these are not formed. The act of drawing and orientation does promote lamellae crystallisation (see Figure 4.30), the sizes of which are small compared with the diameter of the fibre. A typical stress–strain curve for an undrawn fibre is shown in Figure 4.42. The

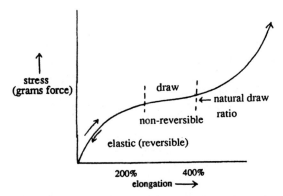

Figure 4.42 Typical stress–strain curve for an undrawn fibre.

horizontal section is where the 'tie' molecules between the lamellae are being orientated; this is non-reversible.

The extent of draw affects the tenacity of the fibre; in fibres for apparel (i.e. soft and flexible) a low draw is used, whereas fibres with high strength (e.g. nylon for tyre cords) require a high draw ratio. Originally this led to irreversible changes in finished garments (e.g. baggy trousers, kneed stockings), but advances in knitting technology and elastomeric fibres have eliminated this. For dimensional stability for water washing, PET fibres, after drawing, are given a steam treatment for ~ 5 min.

With amorphous fibres such as PAN, well-defined crystalline lamellae are not formed (strong dipole–dipole interaction occurs), but the act of drawing still increases the strength.

The drawing of fibres should, in principle, be above T_g, or else brittle fracture may occur. Thus PET and PAN fibres are heated during drawing. In some cases, especially when solution spinning is used, the residual solvent lowers T_g. A similar situation occurs with nylon at high humidity or containing low-MW impurites.

4.9.3   Mechanical strength of fibres

The diameter of monofilament fibres is in the range 10–$50\,\mu$m, but is variable and non-uniform (i.e. not always circular). Therefore the mechanical strength of fibres is quoted against the weight of a length of fibre (i.e. force per unit linear density):

$$1 \text{ tex} = 1 \text{ g per } 1000 \text{ m of fibre}$$

thus if $1\,000\,$m $= 1\,$g, then the fineness is 1 tex; if $1\,000\,$m $= x\,$g, then the fineness is x tex. The higher the tex, the courser the fibre. Originally the definition was 1 denier $= 1\,$g per $9\,000\,$m of fibre, and in order to keep the tex term similar, dtex (mass per $10\,000\,$m) is often used.

Thus the mechanical properties are based upon weight rather than area (e.g. $1\,$N/m$^2 = 1\,$Pa or $1\,$N/mm$^2 = 1\,$MPa). Thus, maximum tensile force (at break)—called tenacity—is expressed in N/tex (or N/dtex, cN/tex, etc.), where $d =$ deci; $c =$ centi and 1 tex $= 10$ dtex, $1\,$N $= 100\,$cN. Another unit sometimes used is gram-force (g-f), where $1\,$N $= 1\,$g-f. Typical values for apparel (textile) and industrial fibres are 0.2 and $3\,$N/tex respectively. Elastic modulus, the initial slope in a stress–strain plot (i.e. $\tan \theta$ in section 4.8.1) is 0.5–$4\,$N/tex for textile fibres and 10–$300\,$N/tex for industrial fibres.

4.9.4 Requirements to be met by textile fibres

Table 4.14 shows the requirements to be met by textile fibres. Many of these requirements are unique to fibres, and these do not include the way in which

Table 4.14   Requirements to be met by textile fibres

Properties of the final product	Corresponding fibre properties
Optical properties	Lustre, fibre surface
Aesthetics	Profile, fibre cross section
Mechanical properties	Modulus of elasticity, tenacity, elongation
Comfort	
Physiological properties of clothing	Moisture absorption, moisture transport
Antistatics	Electrical resistance
Thermal insulation	Heat capacity, porosity, heat conduction
Hand	Textile structure, bending modulus
Feel	Roughness, modulus of elasticity, fineness
Ease of care and washability	Wetting, moisture absorption, glass transition point (wet)
Dry cleaning	Polymer (in)solubility, swelling
Soiling characteristics	Zeta potential, adsorption and dissolution of soil components
Fastness	
Mechanical stability	Tenacity, elongation, moduli, abrasion resistance
Dimensional stability	Melting point, glass transition point (wet and dry)
Forming (e.g. pleating)	Thermoplasticity, glass transition temperature (T_g)
Lightfastness	Chemical structure, sensitisers, stabilisers
Lightfastness of dyeing	Polymer–dye interaction, radical lifetime
Specific properties	
Flame resistance	Chemical structure, combustion mechanism
Impermeability to water	Moisture absorption, wetting properties
Water vapour permeability	Surface properties, morphology, yarn structure
Dyeability	Glass transition point during dyeing
Mechanical properties (ropes, material, tyre cord)	Modulus of elasticity, tenacity, elongation, dynamic modulus
Rubber elasticity	Chemical structure, glass transition temperature, morphological structure, domain structure

the textile is constructed, i.e. how the monofilaments are put together (Figure 4.43), and also whether the finished textile article is woven or knitted.

Staple fibres, being of short length (3–12 cm) and not continuous, are twisted together to form a continuous yarn, like wool and cotton.

For apparel fibres, the effect of temperature on the textile is important. Not only has the T_g of the fibre to be low enough to permit ironing, but the T_m must be sufficiently high to prevent melting. Another aspect is the retention of appearance and shape during cleaning. With water washing, the plasticising action of water can reduce T_g, which may cause permanent deformation, either shrinkage or stretching. With dry cleaning, obviously non-solubility and non-swelling in the fluorochlorocarbon or perchloroethylene solvents is necessary; even in these systems, the temperature must be kept below the T_g of the fibre–solvent system.

Recently, textile finishing has improved these cleaning operations, e.g. thermofixing—heat-setting of polyester (see section 4.9.2), felting or shrink-resistance finishing of wool and crease-resistance of cotton.

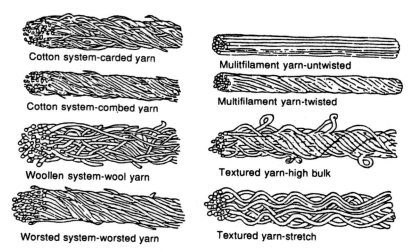

Cotton system-carded yarn

Mulitfilament yarn-untwisted

Cotton system-combed yarn

Multifilament yarn-twisted

Woollen system-wool yarn

Textured yarn-high bulk

Worsted system-worsted yarn

Textured yarn-stretch

Figure 4.43   Schematic description of spun and multifilament yarns.

4.10
Present and future developments in polymer chemistry

The production of commodity polymers (section 4.1.1) is now reaching saturation and a large percentage of these polymers are now manufactured in developing countries, especially in the Middle East and the Pacific Rim.

The technology used for polyolefin (e.g. PE, PP and various copolymers) manufacture is now almost entirely gas-phase using magnesium-based supported titanium catalysts (section 4.3.2.4), the problems of catalyst residues, tacticity, polymer particle shape and size having been overcome. The copolymers obtained from ethylene/α-olefin have been extended to make new grades, e.g. ultra-low density PE (density < 0.9 g/cc). These various grades tend to have specialist markets and may command premium prices.

Another active area of research in polyolefin polymers is the use of dicyclopentadienyl (especially indanyl) complexes of titanium and zirconium together with the activator, polymeric methyl aluminoxane (MAO). Steric control around the metal centre is possible and allows both high stereoregular content polymer from substituted olefin monomers (section 4.4.1), and alternating copolymers (section 4.4.2.2) from mixed olefin feeds.

An area of expansion is in ring-opening metathesis polymerisation (ROMP) of cyclic unsaturated hydrocarbons. Examples include cyclopentene, cyclooctene and norbornene. The general polymerisation reaction scheme is

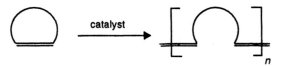

The mechanism is coordination polymerisation (section 4.3.2.4) using Ziegler-type catalysts, i.e. aluminium alkyl chloride activators together with Group VI

and VII metal halides (e.g. chlorides of molybdenum, tungsten and rhenium). These polymers have elastomeric characteristics. If the monomer contains two double bonds, e.g. di-cyclopentadiene (DCBD), a three-dimensional cross-linked polymer is obtained. Because these polymerisation reactions are very rapid (complete in a few seconds), the reaction injection moulding (RIM) technique can be used to make large rigid finished articles (section 4.5.1.7).

The engineering polymer field (section 4.1.2) is still growing—especially in automobile applications. Examples include body panels made from glass-fibre/polymer sheets and moulded into shape by compression moulding (sheet moulding composites, SMC), and large bumper and front facias fabricated using RIM technology. Advantages over conventional metallic components include low cost and weight, corrosion resistance and colour coordination. However, there is a slight drawback in that some countries, e.g. Germany, insist that the automobile manufacturers recover the plastic from the expired automobile and attempt to recycle the polymer (or re-convert the polymer back to the original monomer). This commitment has been termed 'cradle-to-grave' responsibility.

The re-cycling of plastics is also becoming an important issue for environmental reasons. Although only approx. 4% of crude oil is converted directly into polymers and the average household waste contains approx. 7% plastics, various National Plastic Councils and Advisory Bodies are aiming for a 25% recycled plastic target (particularly with plastic bottles and containers). Normally when a plastic is recycled the properties are slightly impaired. New research is being carried out into additives to minimise this impairment as well as into ways of using the recovered polymer as a feedstock for the preparation of the original or new monomers.

The small but expanding area of advanced polymeric materials (section 4.1.3, 4.6.2.1, 4.6.4) which can withstand hostile environments and elevated temperatures (> 200°C) has been slightly curtailed. This is because many of these materials have been developed for application in the aerospace and defence industries and funding from these organisations has recently been reduced. Nevertheless, specialised uses and applications have been forthcoming, especially in the field of thermoset composites (section 4.7.2). Several liquid crystalline polymers (section 4.6.5) are also now commercially available.

Very recently there has been considerable activity in the synthesis of polymers which have properties similar to, and therefore can mimic, natural polymers. This includes not only polymeric pharmaceuticals and foodstuffs, but also the concept of 'smart materials'. These are often polymeric materials which can sense external conditions and respond appropriately. The sensing can be done by the polymeric material itself, which can alter shape or colour on temperature or pressure changes, or by fibre optics or attached strain gauges, accelerometers, etc. Again the field is led by the aerospace and defence industries, but the spin-offs from these materials and applications will be enormous.

Bibliography

Individual references have not been given because in recent years several excellent encyclopaedias have appeared. These cover not only the individual polymers themselves, but also the preparature, fabrication and application of the various homopolymers, copolymers, blends and composites. These include

'Encyclopedia of Chemical Technology', 4th edn, Kirk-Othmer, Wiley-Interscience, 1991–.
'Encyclopedia of Polymer Science and Technology', J.I. Kroschwitz (ed.-in-chief), 2nd edn, Wiley-Interscience, 1985–1989.
'Encyclopedia of Material Science and Engineering', M.B. Bever (ed.), Pergamon, 1986–1988.
'Encyclopedia of Engineering Materials', N.P. Cheremisinoff (ed.), Decker, 1988–.
'Ullmann's Encyclopedia of Industrial Chemistry,' 5th edn, VCH, 1985–.
'The International Encyclopedia of Composites', S.M. Lee (ed.), VCH, 1990–.
'Materials Science and Technology. A Comprehensive Treatment', R.W. Cahn, P. Haasen and E.J. Kramer (eds), VCH, 1991–.
'Comprehensive Polymer Science', G. Allen (ed.), Pergamon, 1989.

Other useful books in polymer chemistry include

'Polymer Handbook', J. Brandrup and E.H. Immergut (eds), 3rd edn, Wiley-Interscience, 1989. This is an encyclopaedic work which gives various physical data on most polymers.
'Polymer Yearbook', R.A. Pethrick (ed.), Harwood Academic 1983–. This is an annual series on various topics in polymer science. It also includes recent books, reviews, significant journal articles, symposia, etc. devoted to polymers.

Dyestuffs 5

Peter Gregory

From prehistoric times, man has been fascinated by colour. From the early cavemen, who adorned their walls with coloured representations of animals, through the Egyptian, Greek and Roman eras, right up to the present time, colour has been a constant companion of mankind.

Until the end of the nineteenth century, these colours were all obtained from natural sources. The majority were of vegetable origin: plants, trees and lichen, though some were obtained from insects and molluscs. Over the thousands of years that natural dyes have been used, it is significant that only a dozen or so proved to be of any practical use, reflecting the instability of nature's dyes. Today, the number of synthetic organic colourants exceeds 4 000.

5.1.1 Natural dyes[1]

In terms of numbers, the yellow dyes comprised the largest group of natural dyes, but they were technically inferior to the reds, blues and blacks, having lower tinctorial strength (i.e. they were only weakly coloured) and poor fastness properties, especially light fastness (i.e. they soon faded). In contrast, the red and blue dyes had good properties, even by modern standards. Natural yellow dyes are based on chromogens* (mainly flavones (1), chalcones (2) and polyenes (3)) that are relatively unstable and which have been completely superseded by superior synthetic yellow chromogens. However, the an-

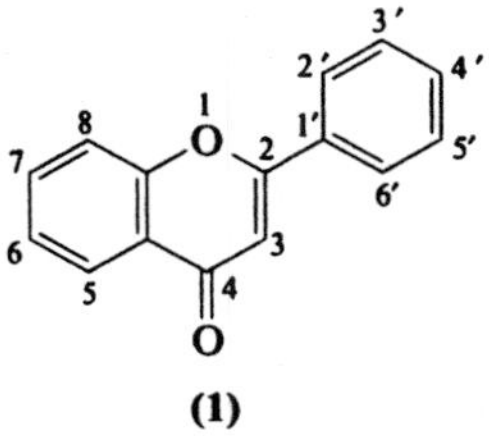

$$RO_2C\{CH=CR\}_n CO_2R$$
$$n = 7/9$$
$$R = H/Me$$

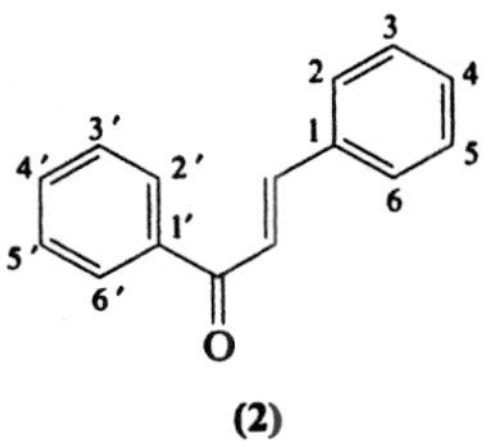

(3)

(1)   (2)

---

* Chromogen is the term used to describe that complete arrangement of atoms that gives rise to the observed colour. The term chromophore describes the various chemical units (building blocks) from which the chromogen is built.

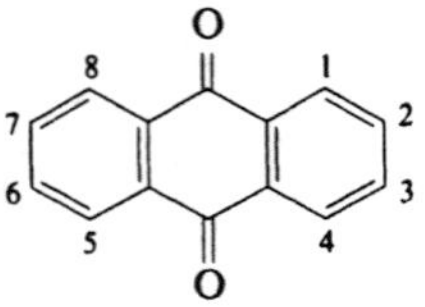

(4) 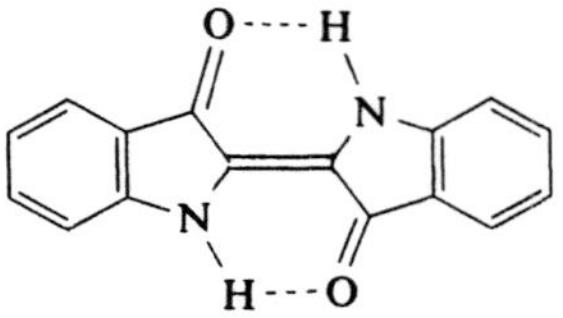   (5)

thraquinone (4) and indigoid (5) chromogens found in the natural red and blue dyes respectively still form the basis of many of the modern synthetic dyes, especially the anthraquinone derivatives.

Some of the more important natural dyes are shown in Table 5.1.

5.1.2   Synthetic dyes

It was W.H. Perkin, whilst working with Hofmann, who discovered the first synthetic dye, mauveine, in 1856, and thereby started the 'dyestuff revolution'. However, he was not intentionally working towards preparing synthetic

Table 5.1   Some important natural dyes

Colour	Class	Typical dyes	Structure (name)	Source
Yellow	Flavone	Weld	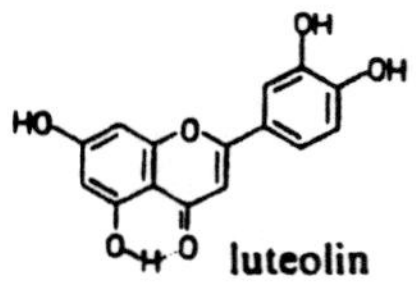 luteolin	Seeds, stems and leaves of the *Reseda luteola* L plant (Dyer's Rocket)
	Flavonol	Quercitron	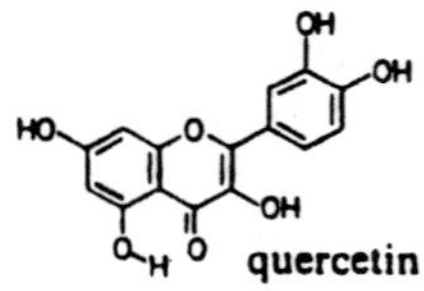 quercetin	Bark of North American can oak, *Quercus tinctoria nigra*
	Chalcone	Safflower	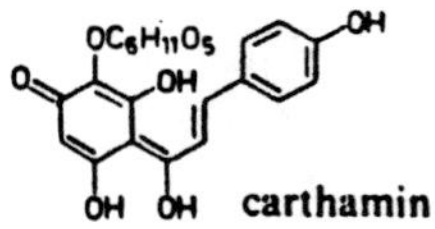 carthamin	Dried petals *of Carthamus tinctorius* (Dyer's Thistle)
	Polyene	Saffron	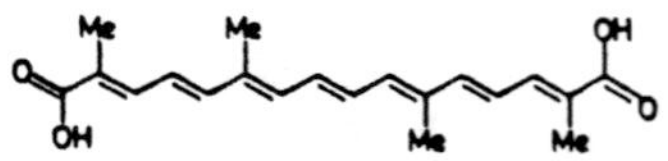 crocetin	Stigmas of *Crocus sativus* (4 000 required to give 25 g of dye)

Table 5.1 Contd.

Colour	Class	Typical dyes	Structure (name)	Source
Red	Anthraquinone	Kermes	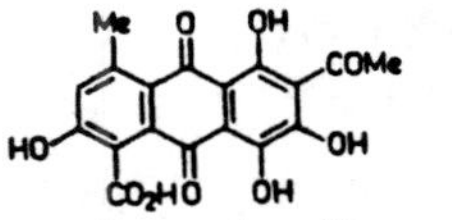 kermesic acid	Female scale insects, *Coccus ilicis*, which infect the Kermes oak
	Anthraquinone	Cochineal	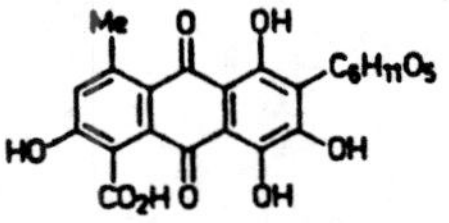 carminic acid	Female insect, *Coccus cacti*, which lives on cactus plants of the prickly pear family found in Mexico (200 000 → 1 kg of dye)
	Anthraquinone	Madder or alizarin	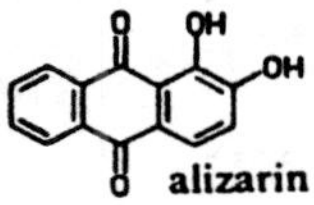 alizarin	Roots of the *Rubia tinctorum* plant. Root was known as 'alizari', hence alizarin
	Anthraquinone	Turkey Red	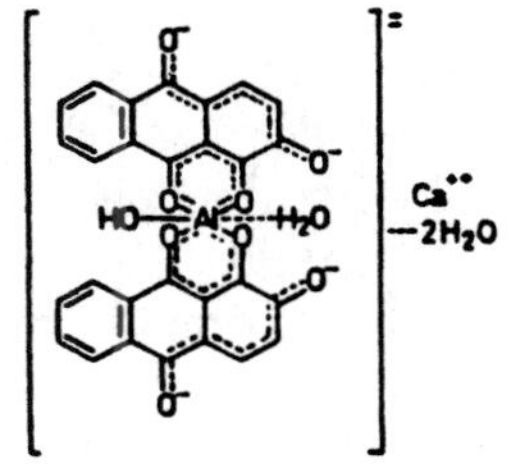	
Purple	Indigoid	Tyrian Purple	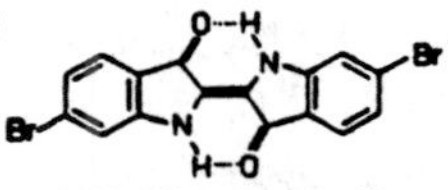 6,6'-dibromoindigo	Mollusc (i.e. shellfish) usually *Murex brandaris* plentiful in the Mediterranean
Blue	Indigoid	Woad; indigo	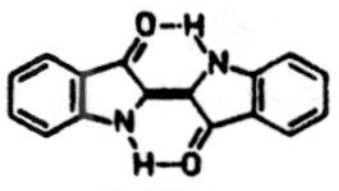 indigo	Leaves of indigo plant, *Indigofera tinctoria* L
Black	Chroman	Logwood	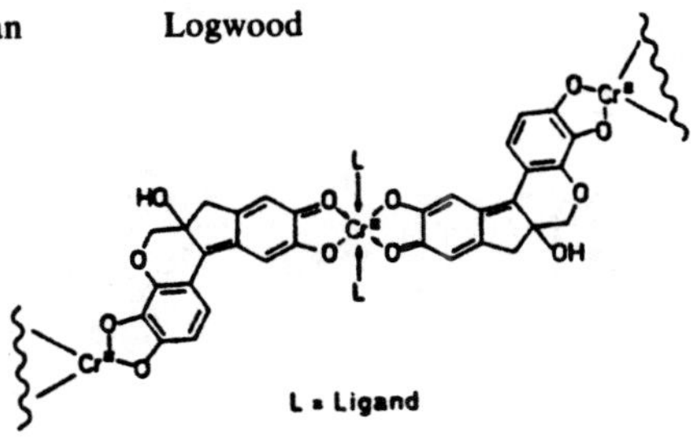 L = Ligand	Heartwood of the tree *Haematoxylon campechiancum* L found in Central America

dyestuffs but towards quinine, the antimalarial drug. By consideration of the molecular formulae of allyltoluidine (6) and quinine (7) he had arrived at the relationship shown in equation 5.1 and thus attempted

$$2C_{10}H_{13}N + 3[O] \longrightarrow C_{20}H_{24}N_2O_2 + H_2O \qquad (5.1)$$

$$\quad (6) \qquad\qquad\qquad\qquad (7)$$

the preparation of quinine by the oxidation of allyltoluidine with potassium dichromate in sulphuric acid. If Perkin had known the structure of quinine (7a)

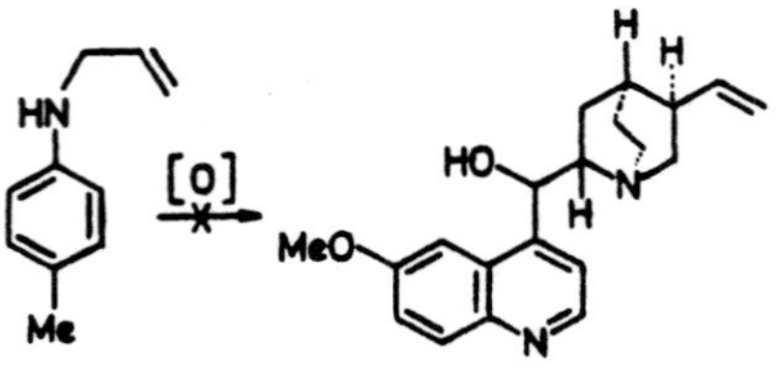

(6a)   (7a)

and allyltoluidine (6a) he would almost certainly have abandoned this route, but in 1856 the structure of benzene was still unknown, never mind that of quinine! Perkin tested his theory but found he obtained a very impure brown powder which did not contain any quinine. He then turned his attention to the simplest aromatic amine, aniline, to determine if the oxidation reaction was a general one. Once again he obtained a very unpromising mixture, this time a black sludge, but on boiling his reaction mixture with ethanol he obtained a striking purple solution which deposited purple crystals on cooling. Perkin recognised that this new compound might serve as a dye, later to be called mauveine (8), and despatched a sample to Pullar's dyehouse in Scotland. The dyers produced a very enthusiastic report about this new dye, for it had superior fastness properties on silk to the available natural dyes.

Perkin had been very fortunate. Not only did his discovery arise from testing an erroneous theory but it also required the presence of substantial toluidine impurities in the starting aniline (equation (5.2)). However, there was no luck associated with the way in which Perkin followed up his astute observation. Once he had established the usefulness of mauveine as a dye, he set about preparing it on a large scale. The problems he faced were

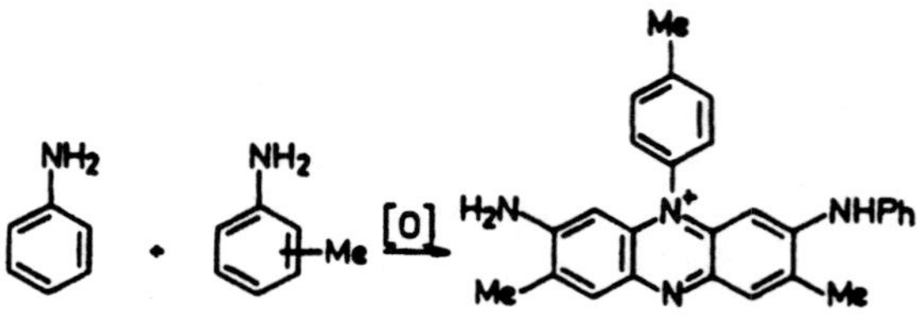

(8)   (5.2)

enormous. As well as having to manufacture the mauveine, he also had to synthesise large quantities of aniline by Hofmann's route, and in the process pioneer the techniques of large-scale synthesis.

Perkin solved all these problems in a surprisingly short time and, with the help of his family, was soon manufacturing and selling mauveine. As a dye for silk, mauveine was an instant success. However, Perkin's dye increased greatly in importance when he discovered, with others, a method for dyeing wool with mauveine using tannic acid. Perkin's remarkable achievements have since earned him the title of 'founder of the dyestuffs industry', and justifiably so!

Perkin's success attracted many competent chemists to the dyestuffs industry, so that the nineteenth century saw the discovery of the major dye chromogens. Indeed, apart from one or two notable exceptions, all the dye types used today were discovered in the nineteenth century.[2] The introduction of the man-made fibres, nylon, polyester and polyacrylonitrile, during the period 1930–1950, produced the next major challenge. The discovery of reactive dyes by Rattee and Stephen in 1954 heralded a major breakthrough in the dyeing of cotton, and intensive research into reactive dyes followed over the next two decades and, indeed, is still continuing today. The oil crisis in the early 1970s, which resulted in a steep increase in the prices of raw materials for dyes, created a drive for more cost-effective dyes, both by improving the efficiency of the manufacturing processes and by replacing tinctorially weak chromogens, such as anthraquinone, with tinctorially stronger chromogens, such as azo, triphendioxazine and benzodifuranone. These themes are still important and ongoing, as are the current themes of product safety, quality and protection of the environment. There is also considerable activity in dyes for high-technology applications, especially for the electronics and reprographics industries.

There are two main ways of classifying dyes: by chemical structure and by their usage or application.

**5.2
Classification and
nomenclature of dyes**

5.2.1 Chemical classification of dyes

The Colour Index[3] already classifies dyes (and pigments) by chemical structure. To repeat this classification here would be rather pointless for several reasons. First, the reader can consult the Colour Index classification directly. Second, the Colour Index classification includes pigments as well as dyes, while this chapter is confined to dyes only. Thus, important pigment* classes such as quinacridones are not included. Third, the classification adopted in this section, while maintaining the backbone of the Colour Index classifica-

---

* Unlike dyes, pigments do not pass through a solution phase during application to the substrate; they remain insoluble.

Table 5.2   Classification of dyes by chemical structure

Azo dyes
Anthraquinone dyes
Phthalocyanines
Indigoid dyes
Benzodifuranone dyes
Oxazine dyes
Polymethine and related dyes
Di- and tri-aryl carbonium and related dyes
Polycyclic aromatic carbonyl dyes
Quinophthalone dyes
Sulphur dyes
Nitro and nitroso dyes
Miscellaneous dyes

tion, is intended to progress beyond the latter classification by simplifying and updating it. Finally, the classification is chosen to highlight some of the more recent discoveries in dye chemistry. Table 5.2 summarises the classification of dyes by chemical structure.[4]

5.2.2   Usage classification of dyes

Only the more important usage applications are discussed briefly. Less important ones are simply listed.

5.2.2.1   Acid dyes.   Acid dyes are chiefly used for dyeing wool and nylon. They are anionic and are applied from an acid dyebath. These dyes belong mainly to the azo and anthraquinone classes.

5.2.2.2   Basic dyes.   The basic dyes are cationic dyes, often brilliant in shade and of very high tinctorial strength, but having poor fastness properties on cotton and paper. Modified basic dyes of the polymethine class, such as (di)-azacarbocyanines and diazahemicyanines, are used extensively for dyeing polyacrylonitrile. They exhibit excellent fastness properties on this fibre.

5.2.2.3   Direct dyes.   All dyes of this class are anionic dyes having affinity for cellulosic fibres when applied from an aqueous dyebath containing an electrolyte. Certain direct dyes are extensively employed in dyeing paper, leather, bast fibres and other substrates.

5.2.2.4   Disperse dyes.   This class comprises dyes whose molecules are devoid of the common solubilising groups but which have a minute solubility in water and when milled to extremely fine particle size will dye secondary acetate. Disperse dyes have been developed for polyester fibres, the dyeing of which may require the presence of swelling agents and the use of pressure to achieve higher temperatures (see later). Disperse dyes are predominantly azo, anthraquinone and benzodifuranone.

5.2.2.5 *Pigments.* Synthetic organic pigments are inert, stable, coloured substances insoluble in water and in organic solvents such as the oils commonly used as vehicles in paint manufacture. They are used also in the mass coloration of plastics. Physical form is of paramount importance in ensuring maximum colouring power and reflectance.

5.2.2.6 *Reactive dyes.* A reactive dye may be defined as a coloured compound possessing a suitable group capable of forming a covalent bond between a carbon atom of the dye ion or molecule and an oxygen, nitrogen or sulphur atom of a hydroxy, an amino or a thiol group respectively of the substrate. The establishment of a chemical bond between dye and substrate results in improved wash fastness.

5.2.2.7 *Solvent dyes.* Solvent dyes are so called because of their solubility in organic solvents including esters, ethers, hydrocarbons, oils, fats and waxes. They are synthetic organic dyes whose molecules contain no water-solubilising groups such as sulphonic acid or sulphonate ion groups.

5.2.2.8 *Sulphur dyes.* Sulphur dyes are of indeterminate structure and are derived from the sulphurisation at elevated temperatures of certain aromatic compounds, e.g. 4-aminophenol. They are water-insoluble but are rendered soluble for dyeing purposes by sulphide reduction (vatting), reoxidation occurring on the fibre on contract with air. They are all vat dyes.

5.2.2.9 *Vat dyes.* This group includes dyes derived from anthraquinone, indigo, or thioindigo, giving dyeings with excellent all-round fastness properties. Insoluble in water, they are reduced to the soluble, leuco-form with sodium dithionite in the presence of sodium hydroxide. The true colour is regenerated on the dyeings by exposure to air or by the action of an oxidizing agent. The solubilised vat dyes are classed under vat dyes since they are the soluble sodium salts of the sulphuric esters of the reduced forms of the parent vat dyes. They are applied directly to the fibre where they are regenerated by the use of an oxidising agent, normally atmospheric oxygen.

The less important usage applications include azoic; fluorescent brighteners; food, drug and cosmetic; mordant; natural; and oxidation bases.

5.2.3 *Nomenclature of dyes*

Dyes are named either by their commercial trade name or by their Colour Index (CI) name.

The commercial names of dyes are usually made up of three parts. The first is a trademark used by the particular manufacturer to designate both the manufacturer and the class of dye, the second is the colour, and the third is a series of letters and numbers used as a code by the manufacturer to define more precisely the hue, and also to indicate important properties which the dye possesses. The code letters used by different manufacturers are not standardised. The most common letters used to designate the hue are R for reddish, B for bluish and G for greenish shades. Some of the more important letters used to denote the dyeings and fastness properties of dyes are W for wash fastness and E for exhaust dyes. For solvent and disperse dyes, the heat fastness of the dye is denoted by letters A, B, C or D, A being the lowest level of heat fastness and D being the highest. In reactive dyes for cotton, M denotes a warm (c. 40 °C) dyeing dye and H a hot (c. 80 °C) dyeing dye. The following example illustrates the use of these letters:

DISPERSOL Yellow B-6G

DISPERSOL is the ICI tradename for its range of disperse dyes for polyester. Therefore, it reveals the manufacturer and the usage. Yellow denotes the main hue of the dye. B denotes its heat fastness, i.e. rather low, and 6G denotes that it is six steps of green away from a neutral yellow, i.e. it is a lemon yellow.

The CI name for a dye is derived from the application class to which the dye belongs, the shade or hue of the dye, and a sequential number, e.g. CI Acid Yellow 3, CI Acid Red 266, CI Basic Blue 41, and CI Vat Black 7. A five-digit CI number is assigned to a dye when its chemical structure has been made known by the manufacturer.

5.3
Colour and constitution

This is a complex topic which cannot be covered adequately in a review such as this. However, good accounts[5,6] do exist and these should be consulted for a comprehensive treatment.

There are two important effects which determine both the colour (hue) and colour strength of dyes: electronic effects and steric effects. These are considered briefly with respect to the four most important dye types: azo, anthraquinone, indigo and phthalocyanine.

5.3.1 Azo and anthraquinone dyes

5.3.1.1 *Electronic effects.* To generate colour in both azo and anthraquinone dyes it is necessary to introduce an electron-donating group containing a $2p_z$ lone pair of electrons, such as amino or hydroxy, capable of delocalisation into the π-system of azobenzene and 9,10-anthraquinone. This generates a new, intense $\pi-\pi^*$ charge transfer band in the visible region of the spectrum.

The colour depends upon the strength, position and number of the electron-

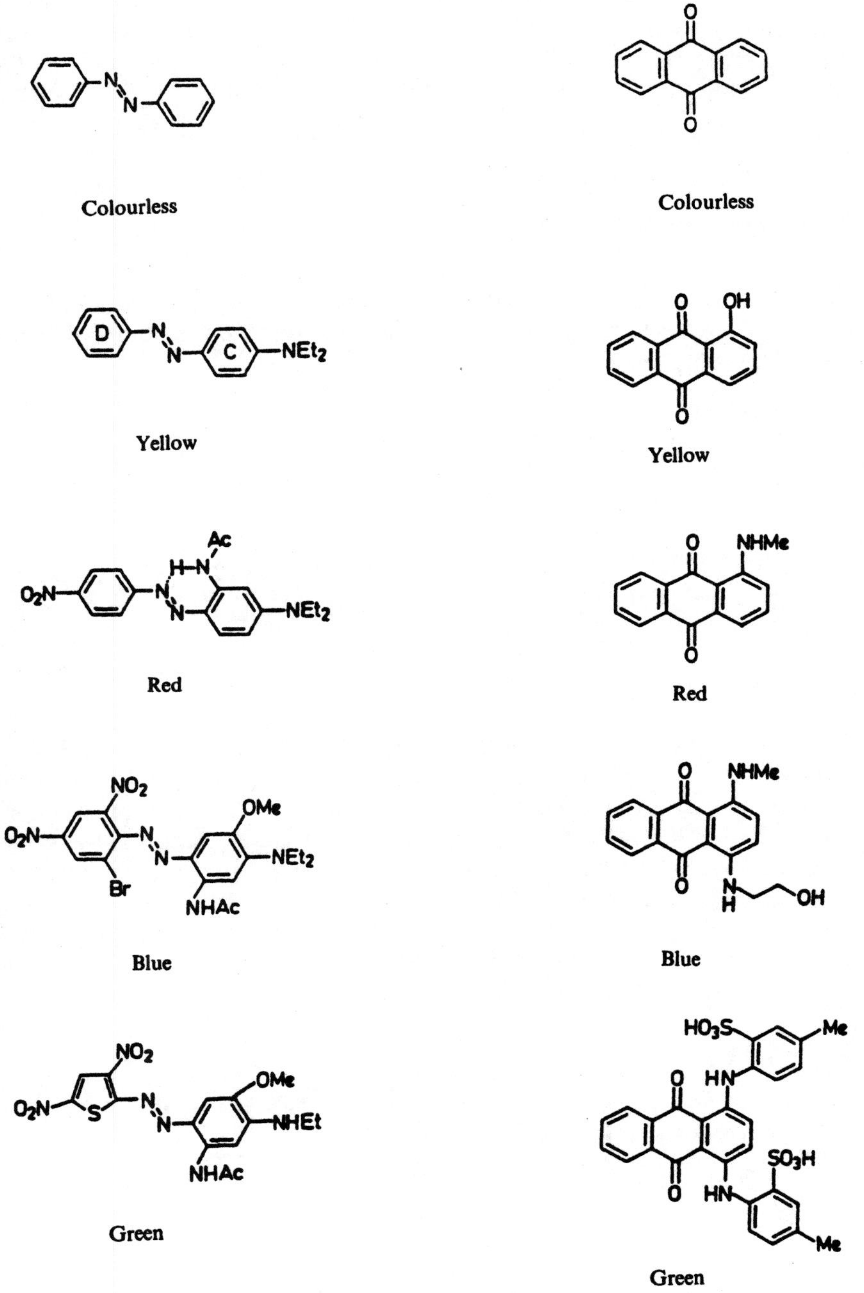

Figure 5.1   Principles of colour generation in azo and anthraquinone dyes.

donating groups. In azo dyes, the nature, position and number of electron-withdrawing groups is also important. (In anthraquinone dyes, the carbonyl groups are the electron-withdrawing groups.) Relative strengths of common electron-donating and electron-withdrawing groups are:

$$O^- > NHAr > N(alkyl)_2 > NHalkyl > NH_2 > NHAc > OR > CH_3$$

$$NO_2 > CN > SO_2R > CO_2R > Hal$$

The examples shown in Figure 5.1 illustrate these principles.

5.3.1.2 Steric effects. Steric effects always lower the tinctorial strength of a dye. The colour shift may be bathochromic or hypsochromic but is predictable.[7] Normally it is hypsochromic.

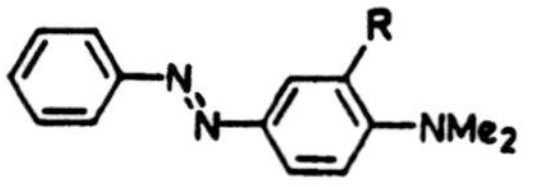

	λ	ε
R = H	408	28,250
R = Me	375	18,200

5.3.2 Indigo

The chromogen of indigo (5) is the sub-unit (9). The fact that such a simple arrangement of atoms produces a blue colour is unusual (*cf.* azo and anthraquinone blues) and led to this type of molecule being termed an H-chromophore[8] (because the shape of the unit resembles a capital H). Because the fixed benzene rings in indigo have only a secondary role on colour, substituents on those rings have only a minor effect on the colour. For example, 6,6'-dibromoindigo (10) is purple (Tyrian Purple). In contrast, replacing the groups of the H-chromophore itself causes a more profound effect on the colour. Thus, thioindigo (11) is red.

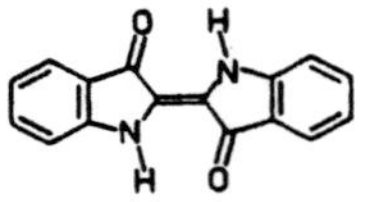

(5)

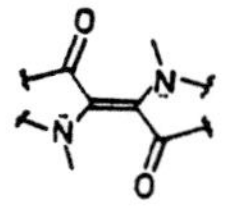

(9)

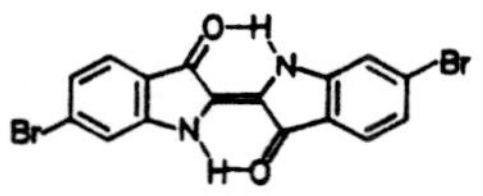

(10)

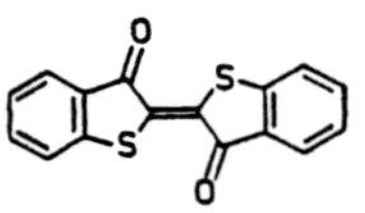

(11)

5.3.3 Phthalocyanines

Phthalocyanines are synthetic analogues of the natural pigments chlorophyll
(12) and haemin (13) based on porphyrins.

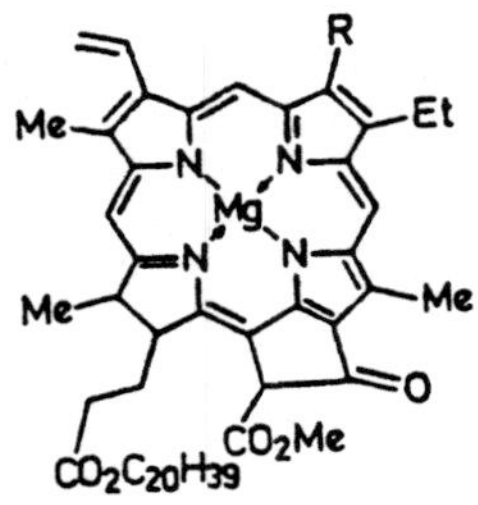

R = Me   (chlorophyll a)
R = CHO (chlorophyll b)

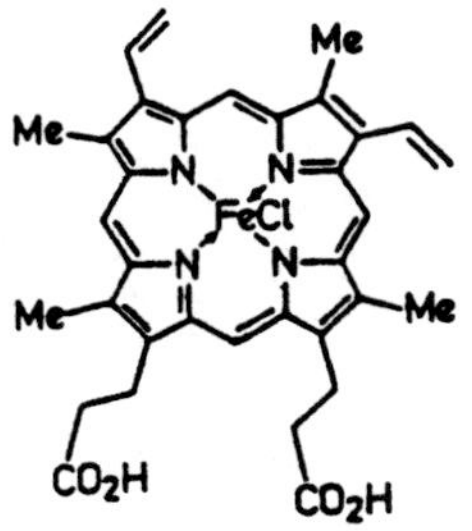

(12)

(13)

In both phthalocyanines and porphyrins the chromogen is the cyclic 16-atom,
18 π-electron aromatic pathway[9] indicated in bold in the structures (14) and
(15). As in indigo, the fused benzene rings have little influence on the colour,
and therefore phthalocyanine dyes (and pigments) are restricted to blue and
green colourants, and near infrared absorbers.[10]

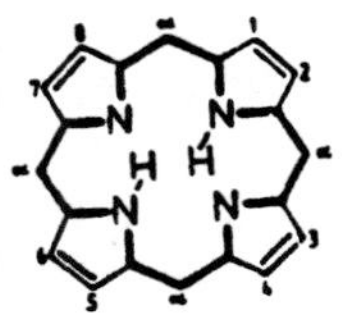

(14)

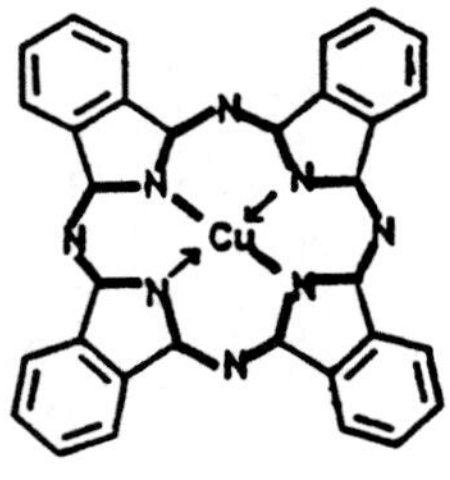

(15)

5.3.4 Theoretical calculations

Rapid advances in computer hardware and software now allow ready calculation of the colour and strength of dyes using molecular orbital (MO) theory.
The calculations range from fairly simple π-electron-only models, such as the
PPP model, for planar molecules, to sophisticated all valence electron models
such as CINDO and *ab initio*.[6]

The scale and growth of the dyes industry is inextricably linked to those of the
textile industry. World textile production has grown steadily to an estimated
35 million tonnes in 1990.[11,12] The two major textile fibres are cotton (the
largest) and polyester. Consequently, dye manufacturers tend to concentrate

**5.4
World production**

Table 5.3 World dyestuff production estimates

Year	Production (tonnes $\times 10^3$)					
	Western Europe	USA	USSR, Eastern Europe, China	Japan	Others	Total
1938	110	37	35		28	210
1948		110				
1958	112	80	127		27	346
1966	191	130		49		
1974	300	138	200*	68	44	750*
1990*	350	150	250*	90	75	915*

* Estimated.

their efforts on producing dyes for these two fibres. The estimated world production of dyes in 1990 was just under 1 million tonnes[11,12] (Table 5.3). This figure is significantly smaller than that for textile fibre because a little dye goes a long way. For example, 1 tonne of dye is sufficient to colour 16 650 cars or 42 000 suites.[12]

Perkin, an Englishman, working under a German professor, Hoffman, discovered the first synthetic dye, and even today the geographical focus of dye production lies in Germany (BASF, Bayer, Hoechst), the UK (ICI) and Switzerland (Ciba-Geigy, Sandoz) (Table 5.4).

Production of dyes in the USA has remained much more fragmented than that of European countries and all the European majors have a significant manufacturing presence in the USA, gradually replacing exports from Europe.

A historically fragmented Japanese industry, with rapid growth to meet internal demand, changed in 1979 with an agreement by its five major

Table 5.4 West European dyestuff production estimates, 1974 and (1991)

Country	Company	Production (tonnes $\times 10^3$)	Value (US$ $\times 10^6$)
West Germany		148 (176)	976 (1 936)
	Bayer BASF Hoechst		
UK		54 (60)	268 (612)
	ICI		
Switzerland		30 (35)	370 (735)
	Ciba-Geigy Sandoz		
France		31 (34)	147 (350)
Italy		15 (17)	62 (136)
Spain		15 (18)	42 (108)
Belgium		6 (7)	23 (52)
Rest		1 (3)	4 (22)
Total		300 (350)	1 892 (3 981)

Table 5.5 USA and Japanese production (tonnes) of colourants, 1980 and (1990 estimates)

Class	Japan	USA
Acid	2 300 (2 400)	11 600 (12 000)
Basic	5 000 (4 500)	6 600 (6 300)
Direct	3 200 (3 300)	14 000 (14 400)
Disperse	13 600 (15 500)	21 000 (24 000)
Reactive	4 700 (6 000)	2 600 (5 000)
Fluorescent brightener	7 500 (7 750)	17 000 (18 500)
Mordant	1 500 (1 200)	200 (160)
Solvent	3 000 (3 100)	4 800 (4 900)
Vat	2 400 (2 500)	18 000 (18 500)
Other (azoic, sulphur, food, etc.)	8 200 (8 500)	15 200 (16 000)
Total dyes	51 400 (54 750)	111 000 (119 760)
Organic pigments	21 000 (23 000)	31 000 (32 000)
Total dyes and pigments	72 400 (77 750)	142 000 (157 760)

manufacturers (Mitsubishi, Sumitomo, Nippon Kayaku, Mitsui Toatsu and Hodogaya) to the formation of a cartel to coordinate research as well as manufacture. Successful operation of this cartel could have profound effects on the international situation by the end of this century.

Comparatively little is known of manufacture in the former Eastern bloc countries, the former USSR and China, although their aim must be to satisfy all their own requirements. The signs are that they need to import technology to hasten the process towards self-sufficiency. They are currently significant importers of dyes. It is noteworthy that these countries were the only ones to announce large-scale investment in new dye-manufacturing capacity for the 1980s. These are a 4 000 tonnes per annum disperse dye plant and a 2 000 tonnes per annum leather dye plant in Russia, both using Montedison/ACNA technology. The Italian group is also reported to be providing know-how for a dyestuff finishing plant to be built in north-east China. Manufacture in India, South America and Mexico is mainly through subsidiaries or partly owned companies associated with the European majors. Within Far Eastern countries, other than Japan, a large number of small manufacturing facilities are evolving, especially in Taiwan and South Korea.[13]

Details of the production of the various colourant classes are available only from the USA[14] and Japan.[15]

Multiplying the sum of the output for these two countries by four permits a very rough estimate of the world manufacture of dyes in 1980, *c.* 850 000 tonnes (Table 5.5).

Table 5.6 shows the usage of dyes by chemical class in the USA. It clearly demonstrates that the three most important dye classes are azo, anthraquinone and phthalocyanine. The dye costs in 1972 ranged from $1.5/kg for the cheapest dyes (sulphur dyes) to $6.8/kg for the most expensive dyes (phthalocyanines), with an average cost of $4.5/kg. The average cost of

Table 5.6   USA dye sales and prices in 1972 and (1990 estimates)

Chemical class	USA sales 1972 (1990) (kg × 10^6)		USA value 1972 (1990) ($ × 10^6)		Price ($/kg)	
Azo	56.4	(65)	290	(975)	5.1	(15)
Anthraquinone	21.8	(17)	125	(340)	5.7	(20)
Phthalocyanine	8	(12)	54.5	(168)	6.8	(14)
Triarylmethane	3.9	(3.8)	20.5	(38)	5.3	(10)
Sulphur (estimates)	6.8	(7.0)	10	(28)	1.5	(4)
Indigoid (estimates)	3.2	(5.0)	5	(30)	1.6	(6)
All others	40	(40)	124	(400)	3.1	(10)
Total	140.1	(149.8)	629	(1979)	4.5(11) (Average)	

colourants in 1984 was $7.7/kg, ranging from low-cost fluorescent brightening agents at $4/kg to reactive dyes at $14.3/kg.[12] The estimated average cost of colourants in 1990 was $11/kg. Thus, a rough estimate for the world market for colourants in 1990 was $10.45 billion (950 000 × $11).

5.5
Major products

The sequence of steps used to make dyes is to convert raw materials into intermediates and intermediates into dyes.

5.5.1   Raw materials

Raw materials are obtained from coal tar and especially petroleum. The major raw materials for dyes are shown in Table 5.7.

5.5.2   Intermediates

These are conveniently divided into primary intermediates and dye intermediates. However, large amounts of inorganic materials are also consumed in both intermediates and dyes manufacture.

5.5.2.1   Inorganic materials.   These include acids (sulphuric, nitric, hydrochloric and phosphoric), bases (caustic soda, caustic potash, soda ash, sodium

Table 5.7   USA production (tonnes) of raw materials[12]

Chemical	1967	1976	1990*
Benzene	977	1426	1800
Toluene	644	998	1200
Naphthalene	108	67	80
Anthracene	94	60	65

* Estimated.

carbonate, ammonia and lime), salts (sodium chloride, sodium nitrite and sodium sulphide) and other substances such as chlorine, bromine, phosphorus chlorides and sulphur chlorides. The important point is that there is a significant usage of at least one inorganic material in every process and the overall tonnage used by, and therefore the cost to, the dye industry is high.

5.5.2.2 The chemistry of intermediates. The chemistry of intermediates may be conveniently divided into two parts: the chemistry of carbocycles, such as benzene and naphthalene, and the chemistry of heterocycles, such as pyridones and thiophenes. The chemistry of carbocyclic intermediates involyes the sequential introduction of groups into the aromatic rings by a variety of steps known as unit processes (Scheme 5.1). The unit processes employed in dye chemistry are listed in Table 5.8.

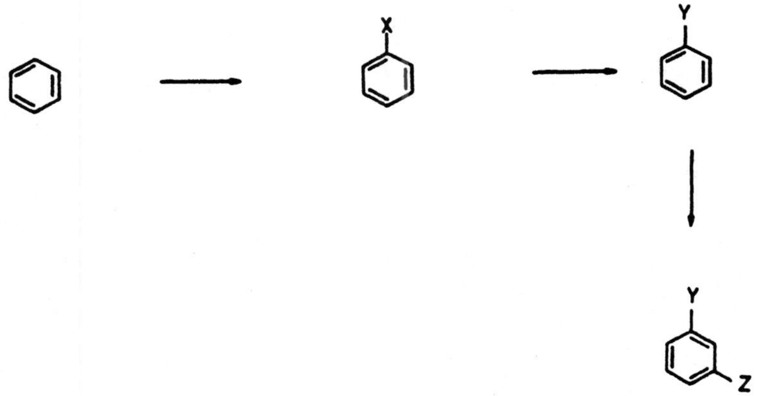

Scheme 5.1.

The substituents are introduced into the aromatic ring by one of two routes, either by electrophilic substitution or by nucleophilic substitution. In general, aromatic rings, because of their inherently high electron density, are much more susceptible to electrophilic attack than to nucleophilic attack. Nucleophilic attack only occurs under forcing conditions unless the aromatic ring already contains a powerful electron-withdrawing group, e.g. NO_2. In this case, nucleophilic attack is greatly facilitated because of the reduced electron density at the ring carbon atoms.

In contrast to carbocyclic chemistry, aromatic heterocycles are built up from acyclic precursors and the ring is functionalised if necessary (equation 5.3).

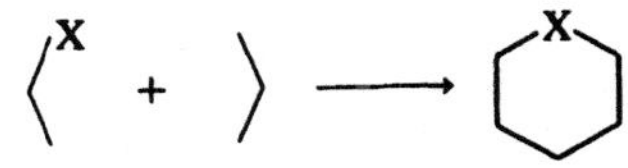

$$(5.3)$$

Table 5.8   Unit processes in dyes manufacture

Process	Primaries (no. of occurrences within 30 identified product manufactures)	Intermediates (common usage)	Colourants (common usage)
Nitration	6	√	
Reduction	8	√	
Sulphonation	4	√	
		(Incl. chlorosulphonation)	
Oxidation	5	√	
Fusion/hydroxylation	3	√	
Amination	3	√	
		(Incl. Bucherer reaction)	
Alkylation	2	√	
Halogenation	2	√	√
Hydrolysis	2	√	√
Condensation	1	√	
Alkoxylation		√	
Esterification		√	
Carboxylation		√	
Acylation		√	√
Phosgenation		√	√
Diazotisation		√	√
Coupling (azo)		√	√

5.5.2.3   Primary intermediates.   Primary intermediates are chemicals that are manufactured on a very large scale because they are used in several industries, of which the dyes industry is just one. Some of the more important primary intermediates are shown in Table 5.9.

Aniline.   This is prepared by nitration of benzene with nitric acid–sulphuric acid to give nitrobenzene, which is reduced to aniline, usually by catalytic hydrogenation (Scheme 5.2).

Table 5.9   Production (tonnes) of primary intermediates

Product	USA	Japan
Phenol	1 167 050	212 173
Phthalic anhydride	371 930	219 943
Aniline	299 737	84 661
Nitrobenzene	278 012	50 961
Salicylic acid	17 749	2 338
p-Nitroaniline	6 558	
p-Chloronitrobenzene	*(2)	20 262
β-Naphthol	*(1)	4 485
Dimethylaniline	*(3)	1 504

*Manufactured (number of producers in parentheses) but quantities not disclosed.

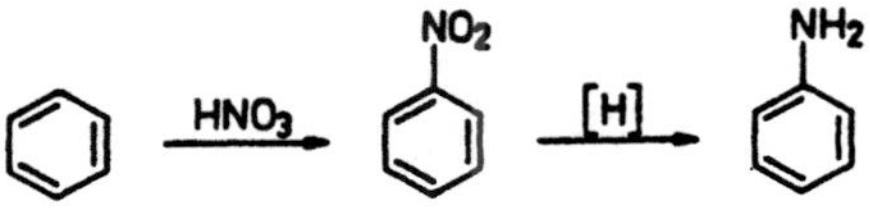

Scheme 5.2.

β-Naphthol.   Naphthalene is sulphonated at high temperature to yield the thermodynamically more stable naphthalene-2-sulphonic acid. (Sulphonation at lower temperatures yields the kinetically controlled 1-isomer.) Caustic fusion causes a nucleophilic displacement of the sulphonic acid group by hydroxy to give β-naphthol (Scheme 5.3).

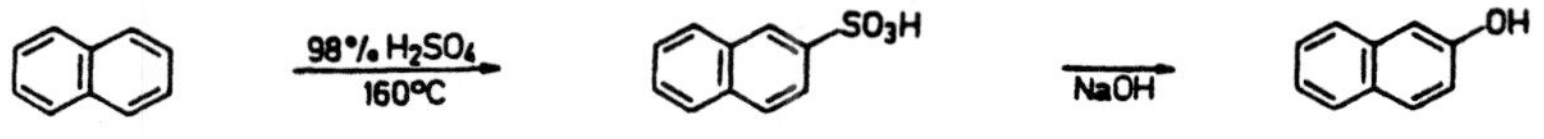

Scheme 5.3.

Phthalic anhydride.   This is prepared by catalysed aerobic oxidation of either naphthalene or *o*-xylene (Scheme 5.4). It is used in large quantities for the production of phthalocyanine dyes and pigments, and for the synthesis of anthraquinone.

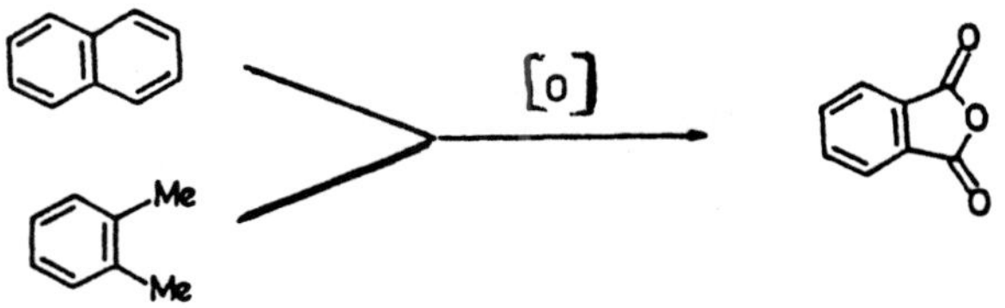

Scheme 5.4.

Phenol.   Phenol is made both by the oxidation of cumene and by nucleophilic substitution of chlorobenzene under forcing conditions (high temperature and pressure) (Scheme 5.5).

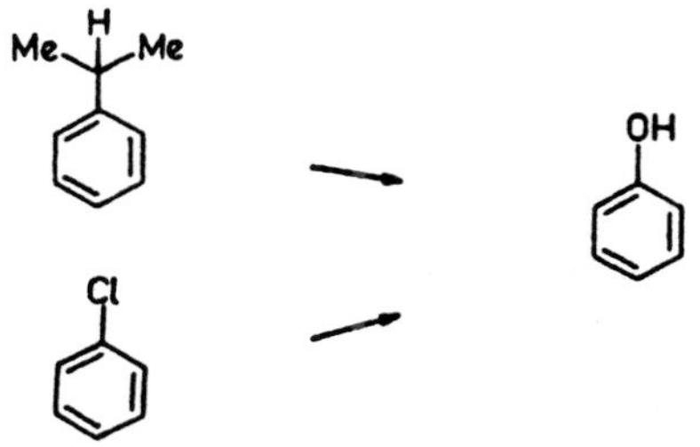

Scheme 5.5.

5.5.2.4 *Dye intermediates.* Products under this heading are of relatively small tonnage and are often made by dye manufacturers themselves for internal use and for sale to other dye manufacturers where competitive interests are compatible.

Carbocyclic.[16] As mentioned earlier these are made by unit processes on benzene and naphthalene.

Benzene intermediates. Aniline is the principle benzene intermediate. The two major uses of aniline intermediates are diazo components and coupling components for azo dyes (Scheme 5.6).

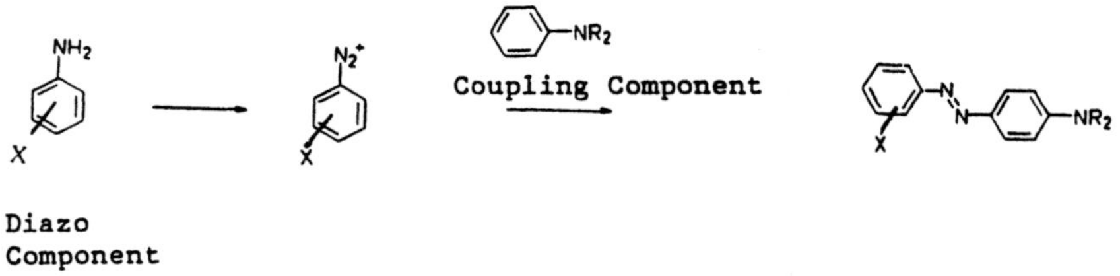

Scheme 5.6.

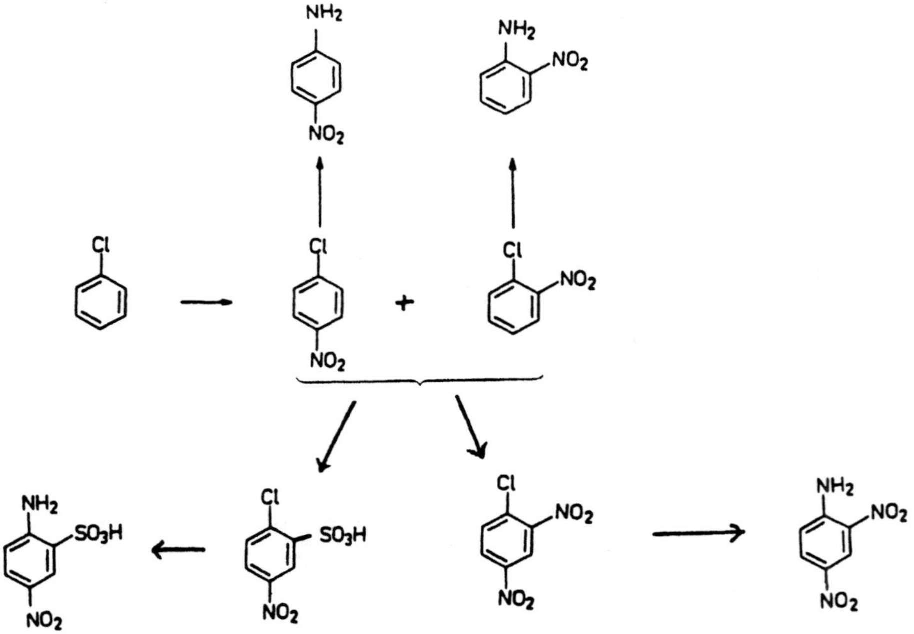

Scheme 5.7.

Aniline-based diazo components contain electron-withdrawing groups such as nitro, halogen, cyano and sulphonic acid. These are introduced by electrophilic attack (equation 5.4). As seen earlier the amino group is normally

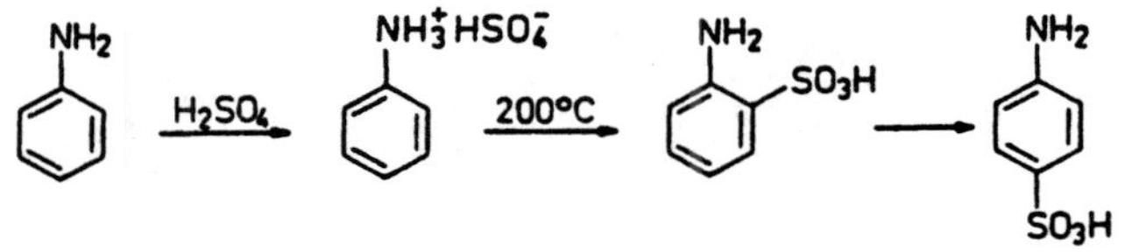

$$(5.4)$$

Sulphanilic Acid

introduced by nitration and reduction, but it may be introduced by nuleophilic substitution of an activated halogen (Scheme 5.7).

Aniline-based coupling components contain further electron-donating groups. Scheme 5.8 shows the synthesis of a fairly complex coupling component using a variety of the unit processes shown in Table 5.8.

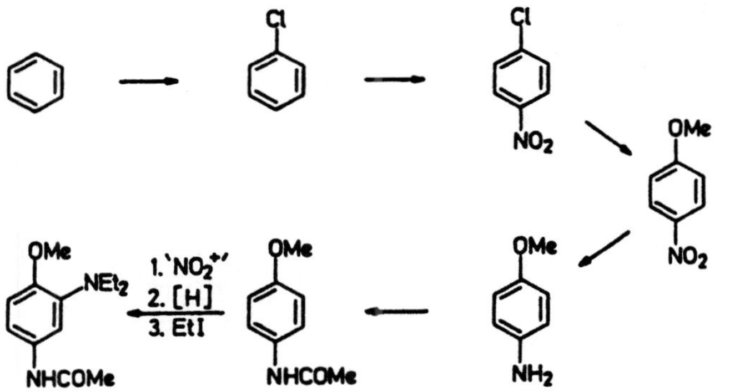

Scheme 5.8. Synthesis of a coupling component used in disperse dyes.

Naphthalene intermediates. As was the case with benzene, naphthalene intermediates can be split into diazo components and, in the case of naphthalene, the more important coupling components.

The most important naphthalene diazo components are 2-naphthylamine-sulphonic acids. These are generally prepared by sulphonation, caustic fusion and amination (Bucherer reaction). The synthesis of 2-naphthylamine-3,6-disulphonic acid (**16**) is typical (Scheme 5.9).

The most important naphthalene coupling components are the aminonaphtholsulphonic acids such as Gamma acid (**17**), J-acid (**18**) and especially H-acid (**19**). The syntheses of H-acid (Scheme 5.10) and of J-acid (Scheme 5.11) illustrate the principles involved.

Noteworthy points are the selective replacement of a 1-sulphonic acid group rather than a 2-sulphonic acid group in the caustic fusion reaction and

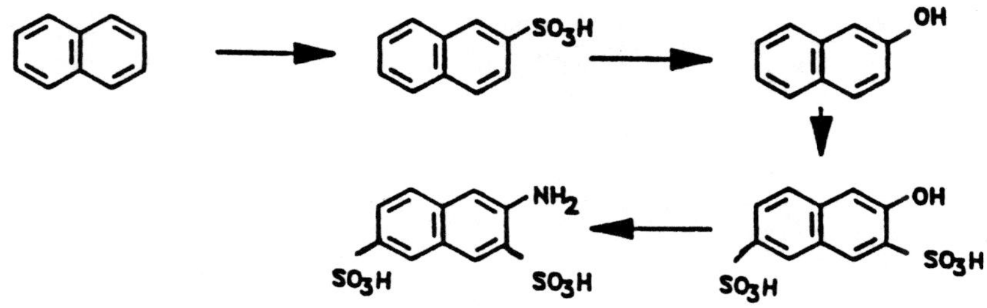

(16)

Scheme 5.9.

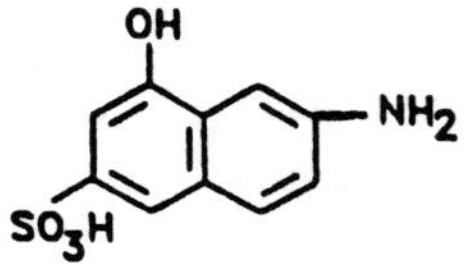

(17)

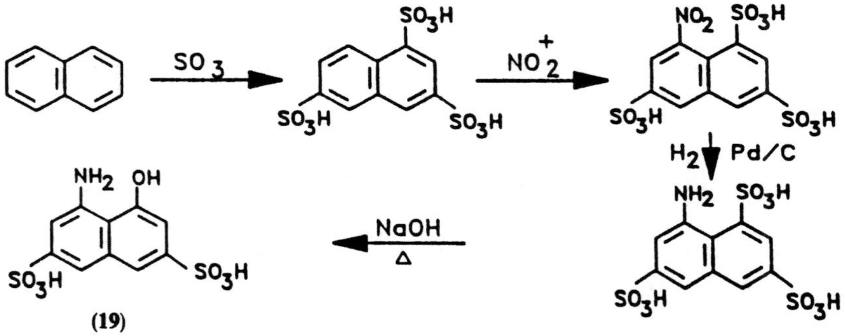

(19)

Scheme 5.10.

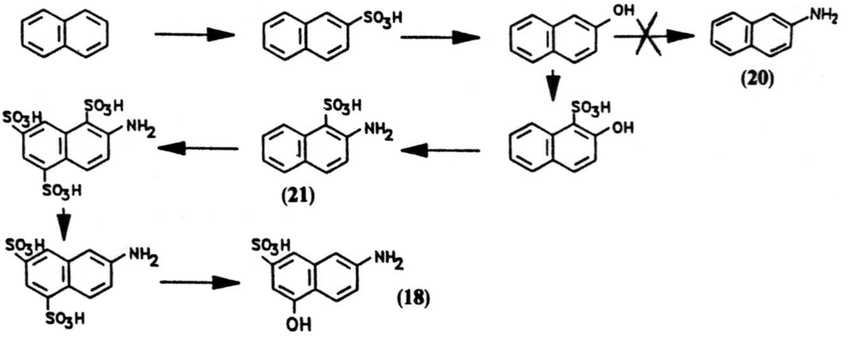

(18)

Scheme 5.11.

the use of the Bucherer reaction to convert a naphthol into a naphthylamine.
Particularly important is the fact that β-naphthylamine (20) is a potent human
carcinogen; it causes bladder cancer. Its use has been banned in dyes. It is
circumvented in the synthesis of J-acid by introducing a sulphonic acid group
prior to the Bucherer reaction. The sulphonic acid group, which is the best
group for conferring water solubility on dyes and intermediates, is also the best
detoxifying group. The presence of just one sulphonic acid group in β-
naphthylamine, as in Tobias acid (21), renders it completely harmless.

The other important naphthalene coupling components are BON-acid (22)
and its arylamides (23). These are important couplers for azo pigments. They
are synthesised as shown in Scheme 5.12.

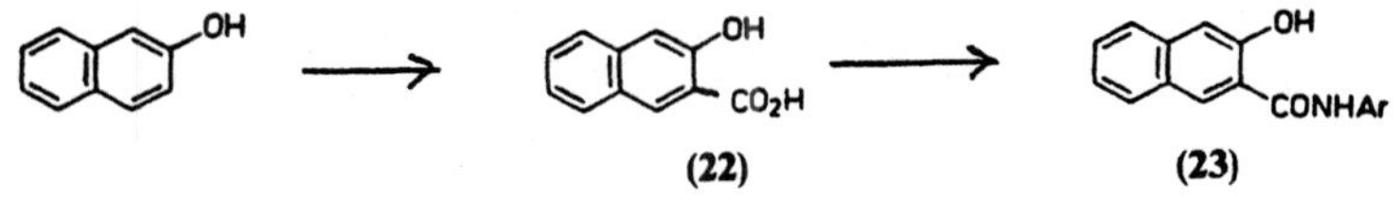

Scheme 5.12.

The only important intermediate from anthracene (24) is 9, 10-an-
thraquinone (4). Anthraquinone is manufactured either by oxidation of anthracene or by the Friedel–Crafts acylation of benzene (Scheme 5.13).

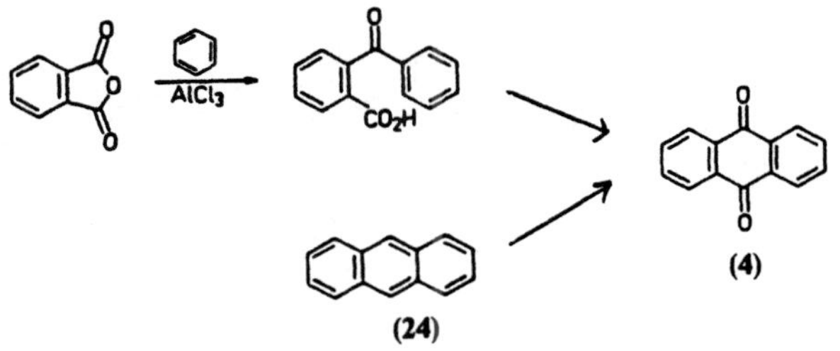

Scheme 5.13.

Aromatic heterocyclic intermediates.[17]   The most important heterocycles
are those with five- or six-membered rings, and these rings may be fused to
other rings, especially a benzene ring. Nitrogen, sulphur and to a lesser extent
oxygen are the most frequently encountered hetero-atoms. They are conveniently divided into two groups: those containing only nitrogen, such as
pyrazolones, indoles, pyridones and triazoles, which are used as coupling
components in azo dyes except for triazoles, and those containing sulphur
(and also optionally nitrogen), such as thiazoles, thiophenes and isothiazoles,
which are used as diazo components in azo dyes. Oxygen is encountered as a
hetero-atom in dyes, such as benzodifuranones and oxazines, rather than in
intermediates.

Triazine is treated separately since it is used as the reactive system in many reactive dyes.

N-*Heterocycles. Pyrazolones.*   Pyrazolones, e.g. (25), are used as coupling components since they couple readily at the 4-position under alkaline conditions to give important azo dyes in the yellow–orange shade area.

The most important synthesis of pyrazolones involves the condensation of a hydrazine with a β-ketoester. Commercially important pyrazolones carry an aryl substituent at the 1-position, mainly because the hydrazine precursors are prepared from readily available and comparatively inexpensive diazonium salts by reduction. In the first step of the synthesis the hydrazine is condensed with the β-ketoester to give a hydrazone; heating with sodium carbonate then effects cyclisation to the pyrazolone. In practice the condensation and cyclisation reactions are usually done in one pot without isolating the hydrazone intermediate. The example in Scheme 5.14 illustrates the condensation of phenylhydrazine with ethyl acetoacetate.

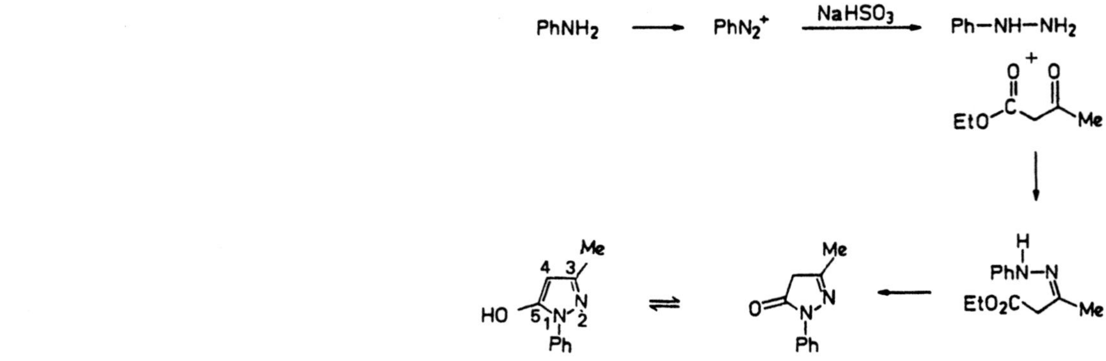

Scheme 5.14.

Pyridones.   Pyridine itself has little importance as a dyestuff intermediate. However, its 2, 6-dihydroxy derivatives have achieved prominence in recent years as coupling components for azo dyes, particularly in the yellow shade area.

The most convenient synthesis of 6-hydroxy-2-pyridones (26) is by the condensation of a β-ketoester, e.g. ethyl acetoacetate, with an active methylene compound, e.g. malonic ester, cyanoacetic ester, and an amine (Scheme 5.15).

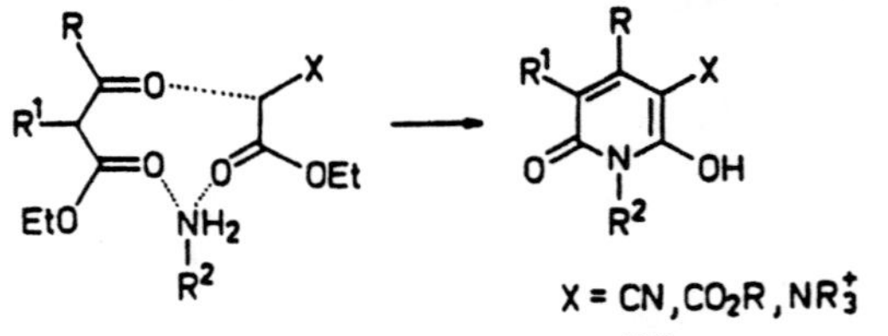

$$X = CN, CO_2R, NR_3^+$$

(26)

Scheme 5.15.

Triazoles. The triazoles are fairly representative of five-membered heterocycles containing three heteroatoms. Heterocycles containing more hetero-
atoms are not generally found in dyestuffs and even triazoles are not of
widespread importance, although they do provide some useful red dyes for
polyacrylonitrile fibres.

The most important triazole is 3-amino-1,2,4-triazole itself, used in the
synthesis of diazahemicyanine dyes. The preparation of 3-amino-1,2,4-
triazole (27) is simple, using readily available and quite inexpensive starting
materials. Thus, cyanamide is reacted with hydrazine to give aminoguanidine
which is then condensed with formic acid. Substituents in the 5-position are
introduced merely by altering the carboxylic acid used.

(27)

Sulphur and sulphur/nitrogen heterocycles. Aminothiazoles. In contrast
to the pyrazolones and pyridones just described, aminothiazoles are used as
diazo components. As such they provide dyes which are more bathochromic
than their benzene analogues. Thus, aminothiazoles are used chiefly to
provide dyes in the red–blue shade areas. The most convenient synthesis of
2-aminothiazoles is by the condensation of thiourea with an α-chlorocarbonyl
compound; for instance, 2-aminothiazole (28) is prepared by condensing
thiourea with α-chloroacetaldehyde—both readily available intermediates.

(28)

Substituents can be introduced into the thiazole ring either by using suitably
substituted precursors or by direct introduction via electrophilic attack. An
interesting example of the latter method is the preparation of 2-amino-
5-nitrothiazole from the nitrate salt of 2-aminothiazole (equation (5.5)).

(5.5)

Aminobenzothiazoles. Aminobenzothiazoles are prepared somewhat differently to the thiazoles. Here the thiazole ring is annelated on to a benzene
ring, usually via an aniline derivative. Thus, 2-amino-6-methoxybenzothiazole
(29) is obtained from *p*-anisidine and thiocyanogen. The thiocyanogen, which
is formed *in situ* from bromine and potassium thiocyanate (the Kaufman
reaction), reacts immediately with *p*-anisidine to give the thiocyanate derivative (30); this spontaneously ring-closes to the aminobenzothiazole (29).

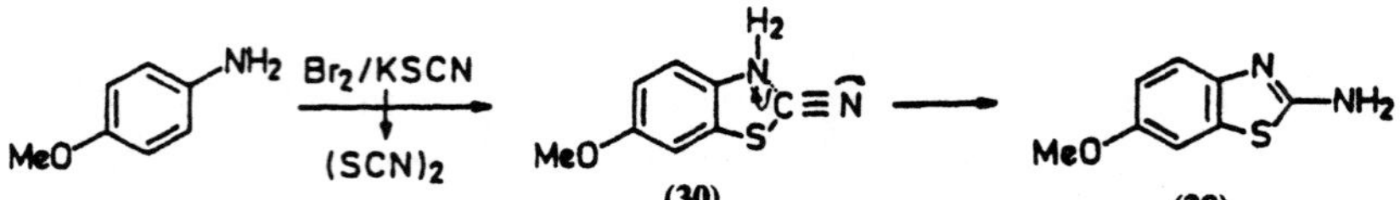

(30)   (29)

Benzoisothiazoles. 5-Aminoisothiazoles are relatively difficult to prepare cheaply but they are used as diazo components in magenta dyes for D2T2 (see later). However, the benzo homologues are easier to prepare and are important diazo components. Aminobenzoisothiazoles give dyes that are even more bathochromic than the corresponding dyes from aminobenzothiazoles. Aminobenzoisothiazoles (31) can be prepared from *o*-nitroanilines by the standard chemical transformation outlined in Scheme 5.16.

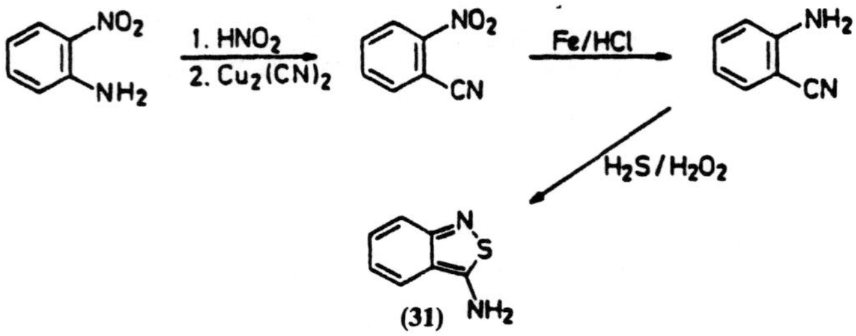

Scheme 5.16.

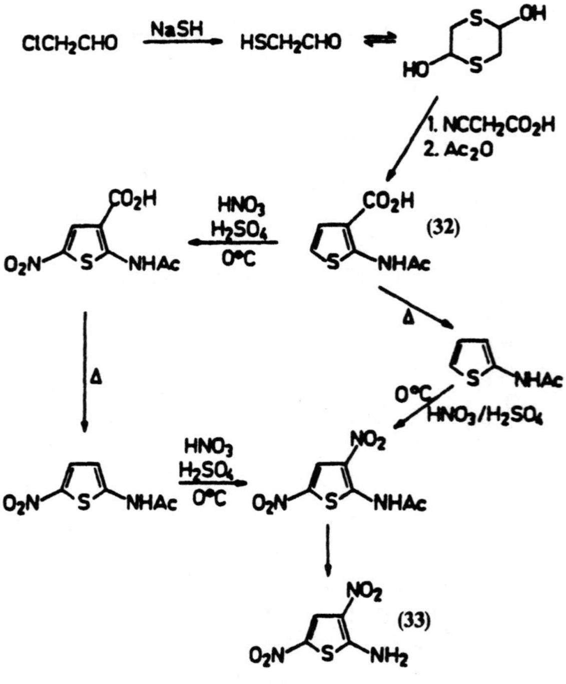

Scheme 5.17.

Thiophenes. The most important thiophenes, i.e. 2-aminothiophenes, are used as diazo components for azo dyes and are capable of producing very bathochromic dyes, e.g. greens, having excellent properties. Part of the reason for their later arrival on the commercial scene was the difficulty encountered in attaining a good synthetic route. However, this problem has been overcome by ICI. The synthetic route is very flexible so that a variety of protected 2-aminothiophenes can be obtained merely by altering the reaction conditions or alternatively by stopping the reaction at an intermediate stage. Thus, the initially formed thiophene (32) can be converted to the useful dinitrothiophene (33) by either of two routes (Scheme 5.17).

Triazines. The most commercially important triazine is 2,4,6-trichloro-*s*-triazine (cyanuric chloride, (34)). Cyanuric chloride has not achieved prominence because of its value as part of a chromogen but because of its use for attaching dyestuffs to cellulose, i.e. as a reactive group. This innovation was first introduced by ICI in 1956 and since then other active halogen compounds have been introduced.

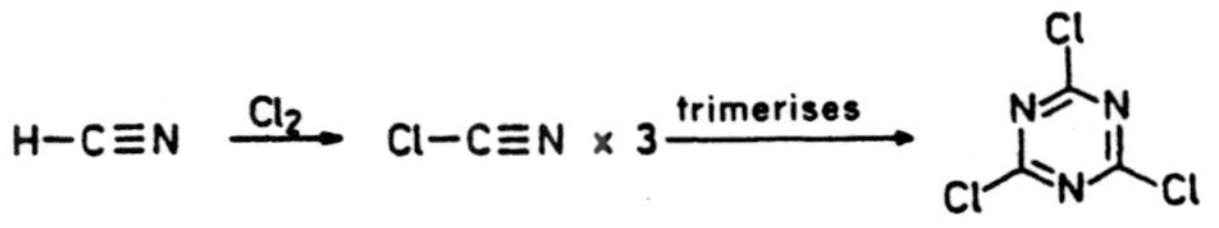

$$H-C\equiv N \xrightarrow{Cl_2} Cl-C\equiv N \ x \ 3 \xrightarrow{trimerises}$$

(34)

5.5.3   Dyes

5.5.3.1   *Azo dyes.*[18]   Azo dyes are the most important class of dye, comprising over 50% of all commercial dyes. Almost without exception, azo dyes are made by diazotisation of a primary aromatic amine followed by coupling of the resultant diazonium salt with an electron-rich nucleophile (Scheme 5.18).

$$ArNH_2 \longrightarrow ArN_2^+ \xrightarrow{EH} Ar-N=N-E+H^+$$

Scheme 5.18.

Formation of the azo dye constitutes the final step in the synthesis. Introduction of substituents is done on the precursor intermediates, the diazo component and the coupling component, normally by electrophilic substitution. This allows tremendous synthetic flexibility in azo dyes.

Azo dyes may be broadly divided into hydroxyazo dyes and aminoazo dyes. Hydroxyazo dyes can exhibit tautomerism[9] (equation (5.6)) and many commercial hydroxyazo dyes exist in the more bathochromic and tinctorially stronger hydrazone form. Important dyes which exist in the hydrazone form are azopyrazolone (35), the azopyridone (36) disperse yellow for polyester, J-acid orange (37) and H-acid red (38) reactive dyes for cotton, and BON-acid (and arylamide) pigments (39).

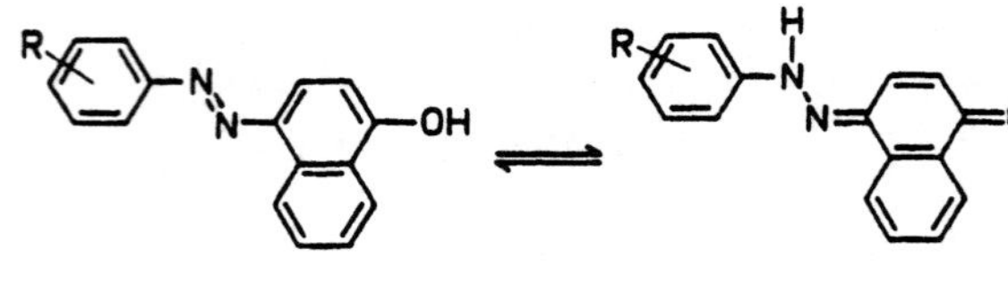

$$(5.6)$$

	Azo		Hydrazone	
(R = H)	λ	ε	λ	ε
	410	25 000	480	35 000

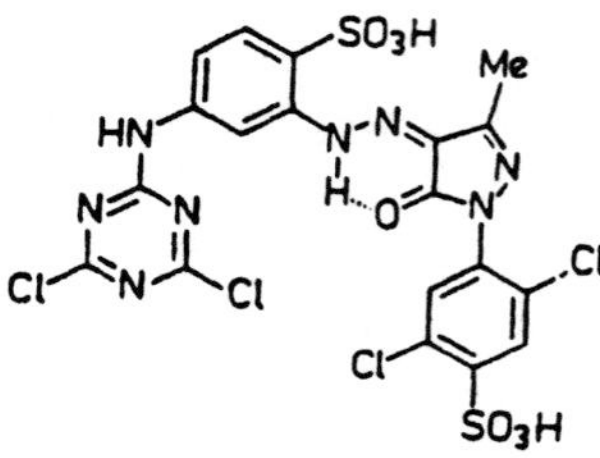

(35)

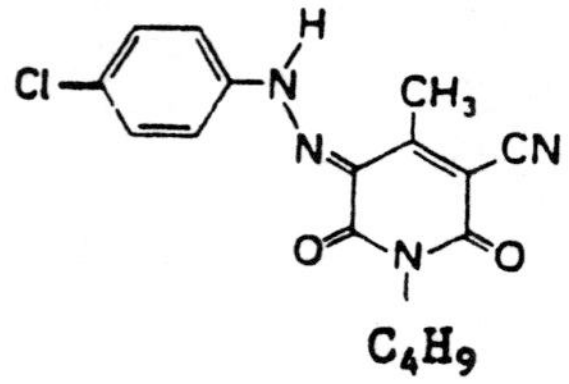

(36)

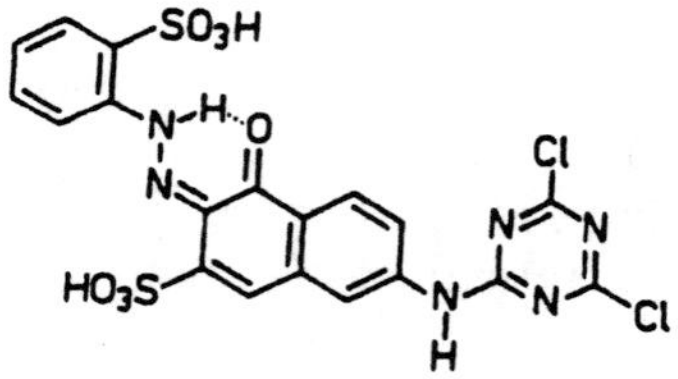

(37)

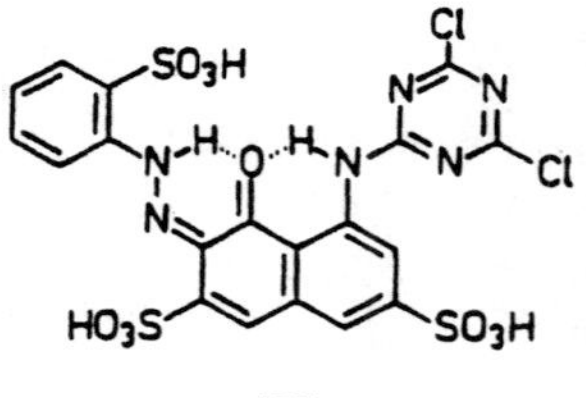

(38)

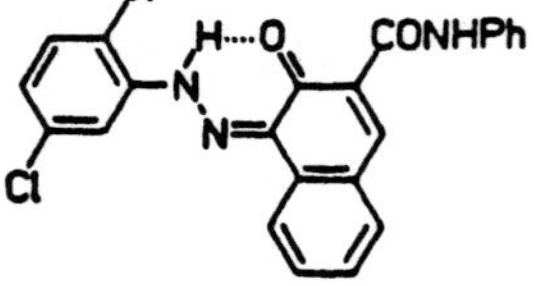

(39)

In contrast, aminoazo dyes exist in the azo form.[19] This is ascribed to the instability of the imine group (C=NH bond energy 147 kcal mol^{-1}) relative to the carbonyl group (C=O bond energy 178 kcal mol^{-1}). Important dyes are the red (40) and blue (41) carbocyclic azo dyes and the green heterocyclic azo dye (42), all for the dyeing of polyester.

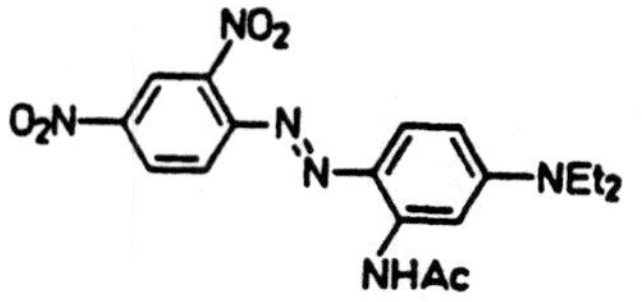

(40)

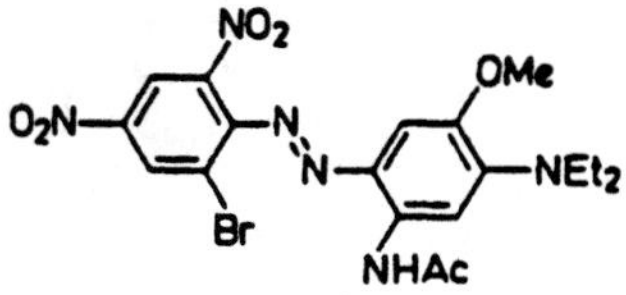

(41)

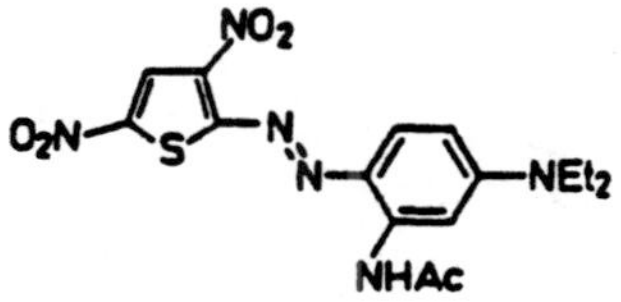

(42)

Certain dis- and tris-azo dyes are also important, such as CI Reactive Black 5 (43).

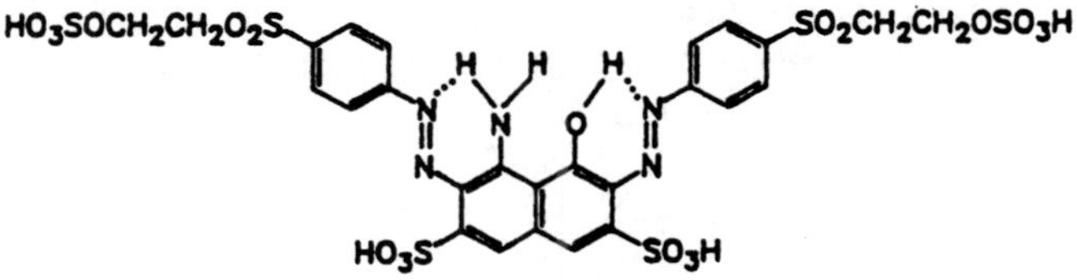

(43)

5.5.3.2 Phthalocyanine dyes.

Phthalocyanines are the next most important class after azo dyes. As was the case for many of the major dye discoveries, phthalocyanines were discovered by accident.[20] In 1928, during the routine manufacture of phthalimide from phthalic anhydride and ammonia (equation (5.7)) it was found that the product contained a blue contaminant. Chemists of Scottish Dyes Ltd, now part of ICI, carried out an independent synthesis of the blue material by passing ammonia gas into molten phthalic anhydride containing iron filings. The importance of the colourant was realised (it was intensely coloured and very stable) and a patent application was filed in the same year.

$$\text{(phthalic anhydride)} + NH_3 \xrightarrow{\Delta} \text{(phthalimide)} + H_2O \quad (5.7)$$

The structure of the blue material was not elucidated until 1934, when Linstead, from the Imperial College of Science and Technology, London, showed it to be the iron complex of (44). Linstead christened the new material phthalocyanine, reflecting both its origin (from phthalic anhydride) and its beautiful blue colour (like cyanine dyes). A year later Robertson, in one of the first uses of X-ray crystallography, confirmed Linstead's structure.

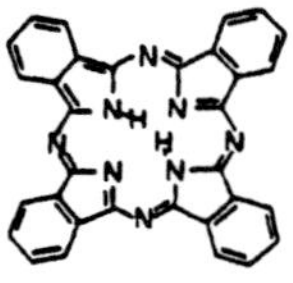

(44)

Further work showed that copper phthalocyanines (45) produced the best dyes and pigments. There are two routes to copper phthalocyanines: the first is the phthalic anhydride–urea process and the second is the phthalonitrile process. In the first process phthalic anhydride is heated with urea, copper(I) chloride and a catalytic amount of ammonium molybdate in a high-boiling solvent. The function of the urea is to act as a source of nitrogen, since the carbonyl group of the urea is lost as carbon dioxide; this has been confirmed by labelling studies. The first step in the transformation is the conversion of phthalic anhydride to phthalimide by the ammonia liberated from the urea. More ammonia then converts phthalimide to isoindoline (46) and finally to the isoindoline (47); the latter has been isolated and identified. However, in the presence of copper(I) chloride, (47) spontaneously tetramerises and is oxidised to phthalocyanine which then forms a complex with the copper.

In the phthalonitrile process phthalonitrile (48) is heated in the presence of a base and cupric and ammonium acetates, with or without a solvent. The mechanism has been less studied than the phthalic anhydride–urea process. However, an intermediate such as (49) is postulated to form before tetramerisation to phthalocyanine occurs. If the reaction is carried out in the absence of copper acetate then metal-free phthalocyanine is obtained. This route is used where high purity is required, as in high-tech applications (see later).

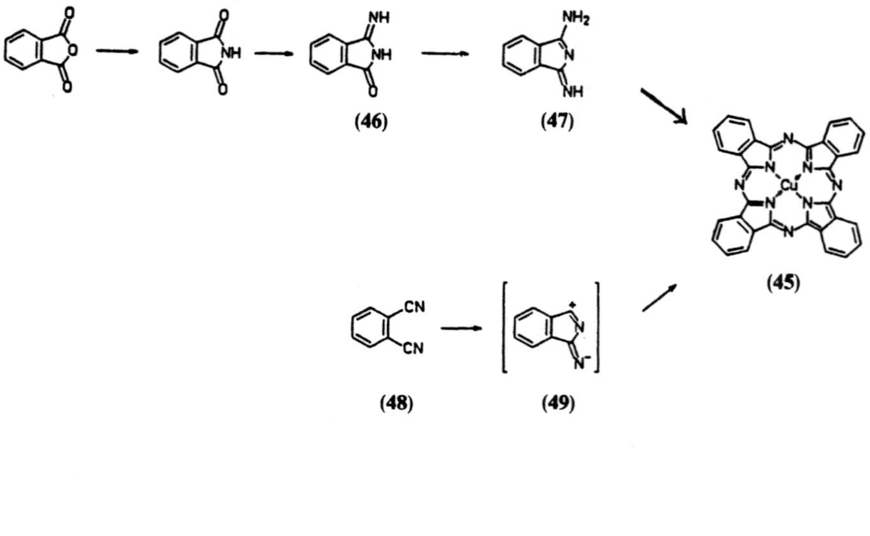

(46)   (47)   (45)

(48)   (49)

As is the case for anthraquinone, the ring is functionalised mainly by chlorination, sulphonation and chlorosulphonation (Scheme 5.19).

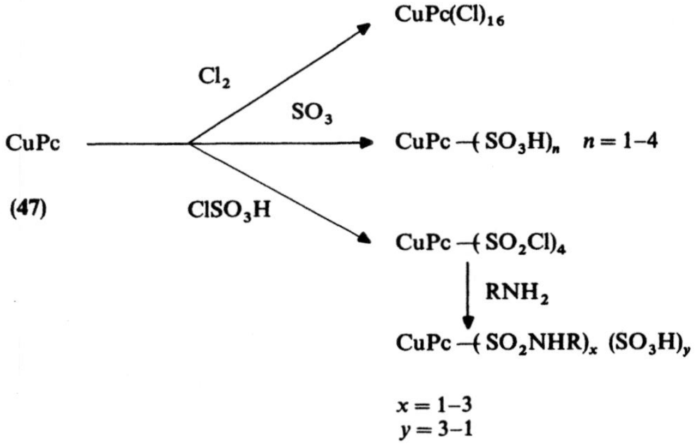

Scheme 5.19.

5.5.3.3 Anthraquinone dyes.

Anthraquinone dyes are probably the next most important class after azo and phthalocyanine but their importance is declining. They are technically good but they are not cost-effective since they have low tinctorial strength and are expensive.

In complete contrast to the azo dyes, anthraquinone dyes are prepared by the stepwise introduction of substituents into the preformed anthraquinone skeleton.[21] The substituents are generally introduced by nucleophilic substitution of a good leaving group (Schemes 5.20 and 5.21) (cf. azo dyes).

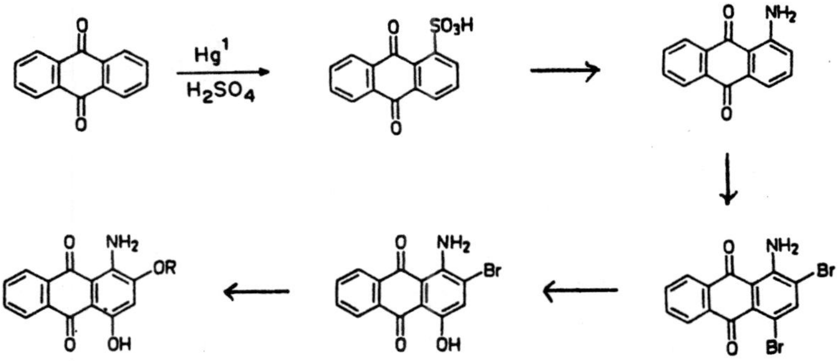

R = Ph

CI Disperse Red 60

Scheme 5.20.

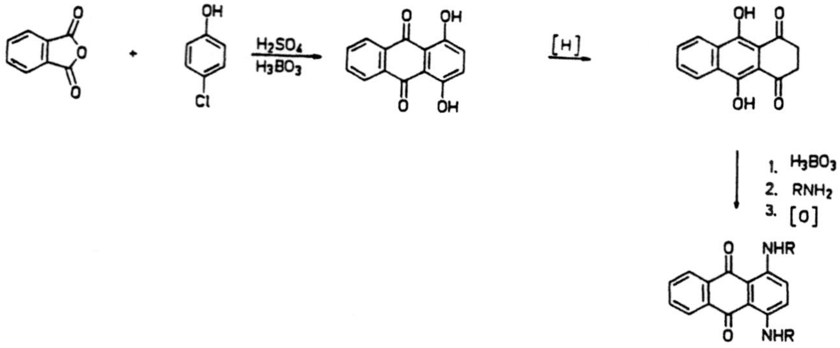

Scheme 5.21.

The synthesis of anthraquinone dyes highlights two very important points. First, the degree of freedom for producing a variety of different structures is restricted. Second, the availability of only eight substitution centres imposes a further restriction on synthetic flexibility. Therefore, there is nowhere near as much synthetic flexibility as exists in azo dyes, and this is one of the drawbacks of anthraquinone dyes.

5.5.3.4   Indigoid dyes.   Although many indigoid dyes have been synthesised, only indigo itself is of any major importance nowadays. Indigo is made by the route shown in Scheme 5.22. In the first step aniline is condensed with formaldehyde–sodium bisulphite and sodium cyanide to give *N*-cyano-

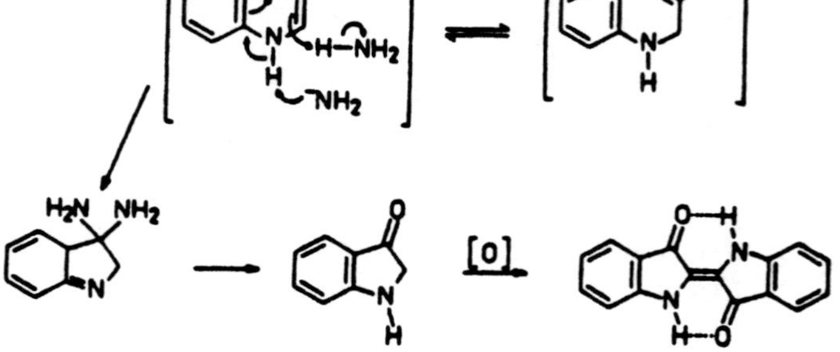

Scheme 5.22.

methylaniline. This compound is then hydrolysed to sodium phenylgly-cinate which is cyclised in molten sodium and potassium hydroxide (containing sodamide) to give indoxyl. The latter is then oxidised in air to indigo.[22]

Functionalisation is normally done prior to the ring synthesis.

Indigo is used for dyeing jeans and denims. It is fashionable because it fades 'on-tone' to give paler shades of blue (see later).

5.5.3.5 Benzodifuranone dyes. Benzodifuranone (BzDF) is one of the very few commercially useful novel chromogens to have been discovered this century. As with several major dye discoveries[2] the BzDF chromogen was discovered by accident. Greenhalgh[23] and co-workers at ICI questioned the pentacenequinone structure (50) assigned by Junek to the intensely red-coloured product obtained from the reaction of *p*-benzoquinone with cyanoacetic acid (Scheme 5.23). (A compound such as (50) should not be coloured owing to the lack of conjugation.) Instead, the red compound was correctly identified as the benzodifuranone (51).

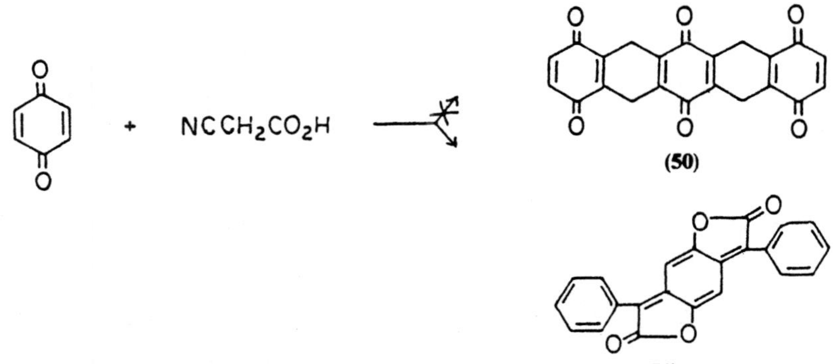

(50)

(51)

$\lambda_{max}^{CHCl_3}$ 446 nm ε_{max} 51 000

Scheme 5.23.

Improved syntheses[24] have been devised for both symmetrical and unsymmetrical BzDF dyes starting from hydroquinone and mandelic acid (52) (Scheme 5.24).

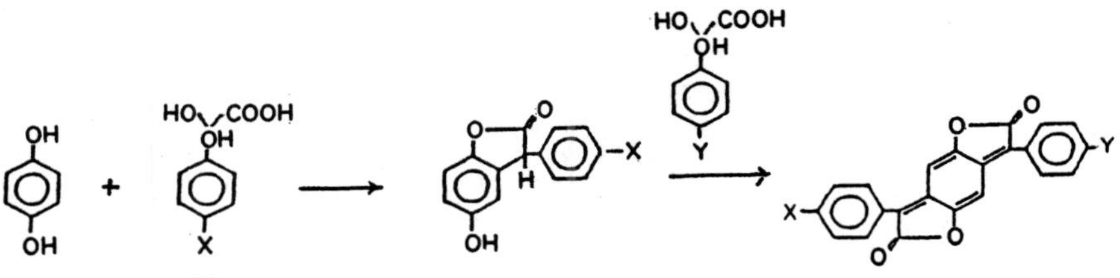

(52)   Scheme 5.24.

By suitable selection of substituents X and Y it is possible for BzDF dyes to span the entire spectrum from yellow through red to blue. The current commercial BzDF dyes are scarlet to bluish-red. They are disperse dyes for polyester, displaying bright, strong colours and excellent wet fastness. Indeed, in recognition of the importance of benzodifuranone dyes, ICI Colours won the Queen's Award to Industry in 1990 and the inventor, Colin Greenhalgh, was awarded the prestigious Perkin Medal in 1992.

5.5.3.6 Oxazine dyes.   The old basic oxazine dyes are prepared by the sequence of reactions shown in Scheme 5.25. These bright blue basic dyes, such as CI Basic Blue 4, having moderate light-fastness, are used for colouring paper and polyacrylonitrile.

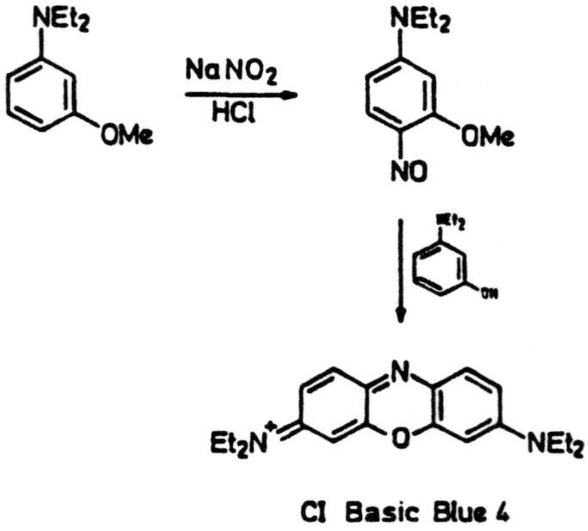

CI Basic Blue 4

Scheme 5.25.

The more recently discovered[25] triphendioxazine reactive blue dyes for cotton are commercially very important. They combine the brightness and durability of anthraquinone dyes with the strength and economy of azo dyes. A typical synthesis is shown in Scheme 5.26.

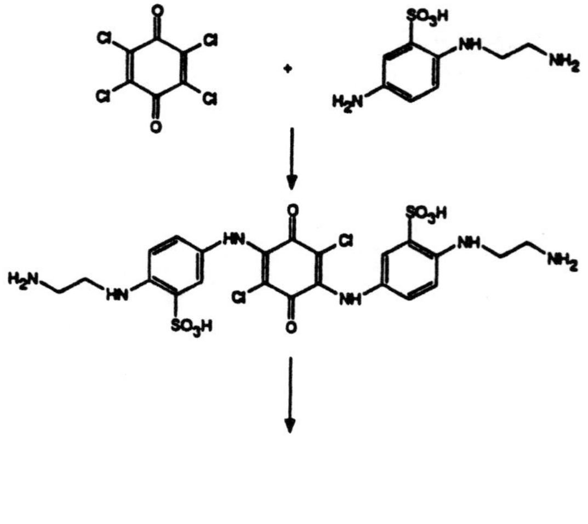

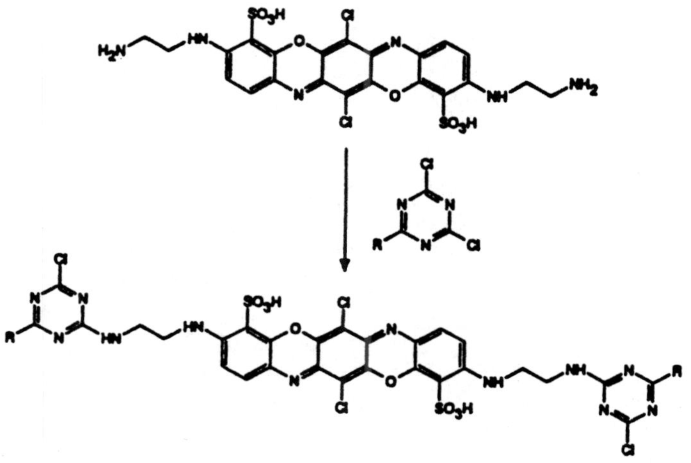

Scheme 5.26.

Other dye types include sulphur dyes,[26] e.g. Sulphur Black (equation (5.8)), polycyclic aromatic carbonyl dyes,[27] such as CALEDON Jade Green (53), polymethine and related dyes,[28] such as CI Basic Blue 41 (54), di- and tri-arylcarbonium dyes,[29] such as Malachite Green (55), quinophthalone dyes,[30] such as CI Disperse Yellow 54 (56), and nitro and nitroso dyes,[31] such as CI Disperse Yellow 14 (57).

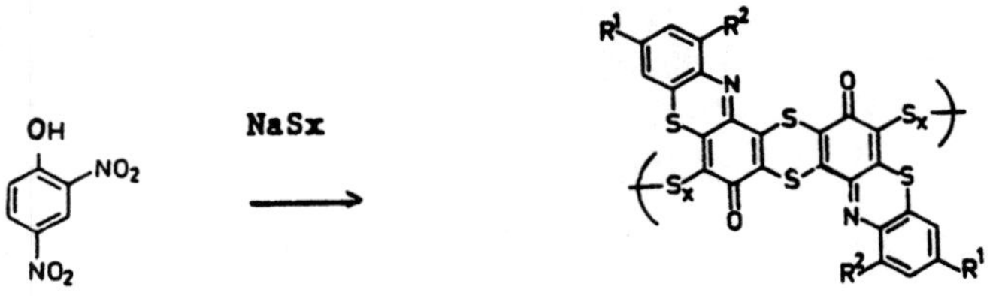

$$(5.8)$$

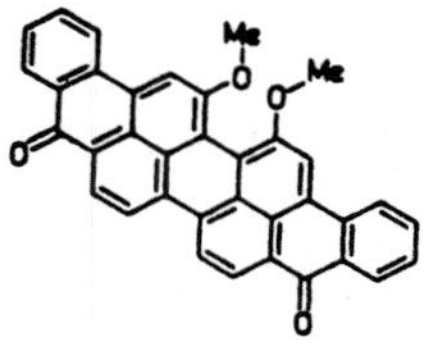

(53)

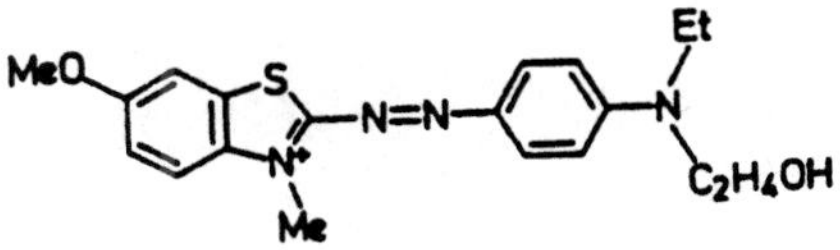

(54)

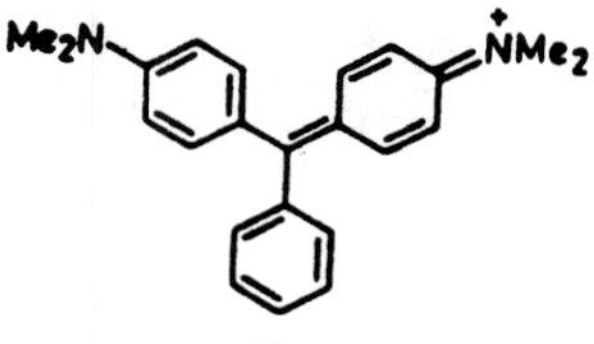

(55)

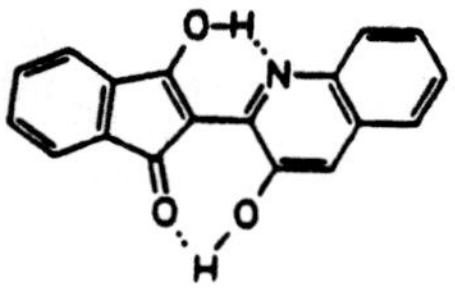

(56)

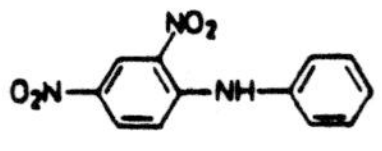

(57)

5.6
Application and fastness properties[32]

In addition to being intensely coloured, a dye must possess two other features: it must be capable of being applied to a substrate, usually a textile fibre; and once on the fibre it must display good fastness properties such as wet fastness (it must not wash out during laundering) and light fastness (it must not fade in sunlight).

The two key properties of fibres that determine which dyes to use and the dyeing conditions to employ are (i) their hydrophobic–hydrophilic nature and (ii) the openness–compactness of the fibre. The water imbibition and density, respectively, give a good indication of these properties (Tables 5.10 and 5.11).

Hydrophilic fibres such as viscose and cotton are dyed with hydrophilic dyes. These are generally sulphonated water-soluble dyes. Hydrophobic fibres such as polyester are dyed with hydrophobic dyes. These are dyes devoid of polar, water-solubilising groups such as sulphonic acid, carboxylic acid or quaternary ammonium.

Dyes easily enter open-structured (low-density) fibres such as cotton but are also easily removed, e.g. by washing treatments. In contrast, tightly packed fibres such as polyester are difficult to dye and special techniques are required to swell (open up) the fibre temporarily. This is normally accomplished by increasing the dyeing temperature.

Most dyeing is done from water, either from solution using water-soluble dyes or from dispersions of water-insoluble dyes. The dyeing temperature is

Table 5.10 Water imbibition of some common fibres

Fibre	Water imbibition (%)
Viscose	100
Acetate	25
Triacetate	10
Polyamide	11–13
Polyacrylonitrile	8
Polyester	3
Polypropylene	0

Table 5.11 Densities of some important fibres

Fibre	Density
Cotton, dry (wet)	1.05 (1.55)
Nylon	1.14
Polyacrylonitrile	1.17
Polyester	1.78

normally 90–100 °C but can be as low as 30 °C (warm-dyeing reactive dyes) and as high as 130 °C (pressurised disperse dyeing of polyester).

Once the dye is on the fibre, it is retained there by one of five basic dye-fibre interactions:

Physical interactions
1. Physical adsorption
2. Solid solution
3. Insoluble aggregates within the fibre

Chemically bound
4. Ionic bonds
5. Covalent bonds

5.6.1   Physical adsorption

The best example of this type of interaction is provided by direct dyes for cotton.

Direct dyes are long, planar, polyazo dyes such as Congo Red (**58**), derived from benzidine and 1-naphthylamine, in which the distance between the azo groups (10·8 Å) is similar to the distance between the cellobiose units (**59**) of cotton. This allows the dye molecules to fit closely on the cellulose polymer, thus maximising the effect of both intermolecular hydrogen-bonding and Van der Waals forces, and the dye is therefore anchored to the fibre. Since Van der Waals forces and hydrogen-bonding interactions are relatively weak, the dyes are not firmly bound to the fibre. Consequently, direct dyes on cotton have the poorest wet fastness properties of all the dyes in this section. However, because they are cheap, they are used widely for colouring paper.

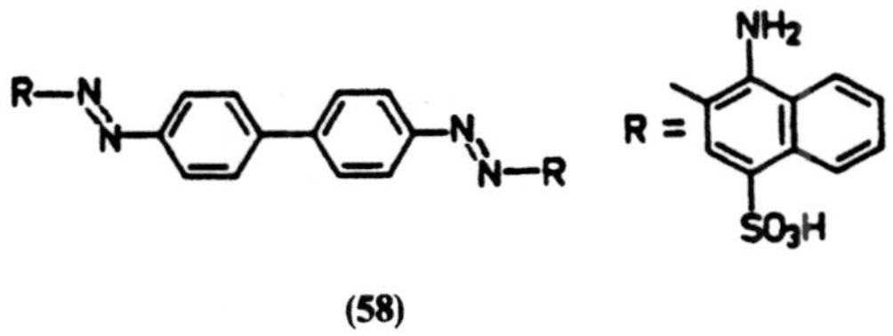

(**58**) 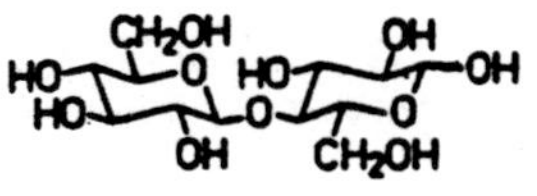   cellobiose unit   (**59**)

5.6.2   Solid solutions

In contrast to cotton, polyester fibres have a tightly packed structure and even small molecules such as water have difficulty in penetrating the fibre under normal conditions. Special dyeing techniques have been developed therefore to allow the entry of dye molecules into the fibre. These techniques were devised to make the fibre temporarily accessible to the dye molecules by loosening or opening its structure; after dyeing, the fibre reverts back to its closely packed state thus trapping the dye molecules within the polymer structure. Even with these techniques only small dye molecules will dye polyester; large dye molecules such as copper phthalocyanine do not. The monoazo dye (60) and the anthraquinone dye (61) are typical dyes for polyester.

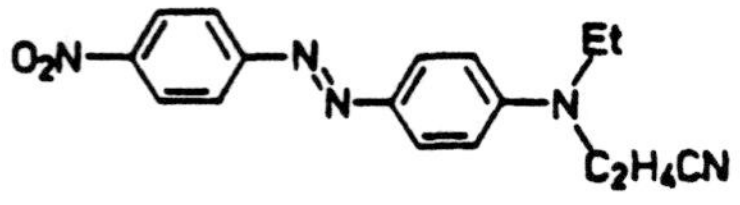

(60) 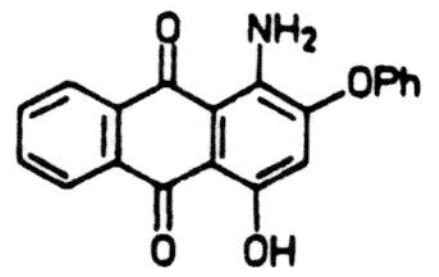   (61)

An additional requirement of dyes for polyester is that they should have only a very low (i.e. $< 1\%$) solubility in water. Thus, the dyes are devoid of water solubilising groups such as $-SO_3H$, $-CO_2H$ and $-NR_3^+$. They are normally supplied and used as aqueous dispersions, and hence the name disperse dyes.

There are three main techniques that have been developed for temporarily opening up the fibre structure:

(i) Dyeing at 95–100 °C with the addition to the dyebath of chemicals called 'carriers';

(ii) Dyeing without carriers at temperatures of 120–130 °C: this necessitates the use of pressurised dyeing vessels, and is the most widely used method; and

(iii) Pad-dry heat methods involving the use of temperatures of 180–220 °C for short periods of time.

The fixation of disperse dyes in polyester fibres is believed to be mainly one of a solution of the dye in the fibre, i.e. a solid solution. There is also the possibility of hydrogen-bonding between the carbonyl groups in the polymer and amino and hydroxy groups in the dye. The dyes have very good fastness to washing treatments.

5.6.3   Insoluble aggregates within the fibre

All these dyes work on the principle of producing water-insoluble aggregates of dye which are larger than the fibre pores within the fibre structure—hence the dye is firmly trapped within the fibre structure (Scheme 5.27).

The most familiar examples of this type are the vat dyes. Vat dyes are water-insoluble dyes that have no affinity for cotton but which can be reduced to a water-soluble leuco form which does have affinity for cotton. The leuco form therefore 'dyes' the fibre, and then atmospheric oxidation regenerates the original water-insoluble vat dye within the cotton; aggregation is usually effected by a soaping treatment.

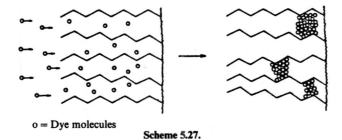

o = Dye molecules

Scheme 5.27.

The chemical reaction involved in converting a vat dye into its leuco form is simply the reduction of one or more of the carbonyl groups to phenolic groups. The practical reducing agent is alkaline sodium hydrosulphite (hydros). Thus the water-soluble sodium salts of the phenolic groups are formed and it is this anionic species which has high fibre affinity. The simplified vatting process is depicted in equation (5.9).

$$\text{water insoluble vat dye} \xrightarrow[\text{NaOH}]{\text{Na}_2\text{S}_2\text{O}_4} \text{soluble leuco form (O}^-\text{Na}^+\text{)} \xrightarrow{[\text{O}]} \text{regenerated dye trapped in fibre} \tag{5.9}$$

Azoic dyes also fall into this category.

5.6.4 Ionic bonds

There are two distinct types of ionic dye-fibre bonding depending upon whether the dye is anionic or cationic, and both are employed.

The best example of negatively charged (anionic) dyes being attached to positively charged (cationic) fibres is acid dyes for nylon, wool and silk. In a weakly acidic dyebath such as dilute aqueous acetic acid the amino groups in the fibre are protonated. During dyeing the negatively charged dye displaces the colourless counter-ion and is retained in the fibre by the ionic bonding between the anionic dye and the cationic ammonium group (equation (5.10)).

$$\begin{array}{c} \text{CO}_2\text{H} \\ \vdots \\ \text{NH}_3^+ \cdot \text{A}^- \end{array} \xrightarrow{\text{Dye}-\text{SO}_3^-} \begin{array}{c} \text{CO}_2\text{H} \\ \vdots \\ \text{NH}_3^+ \cdot \text{O}_3^-\text{S}-\text{Dye} \end{array} \tag{5.10}$$

The reverse case of a positively charged dye and a negatively charged fibre is provided by cationic dyes for polyacrylonitrile. As is the case with anionic dyes for nylon and wool, cationic dyes are applied to acrylic fibres from a weakly acid dyebath. It is thought that the dye is first adsorbed on to the surface: heat then causes it to diffuse into the fibre interior where it becomes anchored to an anionic centre (equation (5.11)).

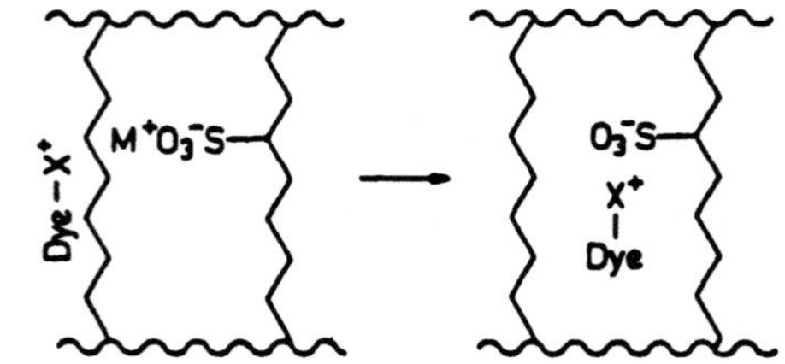

$$\text{(5.11)}$$

The wet-fastness properties of dyes bound by ionic forces are better than those of dyes bound by physical adsorption, e.g. direct dyes for cotton, but inferior to those which form insoluble aggregates in the fibre, e.g. vat dyes.

5.6.5 Covalent bonds

The first practical success of binding a dye to cotton by a covalent chemical bond achieved by Rattee and Stephen (ICI) in 1954 was a major breakthrough: it enabled for the very first time the production of bright, wet-fast dyeings on cotton. (Direct dyes and vat dyes are dull.) The chlorotriazine reactive group first used by Rattee and Stephen is still the most prevalent today. Dyeing is done under alkaline conditions to ionise the cellulose hydroxy groups. The ionised cellulose is a good nucleophile (far better than OH^-) which displaces an active chlorine from the triazine ring, forming an ether bond (equation (5.12)).

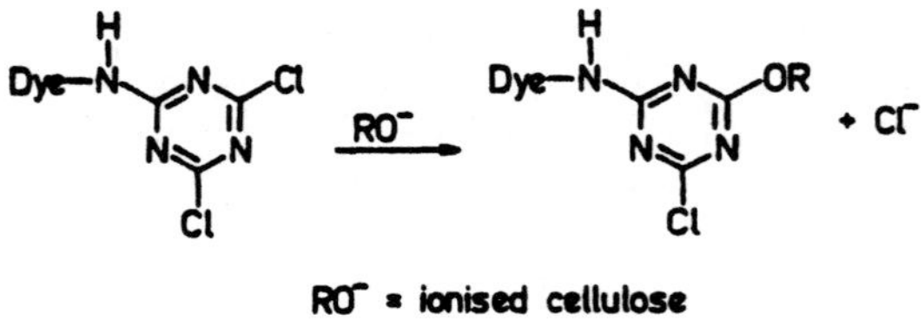

$$\text{(5.12)}$$

RO^- = ionised cellulose

The majority of reactive dyes utilise the nucleophilic displacement reaction generalised in equation (5.12). Basilene (BASF), Cibacron (Ciba-Geigy), Drimarene P (Bayer) and Procion (ICI) ranges are based on 1,3,5-triazines (34), whereas the principal Drimarene range and the Levafix range (Bayer) are based on pyrimidines. Hoechst chemists invented a different type of reactive system, a vinylsulphone group, which reacts with the cellulose by Michael addition (equation (5.13)).

$$\text{Dye}-\text{SO}_2\text{CH}_2\text{CH}_2\text{OSO}_3\text{H} \xrightarrow{\text{OH}^-} \text{Dye}-\text{SO}_2\text{CH}=\text{CH}_2 \xrightarrow{\text{Cell}-\text{O}^-} \text{Dye}-\text{SO}_2\text{CH}_2\text{CH}_2\text{O}-\text{Cell}$$

$$(5.13)$$

Recently, dyes have been marketed which contain two different reactive groups. Sumitomo were the first company to market these dyes, Sumifix Supra dyes, containing both a monochlorotriazinyl group and a vinylsulphone group. Ciba-Geigy followed with the Cibacron C dyes, based on a mono-fluorotriazinyl group and a vinylsulphone group. These hetero-functional reactive dyes are more robust than conventional reactive dyes, being less sensitive to changes in temperature, pH, electrolyte, etc.

5.6.6   Light fastness

The major photofading pathway is oxidation and the main causative agent is singlet oxygen. In hydroxyazo dyes, it is the hydrazone form that is normally attacked, the final products being a quinone and an aryl diazonium salt (Scheme 5.28).

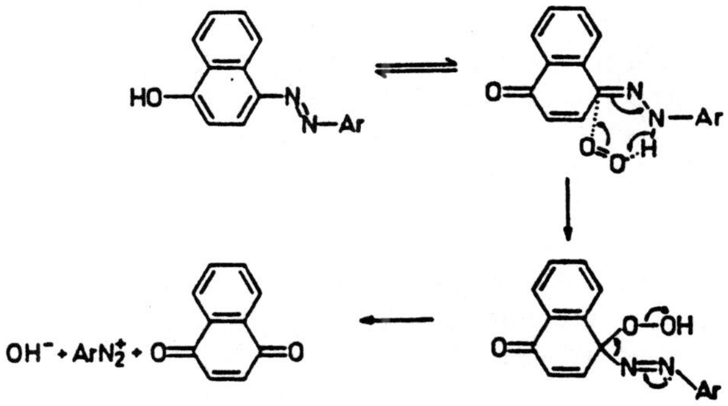

Scheme 5.28.

In aminoazo and anthraquinone dyes, the amine nitrogen atom is attacked resulting, ultimately, in dealkylation (Scheme 5.29). Thus, photofading pro-

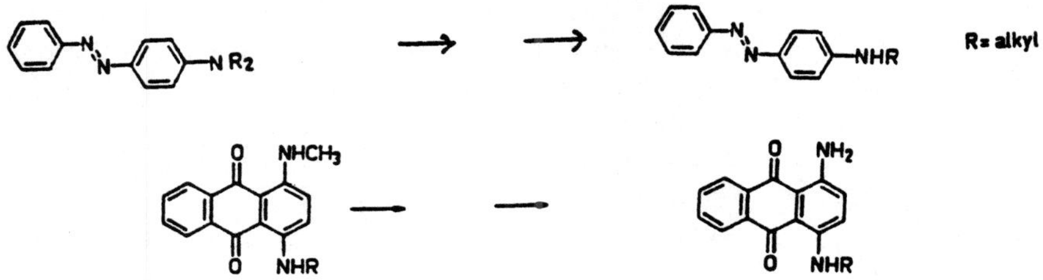

Scheme 5.29.

duces a more hypsochromic product. Two ways of improving the light fastness is by reducing the basicity of the amine nitrogen atoms (e.g. by incorporating electron-withdrawing groups in the alkyl chain, such as β-cyanoethyl in azo dyes) and by avoiding alkylamino groups (e.g. by using NH_2 or NH-aryl, in anthraquinone dyes).

Finally, indigo is unique since its photofading product, isatin (62), is colourless. Thus, indigo fades 'on tone'.

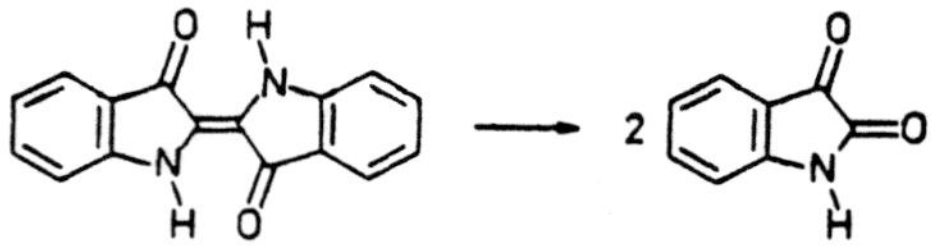

(62)

5.7
Dye manufacture[3,12]

Dyes are manufactured by the processes outlined earlier. Large-tonnage products, such as primary intermediates, the larger-volume dye intermediates such as H-acid, and the large-volume dyes, such as copper phthalocyanine and indigo, are made in dedicated plants. The majority of products, however, are made in multipurpose manufacturing units. The equipment and methods used have changed little from those described in the first edition of this book and this, together with reference 12, should be consulted for further details. The only real change in intermediates and dyes manufacture over the past decade has been the increasing use of computerisation in the manufacturing process.

5.8
Current and future trends

This section is best considered in terms of those factors affecting toxicity and ecology and those affecting techno-economic aspects, although in some instances these are inter-related.

5.8.1   Toxicity and ecology

5.8.1.1   *Toxicity.*   The toxic nature of some dyes and intermediates has long been recognised. Acute, or short-term, effects are generally well known. They are controlled by keeping the concentration of the chemicals in the workplace atmosphere below prescribed limits and avoiding physical contact with the material. Chronic effects, on the other hand, frequently do not become apparent until after many years of exposure. Statistically higher incidences of benign and malignant tumours, especially in the bladders of workers exposed to certain intermediates and dyes, were recorded in dye-producing countries during the period 1930–1960. The specific compounds involved were: 2-naphthylamine (20), 4-aminobiphenyl (63), benzidine (64), fuchsine (65) and auramine (66). There is considerable evidence that metabolites of these compounds are the actual carcinogenic agents. Strict regulations concerning the handling of known carcinogens have been imposed in most industrial nations. In the USA the regulations caused virtually all the dye companies to discontinue use of the compounds. Other actual or suspected carcinogens, such as N-nitroso compounds, polycyclic hydrocarbons, alkylating agents and other individual compounds, such as the dichromates, should be considered in the wider context of industrial chemistry rather than as dye intermediates.

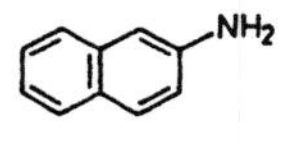

(20) 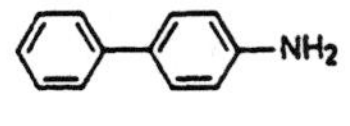   (63) 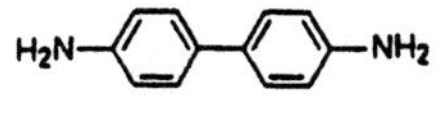   (64)

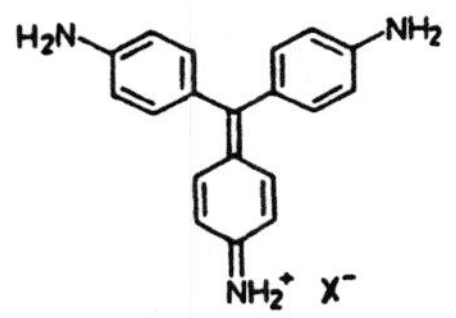

(65) 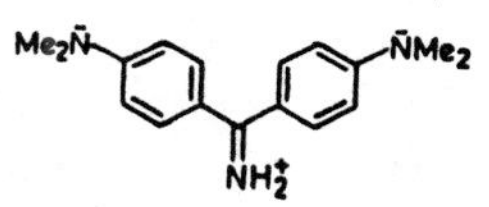   (66)

The positive links between benzidine derivatives and 2-naphthylamine and bladder cancer prompted the introduction of stringent government regulations to minimise such occurrences in the future. Currently, the three major regulatory bodies worldwide are the Commission of the European Communities (CEC) with its European Core Inventory (ECOIN) and European Inventory of Existing Commercial Substances (EINECS) in Europe, the Environmental Protection Agency (EPA) with its Toxic Substances Control Act (TOSCA) in the USA, and the Ministry of International Trade and Industry (MITI) in Japan. Each of these bodies has its own set of data and testing protocols for registration of a new chemical substance.

Obtaining all the data for a full registration can be time-consuming and costly. In 1989 it cost approximately $150 000 and took about a year to register a new substance in Europe. In order to expedite the launch of a new chemical and allow further time to complete the toxicological package for full registration, a 'limited announcement' is normally used. This requires only parts of the full toxicological packages, usually acute toxicity and the Ames test. Consequently, it is less expensive ($20 000) and quicker (90 days) than full registration. However a maximum of only 1 tonne of the chemical per year is allowed to be sold in the EC.

5.8.1.2 Ecology. Ecology has assumed increased importance in recent years in line with the general awareness and concern over protecting the environment. Dyes, because of their inherent properties of being intensely coloured, present special problems in effluent discharge; even a very small amount is noticeable. However, the effect is aesthetically displeasing rather than hazardous, e.g. red dyes discharged into rivers and oceans. Of more concern is the discharge of toxic heavy metals such as mercury and chromium. Strategies being implemented to minimise dye and related effluent include designing more environmentally friendly chemicals, more efficient (higher yielding) manufacturing processes and more effective dyes, e.g. reactive dyes having higher fixation.

5.8.2   Techno-economic factors

As mentioned above, the trends in traditional dyes areas are for improved cost-effectiveness and technical excellence. Improved cost-effectiveness is being tackled in two ways: firstly, by replacing the weaker, expensive chromogens with stronger, less expensive chromogens—examples include replacing anthraquinone dyes with heterocyclic azos, benzodifuranones and triphendioxazines; secondly, by continuously seeking better synthetic routes and more efficient manufacturing processes. However, a very important trend that has already begun and is gathering momentum quickly is the diversification of dyes into high technology (high-tech) areas. Comprehensive treatments of the high-tech uses of dyes and pigments exist:[33] only a brief account can be given here.

5.8.2.1   High-tech applications of dyes.   High-tech applications of dyes (and pigments) may be conveniently divided into electronics, reprographics, security and medical. The driving force for the intensive research in these areas is the attractive added value that the novel, highly pure dyes command, together with reasonable volume sales.

Electronics.   Some of the main uses of dyes in electronics are in liquid-crystal displays, lasers, colour filters, non-linear optics, solar cells, electrochromic dyes and the topic discussed further here, optical data storage. Optical data storage is the technology employed in compact discs and optical tape. Its main advantage is the vast amount of data that can be stored per unit area, viz:

  1   optical tape
  $\equiv$ 10^{12} bytes
  $\equiv$ 5 000 magnetic discs
  $\equiv$ 12 km pile of A4 paper
  $\equiv$ 200 acres of rain forest

The principle is very simple (Figure 5.2). The active layer of the optical disc (or tape) is a thin polymer film containing a dissolved infrared absorber tuned

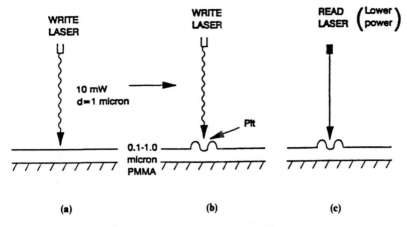

Figure 5.2   Principle of optical data storage.

to an infrared semiconductor laser of 780 nm or 830 nm. The information is written using the laser at full power (*c.* 10 mW). Where the laser radiation hits the disc, a shallow pit is formed by the infrared absorber absorbing the radiation and converting it into sufficient heat to cause melting. (The temperature can reach 300–400 °C.) These pits constitute the recorded data. To read the data a lower-power laser is used so that the energy of the beam is insufficient to cause melting. Instead, a detector records the amount of laser radiation reflected from the disc. Where the disc material is unexposed, there are no pits and there is high reflectivity. In contrast, the reflectivity is low from the data pits.

Phthalocyanines, such as (**67**), prepared by a nucleophilic displacement of the chlorine atoms in the polychloro pigment (**68**), are important infrared absorbers for optical data storage (equation 5.14).

$$CuPc(Cl)_{15/16} \xrightarrow[\Delta]{-S-\bigcirc} CuPc\left(S-\bigcirc\right)_{15/16} \quad (5.14)$$

(**68**)

λ_{max} 770 nm, ε_{max} 180,000 (CHCl$_3$)

(**67**)

Reprographics.   Reprographics, which is by far the biggest of the high-tech areas, comprises three major technologies, electrophotography (photocopiers and laser printers), ink-jet printing and the technology discussed here, dye diffusion thermal transfer (D2T2) printing.

D2T2 is capable of producing photographic-quality full-colour images and its ultimate major use is predicted to be in electronic photography. The basic concept is extremely attractive—an electronic camera replaces the conven-

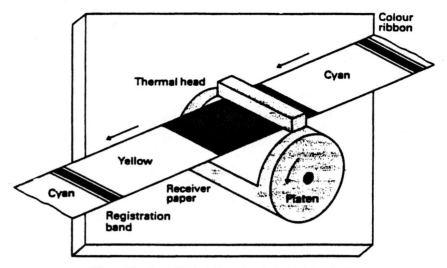

Figure 5.3   Dye diffusion thermal transfer (D2T2) printing.

tional silver halide camera and a magnetic disc replaces the silver halide film. The magnetic disc allows up to 50 pictures to be taken. At any time, the disc can be put into a printer (similar to a video recorder in appearance) which is linked to a television. Each picture can then be viewed on the television screen: if the picture is satisfactory a 'photograph' can be obtained after about a minute by simply pressing a button.

In D2T2, the silver halide film is replaced by a trichromat colour ribbon. It is a three-pass process in which the dyes are transferred to a white receiver paper by a thermal head (Figure 5.3). Although more than a million dyes have been synthesised, these new dyes had to be designed to meet the demanding D2T2 criteria. Azo dyes predominate, such as the pyridone yellow (**69**), the iso-thiazole magenta (**70**) and the thiophene cyan (blue) (**71**).

(69)   (70)

(71)

Phthalocyanines, such as **(72)**, are looking promising in the photodynamic therapy (PDT) treatment of cancer. Here, the chemical is injected into the patient who is then kept in the dark for 48 h, during which time the chemical accumulates preferentially in the cancer cells. Light is then directed on to the cancer cells, e.g. by using an optical fibre. The highly active oxidant, singlet oxygen ($^1O_2$) is then produced (Scheme 5.30), which kills the cancer cells.

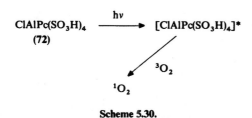

Scheme 5.30.

Security.   Dyes of various types are also being used in security applications. However, for obvious reasons, this cannot be discussed further.

There is no doubt that high-tech applications of dyes will increase rapidly in the future and, although the sales volume will not exceed that of traditional uses, the sales income will become significant.

References

1. 'Organic Chemistry in Colour,' P.F. Gordon and P. Gregory, Springer-Verlag, Berlin, 1983, pp. 1–5.
2. Reference 1, pp. 5–21.
3. *Colour Index*, 3rd edn, The Society of Dyers and Colourists, Bradford, UK, and The American Association of Textile Chemists and Colourists, Lowell, MA, USA, 1971.
4. 'Classification of Dyes by Chemical Structure,' P. Gregory, In 'The Chemistry and Application of Dyes,' D.R. Waring and G. Hallas (eds), Plenum, New York, 1990, pp. 17–47.
5. 'Colour and Constitution of Organic Molecules,' J. Griffiths, Academic Press, London, 1976.
6. Reference 1, pp. 121–158.
7. Reference 1, p. 157.
8. Reference 1, pp. 211–217.
9. Reference 1, pp. 221–226.
10. 'Infrared Absorbing Dyes,' M. Matsuoka, Plenum, New York, 1990; P. Gregory, 'Chemistry of Functional Dyes,' *Proc. 2nd International Conference*, Kobe, Japan, October 1992.
11. A. Calder, 'Dyes in Non-Impact Printing,' *Proc. IS and T's Seventh International Congress on Advances in Non-Impact Printing Technologies*, Portland, OR, USA, October 1991.
12. 'The Manufacture of Organic Colourants and Intermediates,' G. Booth, Society of Dyers and Colourists, Bradford, UK, 1988.
13. Reference 12, p. 9.
14. US International Trade Commission Statistics on Production and Sale of Dyes and Pigments.
15. *Japan Chemical Annual*, 1981.
16. Reference 1, pp. 28–47.
17. Reference 1, pp. 47–56.
18. Reference 1, pp. 57–65.
19. Reference 1, pp. 96–115.
20. Reference 1, pp. 219–221.
21. Reference 1, pp. 66–76.

22. Reference 1, pp. 82–84.
23. C.W. Greenhalgh, J.L. Carey and D.F. Newton, *Dyes and Pigments*, 1980, **1**, 103.
24. C.W. Greenhalgh, N. Hall and D.F. Newton, European Patent 33 583 (1981).
25. B. Parton, British Patent 1 349 513 (1970); A.H.M. Renfrew, *Rev. Progr. Colouration*, 1985, **15**, 15.
26. Reference 1, p. 16; Reference 4, pp. 44, 103.
27. Reference 1, pp. 200–208, 74, 96, 288.
28. Reference 1, pp. 226–242.
29. Reference 1, pp. 242–252.
30. Reference 4, pp. 43, 161.
31. Reference 1, pp. 253–257.
32. Reference 1, pp. 262–304.
33. 'High Technology Applications of Organic Colourants,' P. Gregory, Plenum, New York, 1991.

Bibliography

'Dyestuffs,' E.N. Abrahart, In: 'The Chemical Industry,' C.A. Heaton (ed.), 1st edn, Blackie, Glasgow, 1986, pp. 65–125, and references therein.
'Dyes and Dye Intermediates' in *ECT* 1st edn, Vol. 5, pp. 327–354, G.E. Goheen and J. Werner, General Aniline & Film Corporation, and A. Merz, Americal Cyanamid Company; ' Dyes and Dye Intermediates' in *ECT* 2nd edn, Vol. 7, pp. 462–505, D.W. Bannister and A.D. Olin, Tomas River Chemical Corporation; 'Dyes and Dye Intermediates' in *ECT* 3rd edn, Vol. 8, pp. 152–212, D.W. Bannister, A.D. Olin and H.A. Stingl, Tomas River Chemical Corporation.
'Fibre-Reactive Dyes,' W. Beech, SAF International, New York, 1970.
'The Manufacture of Organic Colourants and Intermediates,' G. Booth, Society of Dyers and Colourists, Bradford, UK, 1988.
'Recent Progress in the Chemistry of Dyes and Pigments,' W. Bradley, The Royal Institute of Chemistry, London, 1958.
'Identification of Dyes on Textile Fibres,' E. Clayton, 2nd edn, Society of Dyers and Colourists, Bradford, UK, 1963.
'The Chemistry and Technology of Naphthalene Compounds,' N. Donaldson, Edward Arnold, London, 1958.
'Light Absorption of Organic Colourants: Theoretical Treatment and Empirical Rules,' J. Fabian and H. Hartmann, Springer-Verlag, Heidelberg, 1980.
'Recent Progress in the Chemistry of Natural and Synthetic Colouring Matters and Related Fields,' T.S. Gore *et al.* Academic Press, New York, 1962.
'Phthalocyanine Compounds,' F.H. Moster and A.L. Thomas, Reinhold, London, 1963.
'Synthetic Dyes in Biology, Medicine and Chemistry,' E. Gurr, Academic Press, London, New York, 1971.
'Diazo and Azo Chemistry,' H. Zollinger, Interscience Publishers, Inc., New York, 1961.
'Fundamentals of the Chemistry and Application of Dyes,' P. Rys and H. Zollinger, Wiley-Interscience, London, 1972.
'Practical Dye Chemistry,' D.R. Waring and G. Hallas (eds), Plenum, New York, 1990.
'Colour Chemistry: The Design and Synthesis of Organic Dyes and Pigments,' A.T. Peters and H.S. Freeman, Elsevier, London, 1991.
'High Technology Applications of Organic Colourants,' P. Gregory, Plenum, New York, 1991.
'Colour Chemistry,' H. Zollinger, VCH, 1987.
'Colour Chemistry,' H. Zollinger, 2nd edn, VCH, 1991.

The sulphur, phosphorus, nitrogen and chlor-alkali industries 6

Stephen Black

6.1
Introduction

Basic chemicals are the orphans of the chemical industry. They are not glamorous, like drugs, and are sometimes not very profitable (and at the very least the profits come in unpredictable cycles of boom and bust). They are not seen or used directly by the general public and so their importance is not often understood. Even within the industry their importance is often insufficiently appreciated. Without them, however, the rest of the industry could not exist.

Basic chemicals occupy the middle ground between raw materials (that is, things that are mined, quarried or pumped from the ground) and end-products. One distinguishing feature of basic chemicals is the scale on which they are manufactured; everything from really big to absolutely enormous. Figure 6.1 shows the top 25 chemicals in the USA market by volume in 1991*, just to give a feel for the sort of chemicals and volumes concerned. Basic chemicals are typically manufactured in plants that produce hundreds of thousands of tonnes of product per year. A plant that produces 100 000 tonnes per year will produce about 12.5 tonnes every hour. It is sometimes difficult to grasp the scale of this; however, if you note that the average driver in the UK uses about one tonne of petrol per year then you are starting to get a feel for the scale.

Another distinguishing and important feature of basic chemicals is their price: most of them are fairly cheap. Again it is helpful to have some feel for what price means in comprehensible terms. Typical selling prices for commodity basic chemicals are anything up to £1000 per tonne. Some comparisons with consumer goods are helpful. Petrol costs about 50 p per litre in UK petrol station forecourts; this is equivalent to about £600–700 per tonne (most of the cost is, however, tax); this can be compared to the naphtha fraction (roughly anything that distils from 100–200 °C) from crude oil which sells for about £100 per tonne when oil prices are about US$ 18 per bbl ('barrel'). The glamour end of the chemical industry is used to completely different ranges of

---

* Adapted from *Chemical and Engineering News.*

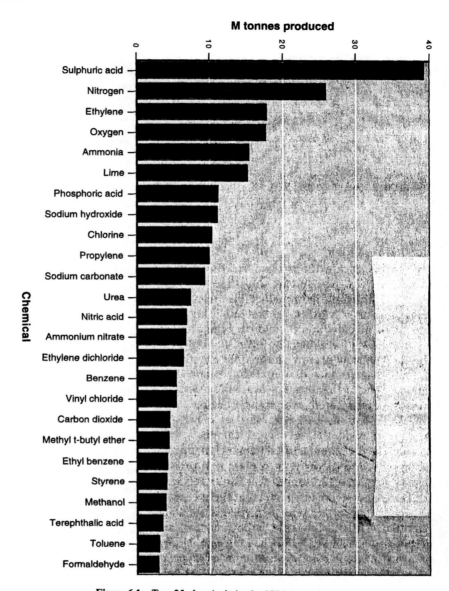

Figure 6.1   Top 25 chemicals in the USA market, 1991.

prices. Aspirin, a 'commodity' pharmaceutical, sells to consumers at a price equating to about £13 000 per tonne; Zantac, the world's biggest grossing pharmaceutical sells to the UK National Health Service (NHS) at about £3 000 000 per tonne but only a few hundred tonnes are produced each year world-wide (both costs include formulation and packing). In contrast, the

USA market alone produces about 10 million tonnes of chlorine and that is only the ninth best selling chemical by weight.

The job of the basic chemical industry is to find economical ways of turning raw materials into useful intermediates. There is little leeway for any company to charge premium prices for its products, so the company that makes the products at the cheapest cost will probably be the most profitable. This situation means that companies must always be on their toes looking for new and more economical ways of making and transforming their raw materials.

It is impossible in any short chapter to deal comprehensively with all the chemicals that might be called basic, so only the key conversions of raw materials to useful chemicals will be considered in detail. Many basic chemicals are the product of oil refining and they already have a chapter to themselves in Volume 1 (chapter 11); this chapter will consider the parts of the industry that put elements other than carbon and hydrogen into chemicals. These are the sulphur, nitrogen, phosphorus and chloro-alkali industries. In combination, these products and the basic products of the petrochemical industry can be combined to produce the myriad of important chemicals that feed the rest of the chemical industry.

This chapter will not delve into the details of plant construction or the inner workings of different processes. Only the basic principles will be covered, along with some indication of the uses of the products. Those who need or want to know details of particular processes (which often differ from manufacturer to manufacturer) can find suitable sources in the bibliography at the end of this chapter.

6.2.1 Introduction

**6.2
Sulphuric acid**

Sulphuric acid is the chemical that is produced in the largest tonnage. Such is its importance as a raw material for other processes that its production was, until recently, considered a reliable indicator of a country's industrial output and the level of its industrial development. It is still used in an incredible variety of different processes and remains one of the most important chemicals produced.

6.2.2 Raw materials

The raw material for sulphuric acid production is elemental sulphur, which can be obtained from several sources. Almost all the elemental sulphur produced is used in the manufacture of sulphuric acid.

The biggest source used to be the direct mining of underground deposits of sulphur by the Frasch process. This involves injection of superheated water and air into the deposits of sulphur via drilling. The resulting aerated liquid sulphur–water–air mixture is buoyant enough to rise to the drill head where it can be separated into its components, and pure sulphur is recovered.

The petrochemical industry now provides more sulphur (from the desulphurisation of oil and gas) than the Frasch process. Some petrochemical deposits contain large quantities of sulphur-containing compounds (25% in some Russian deposits) which must be removed to avoid poisoning the cracking catalysts or indeed the public (though the residual sulphur in some petrol is enough to cause bad hydrogen sulphide smells from some catalytic converters). The process that removes the sulphur creates hydrogen sulphide which is easy to separate from oil and gas and is easily converted to elemental sulphur. In the USA, for example, of about 12 million tonnes of sulphur produced in 1991 about 7.5 million was recovered and only about 3.2 million mined via the Frasch process.

Some sulphur is also produced as a by-product of metal extraction. Many sulphide-containing metal ores are burned as part of the extraction process, giving off sulphur dioxide, which can be recovered and used in the sulphuric acid industry.

One untapped source of sulphur is the coal used in electricity generation. More sulphur is emitted from power stations than is used by the chemical industry. Most of this ends up in the atmosphere where it causes acid rain, though power stations are increasingly having to scrub their flue gases. Unfortunately, there is at present no convenient way to recover this sulphur in a useable form.

6.2.3   The manufacturing process

The production of sulphuric acid has three stages:

1.   The burning of sulphur in air to give sulphur dioxide

$$S + O_2 \longrightarrow SO_2$$

2.   The reaction of sulphur dioxide and oxygen to give sulphur trioxide

$$2SO_2 + O_2 \longrightarrow SO_3$$

3.   The absorption of sulphur trioxide in water to give sulphuric acid.

The first stage is simple, with few of the all too common complications that beset many industrially important 'simple' reactions. Molten sulphur is sprayed into a furnace in a current of dry air at about 1000 °C to produce a gas stream containing about 10% sulphur dioxide. The stream is cooled in a boiler where the energy of the exothermic reaction can be extracted and the temperature brought down to 420 °C.

The second stage is the key to the process. The direct reaction between sulphur dioxide and oxygen to give sulphur trioxide is slow and requires a catalyst. In the old lead chamber process nitrogen dioxide (NO_2) was used for the oxidation—though in practice mixtures of nitrogen oxides were used. This process has now been completely superseded by the contact process, which speeds the direct reaction via a solid-state catalyst of vanadium pentoxide (V_2O_5). The catalyst is normally absorbed on an inert silicate support and lasts for about 20 years.

The reaction between oxygen and sulphur dioxide is exothermic so the equilibrium favours the product at lower temperatures. So much heat is produced by the reaction that it is difficult to achieve good yields in a single-bed reactor: the reactor warms up so much that the reaction goes into reverse. Most plants therefore used several reactor stages (with coolers between the reactors) so that the heat produced does not drive the reaction backwards. The first reactor chamber converts the mixture of oxygen and sulphur dioxide into a stream about 60–65% converted to products at about 600 °C. It is then cooled to about 400 °C and passed into the next layer of catalyst, and so on. After three layers 95–96% of the starting materials have been converted to products (this is near the maximum possible conversion unless sulphur trioxide is removed).

The gas mixture can then be passed into the initial absorption tower where some sulphur trioxide is removed. The remaining gas mixture is then reheated and passed back into a fourth converter which enables overall conversion of up to 99.7% to be achieved.

The final stage involves passing the gas mixture into the final absorption tower where the sulphur trioxide is hydrated with water to give, at the end of the tower, a 98–99% acid solution. If excess sulphur trioxide is used the mixture consists of, effectively, sulphur trioxide dissolved in pure sulphuric acid: this is known as oleum.

6.2.4   Uses of sulphuric acid

A large amount (66% of total production in the peak year) of sulphuric acid is used in the manufacture of phosphoric acid for fertiliser (phosphate rock plus acid gives impure phosphoric acid), though this use is now declining as demand for fertilisers declines. Some more goes into production of ammonium sulphate, a low-grade fertiliser; about 2.2 million tonnes of this were produced in the USA in 1991.

More interesting are some of the more speciality uses of the compound, which are less significant in volume but are—possibly—more important for the rest of the industry. There are many important large-scale industrial syntheses that use sulphuric acid, for example: the manufacture of ethanol from ethene; one of the routes to titanium dioxide (a white pigment important to the paint industry); the production hydrofluoric acid from calcium fluoride (the ultimate source of about 70% of all the fluorine in fluorinated compounds); the production of aluminium sulphate (an important water-treatment chemical); the production of sulphonated surfactants (detergents and many other applications). Many uses are based on the fact that sulphuric acid represents a cheap source of protons. For example, the manufacture of hydrofluoric acid from fluoride-containing minerals involves mixing them with concentrated sulphuric acid in a rotating kiln: the fluoride is protonated to hydrogen fluoride by the acid and the hydrogen fluoride is given off as a gas.

6.3 Phosphorus-containing compounds

6.3.1   Raw materials

All phosphorus-containing compounds in the chemical industry are derived from minerals of the apatite family. These are basically calcium phosphate, though they often contain fluoride as an impurity which must be removed from the product but can then be sold as a source of fluorine. The actual mineral most often used in the preparation of phosphoric acid tends to have the composition $Ca_{10}F_2(PO_4)_6$. Other impurities, such as silica, are common.

6.3.2   Direct production of phosphoric acid

Much of the phosphorus-compound production world-wide involves the direct production of phosphoric acid (H_3PO_4) in an impure form for the fertiliser market, where high purity is not a requirement (except for the removal of poisonous heavy metals). The key reaction used to turn the mineral into fertiliser-grade acid is the reaction of the crushed material with sulphuric acid. The initial stage of this reaction proceeds approximately according to the following equation:

$$Ca_{10}F_2(PO_4)_6 + 7H_2SO_4 \longrightarrow 3\dot{C}a(H_2PO_4)_2 + 7CaSO_4 + 2HF$$

Silica, present as an impurity, reacts with hydrofluoric acid to give silicon tetrafluoride, which is a gas and is collected and converted to hydrofluorosilicic acid, a convenient source of fluorine which accounts for about 30% of all the fluorine produced world-wide. Apart from the recovery of the valuable by-product fluorine, the process also has the problem of how to separate the calcium sulphate from the desired phosphoric acid. There is also the problem of how to encourage the reaction to go to completion, since the calcium sulphate tends to form an insoluble coating on the lumps of apatite. This problem is overcome by crushing the mineral thoroughly and by passing the wet acid through a reactor tank where the conditions are controlled to encourage the formation of optimally sized crystals of calcium sulphate. These can be filtered and the remainder of the mixture recycled.

6.3.3   Elemental phosphorus and derivatives

Elemental phosphorus is produced by using a highly energy-intensive process involving an electric furnace. A mixture of apatite, silica and coke is heated in an electric furnace to about 1400–1500 °C. The silica and apatite react to form a molten mass which reacts with the coke to reduce the phosphate to elemental phosphorus according to the following equation:

$$2Ca_3(PO_4)_2 + 6SiO_2 + 10C \longrightarrow P_4 + 6CaSiO_3 + 10CO$$

Any fluoride present ends up as calcium fluoride. The elemental phosphorus is quite volatile and effectively distils from the melt. The actual plants have some sophisticated extraction and purification stages to make sure they do not lose any of the volatile and toxic compounds up the chimney.

6.3.4   Uses of phosphorus-containing compounds

Much of the crude acid is converted into sodium or ammonium salts before use. In volume terms, phosphoric acid has often come second only to sulphuric acid in sales. The total world production of phosphorus-containing fertiliser has been estimated to be about 40 million tonnes (expressed as phosphorus pentoxide). More than half of this was in the form of ammonium phosphate or compound phosphates (i.e. mixed ammonium, potassium and sodium salts).

Though most of the production of phosphorus-containing compounds goes into fertilisers, the use of phosphates for other purposes is probably more significant. For example, the sodium salt of phosphoric acid (Na_3PO_4) has good dirt- and fat-solving properties and so is widely used in cleaning agents. An important industrial use is for the cleaning of metal surfaces. It is also used to soften water by precipitating calcium and magnesium ions.

In 1988 about 750 000 tonnes of elemental phosphorus was produced, of which 83% went into the production of pure phosphoric acid for use in detergents and foodstuffs. Although the cost of this acid is about 3–4 times that of the impure product from the more direct process, it is costly to purify the cheap product enough for consumer products. The rest of the production of elemental phosphorus went into red phosphorus (used in matches) and various phosphorus halides (PCl_3, PCl_5, $POCl_3$) which are used in agro-chemicals, drugs and some other speciality chemicals such as flame-retardant additives for polymers.

Polyphosphates are made via polyphosphoric acid (PPA) which is produced when phosphoric acid is concentrated beyond a certain point by the loss of water. The compounds consist of chains or rings containing a backbone of P—O bonds. An example of a trimeric species is shown in Figure 6.2.

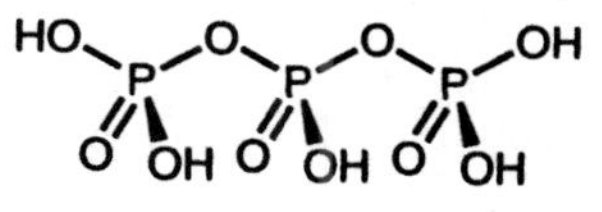

Figure 6.2   Structure of a tripolyphosphoric acid.

Polyphosphoric acid (a mixture of several polymeric species) is a thick syrupy liquid that is easy to transport, and is a good non-oxidising acidic catalyst and a good dehydrating agent and so is widely used in industries needing such properties in bulk. Several important industrial acid catalysts are based on PPA adsorbed on to inert supports.

Polyphosphate salts are also significant. One of the most important polyphosphates is pentasodium tripolyphosphate (often known as STPP). This is used in large quantities as a water softener in domestic washing powder and in other detergents. Polyphosphate salts are also used in the food industry as buffers and as water-retaining agents to improve the storage quality of packaged food.

6.4 Nitrogen-containing compounds

6.4.1   Introduction

Dinitrogen makes up more than three-quarters of the air we breathe, but it is not readily available for further chemical use. Biological transformation of nitrogen into useful chemicals is embarrassing for the chemical industry, since all the effort of all the industry's technologists has been unable to find an easy alternative to this. Leguminous plants can take nitrogen from the air and convert it into ammonia and ammonium-containing products at atmospheric pressure and ambient temperature; despite a hundred years of effort, the chemical industry still needs high temperatures and pressures of hundreds of atmospheres to do the same job. Indeed, until the invention of the Haber process, all nitrogen-containing chemicals came from mineral sources ultimately derived from biological activity.

Essentially all the nitrogen in manufactured chemicals comes from ammonia derived from the Haber-based process. So much ammonia is made (more molecules than any other compound, though because it is a light molecule greater *weights* of other products are produced), and so energy-intensive is the process, that ammonia production alone was estimated to use 3% of the world's energy supply in the mid-1980s.

6.4.2   The Haber process for ammonia synthesis

6.4.2.1   Introduction.   All methods for making ammonia are basically fine-tuned versions of the process developed by Haber, Nernst and Bosch in Germany just before the First World War.

$$N_2 + 3H_2 \rightleftharpoons 2NH_3$$

In principle the reaction between hydrogen and nitrogen is easy; it is exothermic and the equilibrium lies to the right at low temperatures. Unfortunately, nature has bestowed dinitrogen with an inconveniently strong triple bond, enabling the molecule to thumb its nose at thermodynamics. In scientific terms the molecule is kinetically inert, and rather severe reaction conditions are necessary to get reactions to proceed at a respectable rate. A major source of 'fixed' (meaning, paradoxically, 'usefully reactive') nitrogen in nature is lightning, where the intense heat is sufficient to create nitrogen oxides from nitrogen and oxygen.

To get a respectable yield of ammonia in a chemical plant we need to use a catalyst. What Haber discovered—and it won him a Nobel prize—was that some iron compounds were acceptable catalysts. Even with such catalysts extreme pressures (up to 600 atmospheres in early processes) and temperatures (perhaps 400 °C) are necessary.

Pressure drives the equilibrium forward, as four molecules of gas are being transformed into two. Higher temperatures, however, drive the equilibrium the wrong way, though they do make the reaction faster; chosen conditions must be a compromise that gives an acceptable conversion at a reasonable speed. The precise choice will depend on other economic factors and the details of the catalyst. Modern plants have tended to operate at lower pressures and higher temperatures (recycling unconverted material) than the nearer-ideal early plants, since the capital and energy costs have become more significant.

Biological fixation also uses a catalyst which contains molybdenum (or vanadium) and iron embedded in a very large protein, the detailed structure of which eluded chemists until late 1992. How it works is still not understood in detail.

6.4.2.2 Raw materials.

The process requires several inputs: energy, nitrogen and hydrogen. Nitrogen is easy to extract from air, but hydrogen is another problem. Originally it was derived from coal via coke which can be used as a raw material (basically a source of carbon) in steam reforming, where steam is reacted with carbon to give hydrogen, carbon monoxide and carbon dioxide. Now natural gas (mainly methane) is used instead, though other hydrocarbons from oil can also be used. Ammonia plants always include hydrogen-producing plants linked directly to the production of ammonia.

Prior to reforming reactions, sulphur-containing compounds must be removed from the hydrocarbon feedstock as they poison both the reforming catalysts and the Haber catalysts. The first desulphurisation stage involves a cobalt–molybdenum catalyst, which hydrogenates all sulphur-containing compounds to hydrogen sulphide. This can then be removed by reaction with zinc oxide (to give zinc sulphide and water).

The major reforming reactions are typified by the following reactions of methane (which occur over nickel-based catalysts at about 750 °C):

$$CH_4 + H_2O \longrightarrow \underset{\text{Synthesis gas}}{CO + 3H_2}$$

$$CH_4 + 2H_2O \longrightarrow CO_2 + 4H_2$$

Other hydrocarbons undergo similar reactions.

In the secondary reformers, air is injected into the gas stream at about 1100 °C. In addition to the other reactions occurring, the oxygen in the air reacts with hydrogen to give water, leaving a mixture with close to the ideal 3:1 ratio of hydrogen to nitrogen with no contaminating oxygen. Further reactions, however, are necessary to convert more of the carbon monoxide into hydrogen and carbon dioxide via the shift reaction:

$$CO + H_2O \longrightarrow CO_2 + H_2$$

This reaction is carried out at lower temperatures and in two stages (400 °C with an iron catalyst and 220 °C with a copper catalyst) to ensure that conversion is as complete as possible.

In the next stage, carbon dioxide must be removed from the gas mixture, and this is accomplished by reacting the acidic gas with an alkaline solution such as potassium hydroxide and/or mono- or di-ethanolamine.

By this stage there is still too much contamination of the hydrogen–nitrogen mixture by carbon monoxide (which poisons the Haber catalysts), and another step is needed to get the amount of CO down to ppm levels. This step is called methanation and involves the reaction of CO and hydrogen to give methane (i.e. the reverse of some of the reforming steps). The reaction operates at about 325 °C and uses a nickel catalyst.

Now the synthesis gas mixture is ready to go into a Haber reaction.

6.4.2.3 Ammonia production. The common features of all the different varieties of ammonia plant are that the synthesis gas mixture is heated, compressed and passed into a reactor containing a catalyst. The essential equation for the reaction is simple:

$$N_2 + 3H_2 \rightleftharpoons 2NH_3$$

What industry needs to achieve in the process is an acceptable combination of reaction speed and reaction yield. Different compromises have been sought at different times and in different economic circumstances. Early plants plumped for very high pressure (to get the yield up in a one-pass reactor), but many of the most modern plants have accepted much lower one-pass yields at lower pressures and have also opted for lower temperatures to conserve energy. ICI's LCA (leading concept ammonia) process, for example, works at very low pressures (for an ammonia plant) and recycles unreacted gases, paying careful attention to the energy flow in the process so that energy produced at one stage is not wasted by dissolution to atmosphere, but is used for other stages requiring energy input. The LCA process also uses a highly active long-life catalyst which is protected from poisons by very careful purification of the input gases.

In order to ensure the maximum yield in the reactor the synthesis gas is usually cooled as it reaches equilibrium. This can be done by the use of heat exchangers or by the injection of cool gas into the reactor at an appropriate point. The effect of this is to freeze the reaction as near to equilibrium as possible. Since the reaction is exothermic (and the equilibrium is less favourable for ammonia synthesis at higher temperatures) the heat must be carefully controlled in this way to achieve good yields.

The output from the Haber stage will consist of a mixture of ammonia and synthesis gas so the next stage needs to be the separation of the two so that the synthesis gas can be recycled. This is normally accomplished by condensing the ammonia (which is a good deal less volatile than the other components; ammonia boils at about $-40\,°C$).

6.4.2.4 Uses of ammonia. The major use of ammonia is not for the production of nitrogen-containing chemicals for further industry use, but for fertilisers such as urea or ammonium nitrates and phosphates. Fertilisers consume 80% of all the ammonia produced. In the USA in 1991, for example, the following ammonia-derived products were consumed, mostly for fertilisers (amounts in millions of tonnes): urea (4.2); ammonium sulphate (2.2); ammonium nitrate (2.6); diammonium hydrogen phosphate (13.5).

Chemical uses of ammonia are varied. The Solvay process for the manufacture of soda ash uses ammonia, though it does not appear in the final product since it is recycled. A wide variety of processes take in ammonia directly, including the production of cyanides and aromatic nitrogen-containing compounds such as pyridine. The nitrogen in many polymers (such as nylon or acrylics) can be traced back to ammonia, often via nitriles or hydrogen cyanide. Most other processes use nitric acid or salts derived from it as their source of nitrogen. Ammonium nitrate, used as a nitrogen-rich fertiliser, also finds a major use as a bulk explosive.

6.4.3 Nitric acid

6.4.3.1 Production. Much of the nitrogen used by the chemical industry to make other raw materials is not used directly as ammonia; rather, the ammonia is first converted into nitric acid. Nitric acid production consumes about 20% of all the ammonia produced.

The conversion of ammonia to nitric acid is a three-stage process:

1. $$4NH_3 + 5O_2 \longrightarrow 4NO + 6H_2O$$

2. $$2NO + O_2 \longrightarrow 2NO_2$$

3. $$3NO_2 + H_2O \longrightarrow 2HNO_3 + NO$$

The first reaction is catalysed by platinum (in practice platinum–rhodium gauze), as can be observed on the bench with a piece of platinum wire and some concentrated ammonia solution. It might, at first sight, seem that the overall reaction to the acid would be easy; unfortunately, there are complications as nature is a good deal less tidy than chemists and engineers would prefer.

Industrially the first reaction is carried out at about 900 °C in reactors containing platinum–rhodium gauze, the temperature being maintained by the heat produced by the reaction. At these temperatures some important side reactions are also fast. Firstly, the ammonia and air mixture can be oxidised to dinitrogen and water (this reaction tends to happen on the wall of the reaction vessel if it is hot, so it needs to be deliberately cooled). Secondly, the decomposition of the first reaction product, nitric oxide, to dinitrogen and oxygen is promoted by the catalyst. It is therefore important to get the product out of the reactor as fast as possible, though this must be balanced against the need to keep the raw materials in contact with the catalyst long enough for them to react. Thirdly, the product, nitric oxide, reacts with ammonia to give dinitrogen and water, so it is important not to let too much ammonia through the catalyst beds or the result will be wasted raw material that cannot be recovered. Control of these conflicting needs is achieved by careful reactor design and by fine control of temperature and flow-rates through the reactors. The actual contact time is usually about 3×10^{-4}s.

The second and third stages have fewer complications, but both are slow and there are no known—cost-effective—catalysts. Typically, a mixture of air and nitric oxide is passed through a series of cooling condensers where partial oxidation occurs; the reaction is favoured by low temperatures. The nitrogen dioxide is absorbed from the mixture as it is passed down through a large bubble-cap absorption tower; 55–60% nitric acid emerges from the bottom.

This nitric acid cannot be concentrated much by distillation as it forms an azeotrope with water at 68% nitric acid. Nitric acid plants typically employ a tower containing 98% sulphuric acid to give 90% nitric acid from the top of the tower. Near-100% acid can be obtained if necessary by further dehydration with magnesium nitrate.

6.4.3.2 Uses of nitric acid.   About 65% of all the nitric acid produced is reacted with ammonia to make ammonium nitrate; 80% of this is used as fertiliser, the rest as an explosive. The other major use of nitric acid is in organic nitrations. Almost all explosives are ultimately derived from nitric acid (most are nitrate esters—e.g. nitroglycerine—or nitrated aromatics—e.g. trinitrotoluene). Nitration using mixtures of sulphuric and nitric acid is the first step in the synthesis of important nitro- and amino-aromatic intermediates such as aniline (the first step is nitration of an aromatic, then reduction of the nitro group to an amino). Many important dyestuffs and pharmaceuticals are ultimately derived from such reactions, though the quantities involved are small. Polyurethane plastics are built around aromatic isocyanates ultimately derived from nitrated toluene and benzene; this use consumes about 5–10% of nitric acid production.

6.4.4   Urea

6.4.4.1   Production.   One other product of some significance is made directly from ammonia in large quantities: urea (H_2NCONH_2). About 20% of all ammonia is made into urea. It is synthesised by high pressure reaction (typically 200–400 atm and 180–210 °C) of carbon dioxide with ammonia in a two-stage reaction:

1. $CO_2 + 2NH_3 \longrightarrow NH_2CO_2^- NH_4^+$ (ammonium carbamate)

2. $NH_2CO_2^- NH_4^+ \longrightarrow NH_2CONH_2 + H_2O$

The high pressure reaction achieves about 60% conversion of carbon dioxide to the carbamate (stage 1) and the resulting mixture is then passed into low-pressure decomposers to allow for the conversion to urea. Unreacted material is passed back to the start of the high-pressure stage of the process as this greatly improves overall plant efficiency. The solution remaining after the second stage can either be used directly as a liquid nitrogenous fertiliser or concentrated to give solid urea of 99.7% purity.

6.4.4.2   Uses.   The high nitrogen content of urea makes it another useful nitrogenous fertiliser, and this accounts for the vast majority of the market for the compound. Other uses are significant but use only about 10% of all the urea produced. The biggest other use is for resins (melamine–formaldehyde and urea–formaldehyde) which are used, for example, in plywood adhesives and Formica surfaces.

6.5.1   Introduction

**6.5
Chlor-alkali and related compounds**

Historically the bulk chemical industry was built on chlor-alkali and related processes. The segment is normally taken to include the production of chlorine gas, caustic soda (sodium hydroxide), soda-ash (derivatives of sodium carbonate in various forms) and, for convenience, lime-based products.

Soda-ash and sodium hydroxide have competed with each other as the major source of alkali ever since viable processes were discovered for both. The peculiar economics of electrolytic processes mean that you have to make chlorine and caustic soda together in a fixed ratio whatever the relative demand for the two totally different types of product, and this causes swings in the price of caustic soda which can render soda-ash more or less favourable as an alkali.

Both chlorine/caustic soda and soda-ash production are dependent on cheap readily available supplies of raw materials. Chlorine/caustic soda requires a ready supply of cheap brine and electricity; soda-ash requires brine, limestone and lots of energy. Soda-ash plants are only profitable if their raw materials do not have to be transported far. The availability of such supplies is a major factor in the location of many of the chemical industry's great complexes. ICI's Runcorn and Brunner–Mond's Winnington sites in Cheshire, UK, are classic examples—Cheshire has large salt deposits and an enormous deposit of high-purity limestone nearby.

6.5.2   Lime-based products

One of the key raw materials is lime. Limestone consists mostly of calcium carbonate ($CaCO_3$) laid down over geological time by various marine organisms. High-quality limestones are often good enough to be used directly as calcium carbonate in further reactions. Limestone is usually mined in vast open-cast quarries, many of which will also carry out some processing of the material.

The two key products derived from limestone are quicklime (CaO) and slaked lime ($Ca(OH)_2$). Quicklime is manufactured by the thermal decomposition (1200–1500 °C) of limestone according to the equation:

$$CaCO_3 \longrightarrow CaO + CO_2$$

Typically limestone is crushed and fed into the higher end of a sloping rotating kiln where the decomposition takes place and quicklime is recovered from the end. Most frequently, however, the quicklime is not isolated for further reactions; rather, other compounds are fed in with the lime to give final products at the low end of the kiln. For example, alumina, iron ore and sand can be fed in to give Portland cement. Soda-ash manufacture often adds coke to the limestone which burns to give extra carbon dioxide needed for soda-ash manufacture. Slaked lime—which is more convenient to handle than quicklime—is manufactured by reacting quicklime with water.

About 40% of the output of the lime industry goes into steel-making, where it is used to react with the refractory silica present in iron ore to give a fluid slag which floats to the surface and is easily separated from the liquid metal. Smaller, but still significant, amounts are used in chemical manufacture, pollution control and water treatment.

The most important chemical derived from lime is soda-ash, though in the USA much soda-ash is now mined directly rather than being made synthetically.

6.5.3   Soda-ash

6.5.3.1   The Solvay process.   For a simple basic product the Solvay process appears exceedingly complicated. The basic principle of the reaction is to take salt (NaCl) and calcium carbonate ($CaCO_3$) as inputs and to produce calcium chloride and sodium carbonate as outputs. However, the reactions occurring between input and output are not remotely obvious and involve the use of ammonia and calcium hydroxide as intermediate compounds.

The easiest way to follow the logic of the reaction is via a flow-chart (Figure 6.3) rather than through a list of all the component processes. This flow-chart is vastly simplified over the actual sections used in a plant, so the principles can be followed. The essential principle is that, by carefully controlling the concentration of the components (especially ammonia and salt), sodium bicarbonate can be precipitated from solutions containing salt, carbon dioxide and ammonia. The key to making the process work is controlling the strength of the solutions and the rates of crystallisation.

The essential steps of the process are as follows. Ammonia is absorbed into brine which has previously been purified to reduce the amount of calcium and magnesium ions (which tend to precipitate during the process in all the wrong places, blocking pipe-work). The solution (nominally containing sodium chloride and ammonium hydroxide) is then passed down a tower where it absorbs carbon dioxide (passing up the tower) to form ammonium carbonate at first and later ammonium bicarbonate. By the next stage of the plant sodium chloride and ammonium bicarbonate have metathesised to sodium bicarbonate (which precipitates) and ammonium chloride. Filtration separates the solid bicarbonate from the remaining solution. The bicarbonate is passed to a rotary dryer where it loses water and carbon dioxide to give a fluffy crystalline mass known as light soda-ash which is mostly sodium carbonate. The fluffy

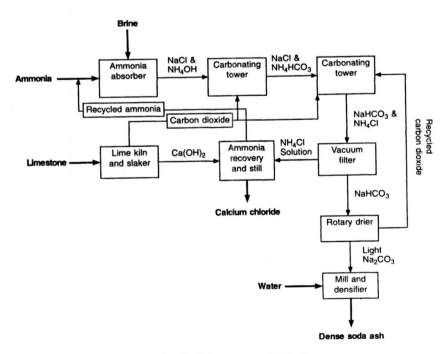

Figure 6.3 The Solvay process: block diagram.

mass is light because the original crystal shape is retained on the loss of carbon dioxide, leaving many voids. It is usually more convenient to make a more dense material and this is achieved by adding water (which causes recrystallisation in a denser form) and further drying.

It is debatable whether the actual chemistry given above is a good description of the process, but it certainly aids understanding. For a detailed understanding, a great deal needs to be known about solubility products of multicomponent systems. The important thing to know is that the system is complex and requires careful control at all parts of the process in order for it to operate effectively.

One disadvantage of the process is the amount of calcium chloride produced. Far more is produced than can be used, so much of the production is simply dumped (it is not a particularly noxious or nasty product). It would be advantageous to use all the input material in this process, for example producing hydrogen chloride from the chloride. In the early 1980s Hüls developed a process (using organic amines to replace the ammonia–calcium hydroxide), but it has had little impact as it is considerably more complicated and there is no economic incentive to use it.

6.5.3.2   *Uses of soda-ash.*   Of all soda-ash, 50% is sold to the glassmaking industry as it is a primary raw material for glass manufacture. The fortunes of the industry are therefore strongly tied to glass demand. Soda-ash also competes directly with sodium hydroxide as an alkali in many chemical processes. Sodium silicates are another important class of chemicals derived from soda-ash by reaction with silica at 1200–1400 °C. Silica-gel is a fine sodium silicate with a large surface area and is used in catalysts, chromatography and as a partial phosphate replacement in detergents and soaps.

6.5.4   *Electrolytic processes for chlorine/caustic soda*

6.5.4.1   *Introduction.*   Both chlorine and caustic soda have, at various times in the history of the chemical industry, been greatly in demand, but unfortunately for operators of electrochemical plants, not always at the same time. Chlorine has been valued as a bleach, or a raw material for the production of bleaching powder, as a disinfectant in water supplies and as a raw material for plastics and solvents manufacture.* Caustic soda has been used in the production of soda-ash, soap, textiles, and as a very important raw material in an incredible variety of chemical processes.

All the electrolytic processes have in common the electrolysis of salt to give chlorine and sodium hydroxide. The vast majority of production electrolyses a

---

* Greenpeace has described chlorine as a useless by-product of the production of caustic soda. However, in the past chlorine demand has often outstripped demand for caustic soda, and it looks like doing so again, at least for a short time.

solution of salt, but there are some significant plants that electrolyse molten salt to give liquid sodium and chlorine. These are used by industries that need the liquid sodium, mainly in the production of tetra-alkyl lead petroleum additives, though the petroleum additive companies are diversifying and other uses may appear.

There are essentially three different types of cell used for aqueous electrolysis: mercury cells, diaphragm cells and membrane cells. Membrane cells are really the only technology that is viable for new capacity in modern plants, but a large amount of old capacity still exists and many companies have not found it economical to replace even their mercury cells, despite the environmental implications. ICI, for example, though it is a leader in the supply of membrane-technology plants, has no membrane capacity installed at its major site in the UK except for the production of potassium hydroxide. Japanese environmental legislation has forced the industry there to move almost entirely to membrane cells, however. The production of chlorine via mercury results in up to 10 g of mercury being emitted per tonne of chlorine produced (actual amounts depend on the stringency of local regulations). Though the amounts are not thought to result in demonstrable harm, the industry is wary of anything that damages its environmental image and many viable mercury cells face being shut down in the near future.

All electrolytic reactions are based on the idea of using electrons as a reagent in chemical reactions. The basic reactions of brine electrolysis can be written as follows:

$$\text{Anode:} \quad 2Cl^- - 2e^- \longrightarrow Cl_2$$

$$\text{Cathode:} \quad 2H_2O + 2e^- \longrightarrow H_2 + 2OH^-$$

The overall reaction is:

$$2Na^+ + 2Cl^- + 2H_2O \longrightarrow 2NaOH + Cl_2 + H_2$$

This reaction has a positive free energy ($\Delta G = 421.7$ kJ/mol at 25 °C) and needs to be driven uphill by electricity.

Like many basic chemical processes, though the reaction appears to be gloriously simple, there are some significant complications. For a start, the reaction products need to be kept apart: hydrogen and chlorine will react explosively if they are allowed to mix. Chlorine reacts with hydroxide to give hypochlorous acid (HOCl) and chloride (both wasting product and creating by-products). The hypochlorous acid and hypochlorite (ClO^-) in turn react to give chlorate (ClO_3^-), protons and more chloride. Hydroxide reacts at the anode to form oxygen, which can contaminate the chlorine. All the reactions reduce efficiency and/or create difficult separation or contamination problems that need to be sorted out before any products can be sold.

The key to understanding the various types of process used for the electrolysis is the way they separate the reaction products. There are basically three types of electrolytic cell for brine electrolysis; though there are many variations of detail among the cells from different manufacturers.

6.5.4.2 The mercury cell.   The mercury cell was first used in Runcorn, UK, in 1892 at a site that is now part of ICI. The original process was known as the Castner–Kellner process, after its inventors, and the names are now commemorated in the name of the site.

A block diagram of a mercury cell is shown in Figure 6.4. The clever idea exploited in it is the use of mercury as the cathode. The use of mercury causes the reaction at the cathode to be slightly different: instead of the direct production of hydroxide ions, sodium ions are reduced to sodium metal in the form of a mercury amalgam. Because amalgamated sodium only reacts slowly with water, there is little problem of direct contamination of the reacting liquors with hydroxide. Since mercury is a liquid, it can flow through the cell continuously and the next part of the reaction can take place elsewhere, so maintaining a continuous overall process. The mercury flows into a chamber known as the denuding chamber where the reaction with water is promoted by graphite electrodes. Denuders are basically short-circuited electrochemical cells that greatly speed up the reaction of amalgam with water. The products are fairly pure hydrogen and sodium hydroxide solution. The mercury is recycled.

In practical terms a continuous flow of mercury used to be obtained in a 'rocking cell'. This was part of the original Castner–Kellner design. More modern cells use the simple device of a slightly sloped base of steel on which a thin film of mercury flows.

The advantages of the mercury cell are the good separation between the products, since they are produced in separate vessels. As a result the caustic

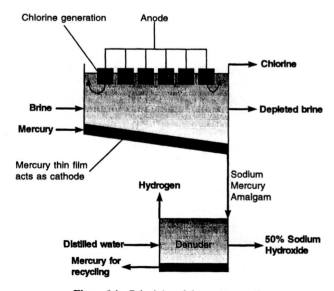

Figure 6.4 Principles of the mercury cell.

soda produced has very low levels of salt. The major disadvantage of the process is that it unavoidably discharges mercury into the environment, and—though the actual amounts in modern plants have not been shown to cause actual harm—this is not something the industry likes to be seen to do. Mercury also demands careful precautions within the plants if it is not to cause problems for the workforce.

6.5.4.3   The diaphragm cell.   A different principle lies behind the diaphragm cell (Figure 6.5). In it the reactants and products are kept apart by an inert diaphragm, usually consisting of asbestos on steel mesh. The liquid pressure is kept high on the anode side (the side where chlorine is produced), in order to maintain a constant flow through the diaphragm into the cathode compartment and thus prevent any back-flow of hydroxide. A disadvantage of the design is that it encourages flow of sodium chloride into the sodium hydroxide-producing side of the system giving, ultimately, an output that contains 15% sodium chloride and only 12% sodium hydroxide by weight. The solution can be concentrated to about 50% sodium hydroxide and only 1% sodium chloride by evaporation, which precipitates sodium chloride. Unfortunately, this may still be much too salty for some applications, and the purification process does waste energy.

6.5.4.4   The membrane cell.   The most modern type of cell is the membrane cell (Figure 6.6). What distinguishes the membrane cell from the diaphragm

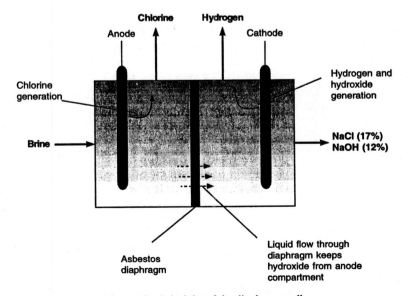

Figure 6.5   Principles of the diaphragm cell.

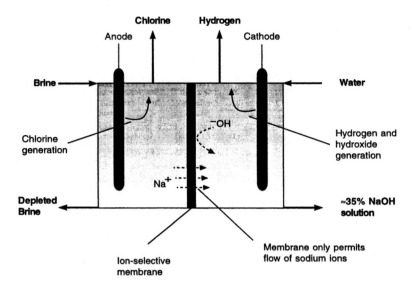

Figure 6.6   Principles of the membrane cell.

cell is the selectivity of the separating material. Membrane cells use Nafion membranes (a DuPont trade name) or similar materials (Asahi glass makes similar membranes), all of which consist of perfluorinated polymers containing sulphonate and/or carboxylate groups. Sophisticated membranes use a sandwich of different types of polymer so that current-carrying capacity, resistance and durability against attack by the surrounding media are maximised. These materials have the ability to transport sodium ions readily, but almost completely inhibit the transport of hydroxide. One consequence of this is that the resulting assemblies have a much lower resistance than diaphragms, giving better current efficiency. One disadvantage of the cells is that the membranes are damaged by some common contaminants of brine such as magnesium and calcium ions, so the brine must be purified to less than 0.1ppm of either. The membranes themselves are fairly expensive, though they may last several years in actual service.

Membrane cells can produce relatively pure (ppm levels of sodium chloride) caustic soda and chlorine at lower costs than the other methods. However, the difference in cost is not so great as to warrant dismantling existing working plants using other technology. There are some other advantages of membrane technology: for example, ICI's engineers pioneered a break from the heavy-engineering tradition of old designs of cell, and produced a modular range of cells that could be easily assembled to make a small or a large chlor-alkali plant. This helped to make smaller-scale local facilities viable.

There are some technical problems that are shared between all the cell types. A particular issue is the material for the electrodes. Anodes used to be made out of graphite, but they wore out, altering the resistance characteristic of the cells and making them less efficient. Later designs have used more robust materials such as titanium and nickel, coated with appropriate electro-active materials. Coatings are often based on platinum-group metals. They tend to wear out at about the same speed as the membranes, but need careful justification as to their economic value. Some membranes help improve current efficiency and are justified when power is costly; in locations with cheap power, the extra expense of a sophisticated membrane coating may not be justified.

6.5.5   The uses of chlorine and sodium hydroxide

Sodium hydroxide has so many chemical uses that it is difficult to classify them conveniently. One of the largest uses is for paper-making, where the treatment of wood requires a strong alkali. In some countries this consumes 20% of production. Another 20% is consumed in the manufacture of inorganic chemicals such as sodium hypochlorite (the bleach and disinfectant). Various organic syntheses consume about another fifth of the production. The production of alumina and soap uses smaller amounts.

Chlorine is widely used in a variety of other products. About a quarter of all production world-wide goes into vinyl chloride, the monomer for making PVC. Between a quarter and a half goes into a variety of other products. Depending on the country, up to 10% goes into water purification. Up to 20% goes into the production of solvents (methylchloroform, trichloroethene, etc.) though many of these are being phased out because of the Montreal Protocol. About 10% world-wide goes into the production of inorganic chlorine-containing compounds. A very significant use in some countries is for the bleaching of wood pulp, though this is another use coming under environmental pressure.

Figure 6.7 summarises many of the uses of chlorine.

6.6.1   The wider world

**6.6
Conclusion**

In a short chapter, it is impossible to go into the detail of every chemical that might be classified as a bulk chemical. The key chemicals that are the first stage from raw materials can be covered, but the dozens of chemicals that lie between these and the chemical products such as polymers with which people tend to be familiar are too many and various to be dealt with in any depth. However, to leave them out is to miss an important lesson about the inter-relationships of the various key chemicals and just how important they are as intermediates. Therefore, the final part of this chapter looks briefly at the synthesis of just one chemical intermediate, methyl methacrylate, to illustrate the wide variety of —often surprising —ways in which the key basic chemicals are combined to produce the myriad variety of more familiar compounds.

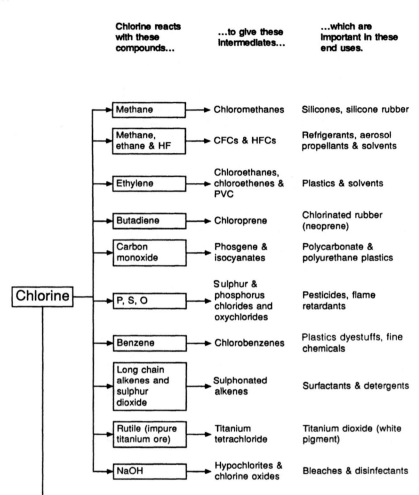

Figure 6.7   Major uses of chlorine.

6.6.2   The synthesis of methyl methacrylate

Methyl methacrylate (CH_2=$CH(CH_3)COOCH_3$, MMA) is the monomer from which the transparent plastics known by their brand names Perspex, Plexiglass and Lucite are made. Though the compound itself contains only oxygen, carbon and hydrogen, both nitrogen-containing and sulphur-containing compounds are used in the synthesis.

Just one of the possible routes to MMA will be examined: the route is known as the acetone–cyanohydrin route. There are alternatives, but none is widely operated worldwide at the moment.

The basic chemicals (and petrochemicals) needed to make MMA are: benzene, propene, ammonia, sulphuric acid, methane, methanol and air. The two key intermediates are hydrogen cyanide and acetone. Figure 6.8 summarises the key transformations.

Cumene (isopropylbenzene) is made from benzene and propene using acid catalysts such as polyphosphoric acid. Cumene can be oxidised to cumene hydroperoxide by a radical chain reaction using air as an oxidising agent. The hydroperoxide can then be decomposed by sulphuric acid to give acetone and

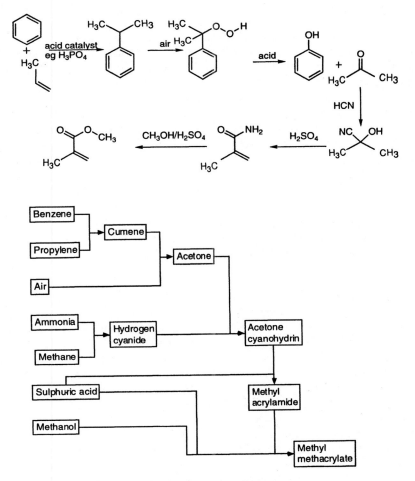

Figure 6.8 The synthesis of methyl methacrylate.

phenol (which is used in resins). About one-third of all the acetone made is consumed by the manufacture of MMA. Since the sulphuric acid used is recycled, this reaction consumes only oxygen from the air in addition to the primary starting materials. The overall reaction can be considered a convenient way to oxidise both starting materials. It is an interesting example of how different parts of the industry are linked together by the drive to use efficient, simple, cheap reactions to elaborate basic raw materials.

The other key intermediate comes from ammonia and methane. Ammonia and hydrocarbons can be oxidised, in a remarkable reaction, over a platinum-based catalyst to give compounds containing cyanide groups. Methane gives hydrogen cyanide and propene gives acrylonitrile (CH_2=CHCN, the polymer of which is the basis for acrylic fibres).

Methanol is made in several different ways. Most, however, is made from mixtures of carbon dioxide, carbon monoxide and hydrogen over various catalysts, one of the most successful of which is the ICI zinc oxide–copper catalyst. The production of MMA is not a major user of methanol, as the product is widely used in many other areas. One rapidly growing one is the synthesis of methyl *tert*-butyl ether (MTBE) which is used in large quantities as an octane enhancer in unleaded petrol.

In the next step, hydrogen cyanide is reacted with acetone to give the cyanohydrin addition product. This is hydrolysed with sulphuric acid in the presence of methanol to give MMA, as Figure 6.8 shows. The route gives a large amount of ammonium sulphate as a by-product, which has often simply been dumped at sea, but which is now mostly recycled. ICI has just brought on stream a new MMA plant with a built-in recycling unit (which doubles the cost of the plant, but saves buying in fresh sulphuric acid) on Teesside.

The cyanohydrin route to MMA illustrates how a few simple products of the basic chemical and petrochemical industries can come together to make much more complex products. Far from being dull and declining, basic chemicals are the foundation on which the rest of the chemical industry is based.

6.7
The future for basic chemicals

The basic chemicals industry is now facing one of the biggest challenges in its history. The main consumer of many of the industry's key products—the agriculture industry—has stopped growing and is severely cutting back its demand for fertilisers. Western farmers have been producing too much food and governments have been cutting subsidies, with the result that less land is being farmed and less fertiliser used. Environmental concerns about the effects of excessive fertiliser run-off have also reduced demand for fertilisers.

Products such as chlorinated compounds have come under threat from environmentalists. Some will be banned under the Montreal Protocol, but others are not harmful and may survive environmentalist pressures. The industry can no longer rely on long-term growth in demand.

The industry may well see increased consolidation as companies swap plants to achieve better economies of scale or better market positions in specific products. This could leave an industry with far fewer players but with a better balance of supply and demand and better profitability. The industry will move more to serving the rest of the chemical industry and less to serving the farming industry.

Another threat is the perceived environmental messiness of many large-scale processes. Despite the relative efficiency of many big plants, the industry has a long way to go to achieve the best environmental standards possible. The drive to increased recycling and the ideal of emission-free plants will be a major factor influencing the development of the industry in the next decade.

Technical developments will not stop. There will be increasing emphasis on plants and processes that do not pollute. Companies will compete on efficiency: those able to produce the best quality products at the cheapest price will prosper. This will require companies to keep investing in technical improvements. New ways of bringing basic chemicals together to form useful intermediates will be found.

There is still much to do in the basic chemicals industry.

A very useful one-volume reference work which covers a vast variety of industrial chemical processes (though it has a disturbing number of typographical and trivial errors) is: 'Survey of Industrial Chemistry,' Philip J. Chenier, Wiley Interscience, 1986.

The key reference source for the details of processes and end uses is: 'Ullmann's Encyclopedia of Industrial Chemistry,' Barbara Elvers *et al.*, eds, 25 vols, VCH.

Up-to-date information about the industry can be found in various journals such as: *Chemistry and Industry* (published every two weeks, good on UK & European news and general industry information); *European Chemical News, Chemical Week* (both good for general industry information; *Chemical and Engineering News* (very good on collected statistics, but biased to USA data).

A useful—though very detailed and technical—guide to some specific manufacturers' processes is: 'Handbook of Chemical Production Processes', Robert A. Meyers, McGraw-Hill, 1986.

Bibliography and recommended reading

7 Agrochemicals

Alan Heaton

Alan Heaton

**7.1
Introduction**

7.1.1 Origins of the industry

There are numerous references throughout history to catastrophic crop losses ranging from the effects of plagues of locusts in Egypt in biblical times to the Irish potato famine, caused by the fungal disease 'potato blight,' in the 1840s. Until the 19th century, except for the occasional 'natural disaster,' traditional methods of agriculture like the crop rotation system had proved adequate in feeding the population, by maintaining soil fertility and controlling the spread of crop diseases. With the drift of workers from the land to the cities, plus the spiralling population, in the late 19th and early 20th centuries, a new approach to agriculture was necessary in order to produce more food. This has resulted in major changes in the pattern of agriculture, some of which have created new problems.

(*a*) *Increased size of units and mechanization.* The size of fields has been increased to cater for the large machinery used for ploughing, sowing and harvesting. Consequent removal of hedgerows has reduced the bird population and hence the level of insect predation. Also, heavy farm machinery can cause damage to the soil structure particularly when operating in wet conditions.

(*b*) *Monoculture.* This has been practised, especially in the United States, for a number of years, and consists of planting vast areas of land with just a single crop like soya beans. It contrasts with nature where diversity is the norm, and if adequate protection measures are not taken may well result in a greater susceptibility to disease and insect attack than if a mixture of crops was grown.

(*c*) *Intensive farming.* This requires the maintenance of soil fertility by application of large amounts of artificial fertilizers. Loss of soil structure may result (see below).

The agrochemical industry developed to assist in providing more food by producing chemicals which maximize yields of crops while minimizing production cost. This may be achieved both by assisting growth by the application of artificial fertilizers and plant growth regulators, and by minimizing the attack of pests on crops through the use of pesticides.

Two ways of providing this assistance may be distinguished. Firstly, artificial fertilizers may be provided which replace the nutrients in the soil extracted by growing plants. Modern intensive farming results in such rapid depletion of nutrients that natural replenishment is inadequate. Therefore to maintain crop growth artificial fertilizers must be added to the soil. Problems resulting from the collapse of soil structure, which is caused by the organic material in the soil being used up, are not so easily resolved. Indeed, some areas in the Midwest of the United States illustrate very clearly the difficulty. This is an important area for growing cereals, and intensive farming over many years has left gigantic dust bowls because the soil now contains virtually no organic material. The second way in which the industry assists in maximizing crop yields is by supplying pesticides, i.e. chemicals which prevent pests from competing for or destroying crops.

Division of agrochemicals into two sections, artificial fertilizers and pesticides, is justified for several other reasons. One of these is the scale on which the chemicals are produced. Fertilizer production is a good example of very large-scale chemical production. Much of it resolves around the manufacture of ammonia—these plants are the giants of the chemical industry, with a single plant easily capable of producing 1000 tonnes per day (see sections 6.3.4 and 6.4.2 of this book and section 11.4.1.2 of Volume 1).

Production of individual pesticides rarely exceeds 1 000 tonnes per year, and is typically hundreds of tonnes.

A second contrast is the nature of the compounds. Those used as fertilizers consist largely of a small number of simple inorganic compounds such as ammonium salts, nitrates and phosphates. Modern pesticides are virtually all organic compounds and the range of chemical structures which they cover is very diverse, encompassing practically all areas of organic chemistry.

Thirdly, the compounds used as fertilizers and their manufacturing processes are now well established, i.e. mature. Therefore the little research that is undertaken is directed mainly at improving process efficiency. New and improved pesticides on the other hand are constantly required. This, coupled with the diversity of structures already referred to, explains why this is one of the most research-intensive areas of the chemical industry, with hundreds of thousands of new compounds being synthesized and screened for biological activity each year.

This chapter will largely concentrate on pesticides, but it is of interest to consider fertilizers briefly.

7.1.2 Artificial fertilizers

In order to appreciate the function of artificial fertilizers it is instructive to look at plant nutrition and some natural cycles which replace nutrients removed from the soil by growing crops.

Plants grow through photosynthesis, the process whereby carbon dioxide and water, in the presence of sunlight and with the aid of chlorophyll, are built up into complex organic compounds such as glucose and cellulose. In addition to carbon, hydrogen and oxygen, many other elements (trace elements) are required in various quantities ranging down to only a few parts per million. Those required in relatively large amounts are termed essential elements: C, H, O, N, P, K, Ca, Mg and S. The trace elements include Fe, B, Mn, Zn, Mo, Cu and Cl. Lack of these elements, however small the quantities required, can have a serious adverse effect on a plant's growth.

Plant nutrition is therefore concerned with ensuring that the growing crop receives adequate supplies of these elements in addition to sunlight and water. The plants themselves participate in a number of natural recycling processes. The carbon and nitrogen cycles are examples and are summarized below.

Carbon cycle.

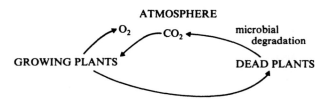

Nitrogen cycle. It is one of the ironies of nature that unlimited supplies of nitrogen are available in the atmosphere, yet this cannot be utilized directly by plants. The nitrogen first has to be fixed, i.e. converted into a form which the plant can use—this means into ammonium salts or nitrates. Atmospheric electrical discharges, i.e. lightning, bring this about and certain bacteria, e.g. *Azotobacter*, can also fix nitrogen. The generalized figures below give an idea of the extent of these inputs.

Nitrogen input to soil
Microbiological fixation	$20 \, \text{kg ha}^{-1} \, \text{y}^{-1}$
Atmospheric electrical discharge	$2 \, \text{kg ha}^{-1} \, \text{y}^{-1}$
Nitrogen loss from the soil	
Growing crops	$25–50 \, \text{kg ha}^{-1} \, \text{y}^{-1}$

Therefore to maintain soil fertility the difference must be made up by the application of artificial fertilizers.

Nitrogen fertilizers are based on the Haber process for producing ammonia from nitrogen and hydrogen. Although this is a very energy-intensive process it is well established, and full benefit has been taken of the economy of scale effect (see section 5.7, Volume 1). The ammonia is then converted into ammonium salts and nitrates:

$$NH_3 \underset{\searrow}{\overset{\nearrow}{\longrightarrow}} \begin{array}{l} (NH_4)_2SO_4,\ (NH_4)_3PO_4 \\ NH_2CONH_2 \\ NO_3^- \to NH_4NO_3,\ NaNO_3 \end{array}$$

(see sections 6.2 to 6.4)

Phosphate fertilizers are derived from rock phosphate and superphosphate which are mined in several areas of the world, e.g. Morocco. Potash fertilizers are again mined directly. Compound fertilizers having specified N : P : K ratio are also produced. Various ratios are formulated for particular applications.

7.1.3   Pesticides

Until the 1930s pesticides were largely inorganic compounds, such as Bordeaux mixture (a copper sulphate/lime mixture) which was used to control fungal attack on grapes, organomercurials as a fungicidal seed dressing, and calcium arsenate for controlling boll weevil on cotton. By today's standards some of these heavy metal salts are exceedingly toxic and dangerous. One or two naturally-occurring organic insecticides had also already been used for quite a number of years. Examples are pyrethrum from the flowers of the chrysanthemum and rotenone from the roots of the derris plant.

Since the 1930s almost all the pesticides which have been developed are organic compounds, with notable examples being DDT, the chlorinated phenoxyacetic acid and bipyridyl herbicides, and glyphosate.

Before considering the characteristics and present position of the industry we must first see the way in which pesticides are subdivided. The word 'pesticide' is a blanket term which covers any chemical, or indeed any agent, which is used for pest control. The pesticide is normally fatal to the pest. Pesticides are therefore divided into groups based on the pest which is being controlled. The three main groups are fungicides (fungal spores), herbicides (weeds) and insecticides (insects). Other groups include nematicides (worms), molluscicides (slugs and snails), acaricides (mites) and rodenticides (rodents). This chapter will only consider the three main groups.

Development of the modern agrochemicals industry really only began with the advent of synthetic organic pesticides in the 1930s. Discovery of the dithiocarbamate fungicides and the insecticidal activity of DDT was followed by the chlorinated phenoxyacetic acid herbicides (2,4-D and MCPA) and hexachlorocyclohexane (BHC). These pesticides are still important products and they have been joined by many others, including the organophosphate insecticides, and triazine, bipyridyl and glyphosate herbicides, to name but a few.

In order to get a feel for the size, importance and characteristics of the industry it is instructive to consider some statistics.

**7.2
Present position of the industry**

Table 7.1 UK pesticide sales, 1991

Total sales £1 172·4 million	UK sales £415·8 million (35·5%)	Fungicides £130·0 million (31·3%)
		Herbicides £206·7 million (49·7%)
		Insecticides £43·4 million (10·4%)
		Others £35·8 million (8·6%)
	Exports £756·6 million (64·5%)	Fungicides £90·9 million (12.0%)
		Herbicides £397·5 million (52·5%)
		Insecticides £230·7 million (30·5%)
		Others £37.5 million (5.0%)

7.2.1 *Financial importance of pesticides*

Tables 7.1 and 7.2 relate respectively to the UK only and to world industries. The sales are further divided amongst the three major groups of pesticides plus other pesticides.[1] (Note that world sales figures are in $US.) UK production probably accounts for about 7% of world sales; the volatility of exchange rates makes a precise judgement difficult.

As far as the UK is concerned, for both home usage and export herbicides are by far the most important group of pesticides. Sales of fungicides in this country are now approximately three times those of insecticides. Reversal of

Table 7.2 World pesticide sales, 1991

Total sales $26 800 million	Fungicides $5 548 million (20·7%)
	Herbicides $11 792 million (44·4%)
	Insecticides $7 745 million (28·9%)
	Others $1 581 million (5.9%)

this situation for exports is clearly apparent. Breakdown of the sales figures into home sales and exports shows that the latter are far more important.

Turning to the world-wide situation, herbicides are again the single most important group of pesticides. However, the sales value of insecticides is not far behind and reflects the severe problems that insects cause in climates warmer than our own temperate one. Not only is there interference with growing crops (e.g. by locusts), but in tropical countries insects are responsible for the transmission of several virulent diseases. Malaria is perhaps the best known of these, and programmes for its eradication have represented a major outlet for insecticides.

7.2.1.1   Geographical usage of pesticides.   If total worldwide pesticide usage is broken down geographically the following figures are obtained for 1991:[1]

	%
Western Europe	31·7
North America	26·0
Far East	22·5
Latin America	7·9
Eastern Europe (including former USSR)	6·3
Rest of World	5·6

Two conclusions may be drawn from these figures.

(i) The most important markets for pesticides are the developed parts of the world, and the North American market is substantial. Hence for any pesticide to be a major sales success it must satisfy the stringent regulations of the EPA (Environmental Protection Agency) in the United States so that it may be marketed there.

(ii) There is a low rate of usage in the underdeveloped parts of the world. This is well illustrated by Africa and the Middle East, whose land area well exceeds that of North America yet whose usage of pesticides is only about one-seventh as great, and by the enormous area covered by the Far East. It is not just coincidence that crop losses in these areas are very much higher than those in North America and Western Europe. Losses during growing, harvesting and storage of crops average about 25% of total production in Europe and America, and about 45% in Africa and the Far East.

7.2.1.2   Pesticide usage by crop.   The 1991 figures[1] were:

	%
Fruit and vegetables	23·9
Cereals	17·0
Others	14·5
Rice	11·8
Cotton	11·0

Maize	9·5
Soybeans	7·2
Sugar beet	3·2
Rape seed	2·0

All the crops listed are grown as food with the exception of cotton. The first two uses in the rank order clearly cover a variety of crops and so the great importance of maize and rice as foodstuffs is evident from the large-scale usage of pesticides—mainly herbicides plus some fungicides—which emphasizes the vast scale on which these crops are grown.

7.2.2. *General characteristics of the agrochemicals industry*

It is convenient to consider the general characteristics of the agrochemicals sector under a number of headings and to emphasize the contrast with some other sectors of the chemical industry. These headings are (i) scale of production, (ii) relative value of products, (iii) research intensiveness, and (iv) convergent nature.

(i)   Scale of production.   This has already been touched upon at the beginning of this chapter and was one of the reasons for dividing agrochemicals into fertilizers and pesticides. Pesticide production is a typical fine-chemical operation and lies generally in the hundreds to a few thousand tonnes per annum range. There are, of course, several exceptions, such as DDT, chlorinated phenoxyacetic acids, bipyridyls and glyphosate. The contrast with fertilizer or petrochemicals production is obvious, and is reflected in the nature of the processes, which are largely batch and much less likely to be automatically (computer-) controlled. Modern large-scale production of petrochemicals represents the other extreme with its large, single-product, computer-controlled, continuously operating plants.

(ii)   Relative value of products.   Pesticides have a relatively high value, selling for between several thousand and several hundred thousand pounds sterling per tonne. Again the contrast with basic petrochemicals, many of which sell for several hundred pounds per tonne, is illuminating. This high value per unit weight coupled with the relatively small scale of production accounts for the fact that plants making pesticides have few restrictions on their location, which may well be in rural areas. Large-scale chemicals production, e.g. of ammonia, is in contrast located at the mouths of river estuaries (see also section 3.3.3, Volume 1).

(iii)   Research intensiveness.   Attention was drawn several times in Volume 1 (see for example Table 4.1, p. 90) to the research-intensive nature of the

chemical industry. In the difficult periods which the industry has faced during the early 1980s, and the late 1980s, and early 1990s, research and development have suffered some cutbacks. The situation prior to this may be considered more typical. For the 1970s, investment in research and development expressed as a percentage of *sales income* was as follows:

	%
Whole chemical industry	5
Pharmaceuticals	10
Agrochemicals	7–8
(Engineering industry	1)

Research and development is clearly an extremely important activity and for agrochemicals (and pharmaceuticals) it has rightly been described as the lifeblood of the industry. Indeed, in the UK agrochemical industry 27% of all employees work in research and development.

The reasons for this great emphasis on research to find new products are: the great range of pests; the ever-increasing need for greater selectivity in pesticide action; problems of resistance to existing pesticides; and finally the requirement to produce the pesticide at an acceptable price. The last-named is an extremely important point; some very efficient new pesticides have been patented but have never found their way on to the market simply because they cannot be made cheaply enough. A price of about £5 per kilogram is a current target.

Space permits consideration of only a limited number of pesticides later in this chapter; nevertheless the extremely diverse collection of chemical structures found among compounds possessing pesticidal activity will soon become apparent.

Some similarities between the pesticide industry and pharmaceuticals have already been indicated in this section. This is not surprising since both types of products are aimed at producing a selective biological effect, i.e. killing or controlling the pest (e.g. an insect or a bacterium) in the presence of other living species which should ideally be totally unaffected. Many of the major companies which are active in one area also have strong interests in the other. Examples are ICI, Ciba-Geigy, Dow, Bayer, and Hoechst. This facilitates the provision of new compounds for screening for biological activity since those synthesized as potential drugs may be passed on for screening for pesticidal activity and vice versa. It is extremely rare to find a compound which is suitable for use as both a drug and as a pesticide. Warfarin is, however, a notable exception, having been used as both an anticoagulant in the treatment of thrombosis and as a rodenticide.

(iv) Convergent nature. The pesticide sector of the chemical industry is described as a convergent industry. This is because it brings together an extremely wide range of chemical products on to a group of closely related

biological targets. Note that here the relationship is biological and economic and not chemical. In contrast, petrochemicals can be described as a divergent sector.

7.2.3   Control of the use of pesticides

The manufacture and use of any chemical product depends on technical, economic and social factors and the inter-relationship, and often the conflicting interplay, between these. It is in the area of pesticides that the influence of social factors has had the greatest impact, with the possible sole exception of pharmaceuticals. Indeed, agrochemicals have been described as the most highly regulated sector of the chemical industry. Since pesticides are by definition toxic (to the pest), it is reasonable that some control should be exercised over their use. In the United States pesticides cannot be marketed until they have been cleared and approved by the EPA, i.e. there is a legal obligation.

The situation regarding control of pesticide usage in the UK changed markedly in 1986. Until then, new pesticides could be freely introduced to the market although in practice the major companies all sought approval from the Ministry of Agriculture, Fisheries and Food (MAFF) through the Pesticides Safety Precautions Scheme (PSPS). It should be noted, though, that in contrast to the situation in the USA, there was no legal obligation to do this—it was purely voluntary. Following pressure from both the agrochemical industry and the environmentalist lobby the government introduced legislation, the Control of Pesticides Regulations 1986, which effectively made the PSPS scheme mandatory. However, it went much further than this, imposing a legal requirement for pesticides to be approved before they can be supplied, stored or moved, and only those with provisional or full approval may be sold or advertised. Furthermore, those who sell, store or use these products are required to be trained first.

A company wishing to introduce a new pesticide must submit full details of all the extensive tests carried out to the regulatory authority. They will then decide whether to approve the product or perhaps to require further tests to be carried out. There is concern that in the UK it is taking 3–4 years for a decision to be reached, eating into the patent life of the product (see section 7.3.3). A full scientific review of the pesticide is required after 10 years.

Further changes are likely after the implementation of EC Registration Directive on Plant Protection Products (Pesticides) on 26 July 1993. This details the European Community procedure for approval for new active ingredients, etc., and should allow products that have been approved in one member state to be freely marketed in others.

Before moving on to look in detail at the discovery and development of new pesticides it is worth while briefly reviewing public concern over the safety of chemicals in, and their effect on, the environment. This topic was touched on

Table 7.3 Requirements for worldwide registration of a pesticide—toxicology studies

1950	1960	1970
Acute toxicity	Acute toxicity	Acute toxicity
30–90 day rat feeding	90-day rat feeding 90-day dog feeding	90-day rat feeding 90-day dog feeding
	2-year rat feeding 1-year dog feeding	2-year rat feeding 2-year dog feeding

		Reproduction:
		  3 rat generations
		Teratogenesis in rodents
		Toxicity to fish, shellfish
		  and birds

briefly in chapter 1. As we shall see, public attitudes have had a major effect on both the time-scale and the cost of developing new pesticides.

Until 1962 there was little interest in, or concern for, the effects of pesticides on the environment. In that year the American marine biologist Rachel Carson published her book *Silent Spring*, in which she drew attention to the damaging effects which one group of insecticides—the organochlorines—had had on the environment. This triggered public interest in the use and safety of chemicals in the environment, and the growing pressure from the environmentalist lobby has resulted in tighter and tighter controls on the use of pesticides and in a few cases in complete banning. It is an area in which emotions have run high and occasionally scientific facts have been ignored. The effect on the development of new pesticides of this public pressure for greater safety is well illustrated in Table 7.3, showing the changing requirements over the years of just one of the many tests which are now routinely carried out.

Registration costs (mainly toxicological and ecological studies) now constitute 25–30% of the total cost of developing a new pesticide. The consequences of this almost exponential rise in the amount of testing carried out will be considered at the end of section 7.3.

The great importance of research directed towards developing new pesticides has already been emphasized. The purpose of the present section is to discuss the way in which this work is carried out and the factors that have influenced and are still influencing the cost, the time-scale and the nature of the testing programme. Effects of public opinion, introduced in the previous section, will be much in evidence.

It cannot be emphasized too strongly that nowadays the primary objective is to produce a pesticide which is selective in its biological action. In this way adverse effects on other living things and the environment will be minimized.

7.3 Discovery and development of new pesticides

As we shall see later, most of the widely publicized problems in the use of just a few pesticides result from a lack of this selectivity. Just one example illustrates why it can be so difficult to achieve the required degree of selectivity, although tremendous advances have been made during the last two decades. Wild oats are a serious pest (i.e. a weed) encountered by farmers growing cereal crops in Britain. Although there would seem to be little difference genetically between wild oats and cultivated oats which are a valuable crop, herbicides are required to kill one and not in any way harm the other—clearly a very difficult task for the research scientists. However, this degree of selectivity has been achieved and suitable herbicides have been marketed for several years now.

7.3.1   Outline schemes

Two approaches to the discovery and development of new pesticides will be considered. First, in order to illustrate how difficult it is to produce an acceptable pesticide, an exercise in listing the properties which the chemical requires will be carried out. A generalized scheme showing the main stages involved will then be considered.

In this illustrative example let us assume that the pesticide is to be an insecticide which will control caterpillars, and that it will act by attacking the caterpillar's stomach. What are its desirable and/or essential properties?

The main properties only would be the following:

(i) Low mammalian toxicity
(ii) No phytotoxicity
(iii) Stability to UV light and resistance to weathering
(iv) No repulsiveness to caterpillar
(v) Resistance to degradation by changes in pH and by plant and caterpillar enzymes
(vi) Penetration of the gut
(vii) No effect on beneficial insects and other forms of life
(viii) (Ideally) inactivation on contact with the soil.

Several of these properties relate directly to the key point of selectivity of action. Thus (i) ensures that accidental ingestion by domestic animals or humans does not have serious consequences; (ii) means that the chemical will not be toxic to the growing crop, e.g. cabbages. Properties (vii) and (viii) are self-explanatory and clearly also influence selectivity. Property (iii) means that the chemical will not be degraded by the UV light in sunshine to a non-pesticidal product, nor will it be, say, washed away by heavy rainfall before it has a chance to do its job. The reasons for the other desirable properties are obvious.

Bearing in mind that this is not a comprehensive list, but merely includes the main properties which the pesticide should have, it can easily be appreciated why it is so difficult to produce pesticides with perfect selectivity—the ultimate objective of the industry.

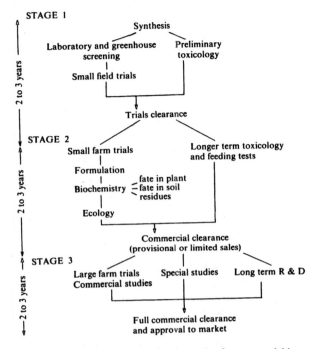

Figure 7.1 Scheme for the development of a new pesticide.

Turning now to the various stages involved in developing and testing a new pesticide, the generalized scheme in Figure 7.1 outlines the main tests which any candidate chemical must undergo before acceptance as a pesticide. Note that the further down the scheme one progresses the more expensive each test becomes, with costs rising very rapidly to exceedingly large figures. For example, world-wide evaluation alone will now cost well in excess of £1 million. Clearly, therefore, if a candidate chemical is not to reach the market-place, for whatever reason, the earlier in the scheme that the problem causing its rejection is detected, the less money will be wasted on testing it.

Let us now consider some of the activities in the scheme in more detail.

The first step is for a team of synthetic organic chemists, having discussed likely structures with their biologist colleagues, to synthesize a large number of compounds which they think will have pesticidal activity. The pool of compounds will be swelled by the efforts of researchers in universities. These are then handed over to the biologists to carry out preliminary screening tests against a representative selection of fungal spores, insects and weeds. This first screening test is carried out at fairly high concentration levels of several hundred ppm. Since a pesticide requiring application at this level would not be commercially viable, if pesticidal activity is found, further tests at levels closer to the commercial are carried out ($\leqslant 100$ ppm).[2] Success here initiates preliminary toxicological studies and small field trials on plots at the company's

Table 7.4   Summary of Figure 7.1

Time	Stage	% of total cost
Year 1 to Year 3	Discovery to trials clearance	20
Year 4 to Year 6	—to limited commercial clearance	30
Year 7 to Year 9	—to full commercial clearance	50
Total 9 Years	Approval	

research station. A successful lead like this will result in many homologues and analogues being synthesized and also tested in order to try to maximize the biological activity. These studies continue for 2–3 years.

Stage 2, which takes at least another two to three years of continuous testing, involves a wide variety of scientific personnel, e.g. toxicologists, agronomists (for assessing the small farm trials), biochemists (studying the degradation of the compound and build-up or otherwise of residues), ecologists (considering the chemical's effect on the environment) and physical chemists (formulation). The last-named studies are very important, since the way in which the active pesticide is formulated—for example as an aqueous spray, an oil-based spray (or emulsion) or impregnated on an inert solid—can have a considerable impact on its activity and hence its rate of application (or concentration in ppm) and finally on its cost. It is an important area of application of colloid and surface chemistry. (However efficient a pesticide, unless it is correctly applied it cannot function properly. There is therefore a considerable effort also devoted to the technology of, and the machinery used for, application of pesticides.)

Finally, Stage 3 covers only a small number of tests but each is very extensive and very expensive. Only at the end of this is the pesticide ready for manufacture and marketing.

A brief summary of the scheme and the apportioning of the total costs between the three stages is given in Table 7.4.

A more detailed breakdown of estimated costs for 1970 is given in Table 7.5, showing that production of one marketable product cost $5.482 million. Note that to the development costs must be added the considerable cost of constructing a plant in which to manufacture the pesticide. An updating of the main costs, the success rate and the time scale follows in section 7.3.2.

It is evident from this section that the discovery and development of a new pesticide requires collaboration between a wide range of scientists—organic and physical chemists, biochemists, biologists, ecologists, toxicologists, etc.—and commercial personnel—market researchers, economists, patent agents etc. Therefore it is very much a large team effort. Equally apparent, even from

Table 7.5 Breakdown of development costs of a pesticide (1970)[3]

Year	Activity	No. of CPDs	Cost per CPD $	Total cost $000s
1	Synthesis	8 000	140	
	Screening (survival rate 1 *in* 100)		65	1640
2	Glasshouse trials	80	5 000	460
	Initial field trials (1 *in* 5)			
3	Field trials	16	5 000	
	Initial toxicology (1 *in* 4)		2 000	112
4	Field evaluation	4	50 000	
	Toxicology		20 000	
	Formulation and process (1 *in* 2)		50 000	480
5	Worldwide evaluation	2	500 000	
6	Toxicology, environment, ecology		350 000	
	Formulation and process		200 000	
	Production		200 000	
	Registration and patent (1 *in* 2)		175 000	2 850
		1		5 482

the 1970 figures, is the long time-scale, the high cost and the considerable risk involved in this business.

7.3.2 Cost, timescale and success rate

In 1970, as Table 7.5 shows, to produce one new marketable pesticide on average a company started with 8000 new compounds and carried out 6 years of continuous testing at a total cost of $5·5 million. Since then, public concern about the safety of chemicals in general and the effect of pesticides (e.g. DDT and 2,4,5-T, or Agent Orange) on the environment in particular has grown considerably. This has resulted in the need for much more extensive and demanding testing procedures. It is particularly evident in the area of toxicology, which Table 7.3 has already clearly demonstrated. Several major consequences of this are apparent.

(i) Lower success rates
(ii) Increased time-scale
(iii) Increased costs
(iv) Reduction in the number of companies carrying out basic research
(v) A fall in the rate of introduction of new pesticides.

(i) Lower success rates. The more tests carried out on the chemical, the more likely it is to fail one and be rejected. It is also somewhat ironic that the success of analytical chemists in developing methods such as gas–liquid

chromatography, which can detect microgram quantities of pesticide residues, has led to the laying down of lower acceptable levels for these residues and as a result the lowering of the success rates for biologically active compounds. Currently the success rate has fallen to about 1 in every 10 000 compounds tested.

(ii) Increased time scale and (ii) increased costs.   These are a direct result of the more extensive testing programme. Current estimates are that it now takes 10 years and costs in excess of £70 million to develop one new pesticide. Of course, this figure includes the cost of testing the other 9 999 compounds which were rejected at some stage.

(iv) Reduction in the number of companies carrying out basic research.   The lowering of the success rate coupled with the increasing cost and time-scale has meant that the pesticide business has become more risky. It is therefore not surprising that the number of companies world-wide carrying out basic research work has fallen from about 50 in 1970 to about 20 at the time of writing. The latter are invariably giant multinational companies which operate in virtually all sectors of the chemical industry and are therefore able to carry the risk of their agrochemical operations. It does appear, however, that the companies that have survived are very competent and are operating profitably in this area.

(v) Fall in the rate of introduction of new compounds.   Although the considerably expanded test programme is one reason for this, there are several others. For example, since most of the compounds having fairly simple structures have already been tested, this means that compounds now being synthesized are structurally more complex. They therefore have longer and more expensive synthetic pathways. For a pesticide to succeed commercially, farmers must buy it, and they will only do this if the price paid for it is more than compensated for by the increase in yield of crops as a result of its pesticidal action. Hence the need to keep the selling price as low as possible. Many perfectly efficient pesticides have had to be rejected purely because their manufacturing cost was too high.

During the 1960s 20–25 new pesticides were registered each year, but in the 1970s and 1980s this had fallen to 10–12, and this falling rate is clearly a matter of increasing concern.

7.3.3   Patents position

As the time taken to test a pesticide before it can be marketed increases (and also the time for regulatory approval), a worrying consequence is that the remaining time for which it is covered by patent protection becomes smaller. It is during this time that the development cost of £50 million or more will be recovered and the product can start to generate profit. According to the figures already given in the previous section, this means that as little as seven years of protected sales will be available before the patent runs out and other companies, who have had no development costs, may start producing the same pesticide and marketing it. There is also always the danger that a competitor may bring out a better, alternative product.

This eroding of the patent-protected marketing life because of the insistence on longer and greater testing means that the chance of a pesticide becoming a commercial success is reduced, and so there is strong pressure by bodies such as the British Agrochemicals Association for extending patent life by a further 4–5 years. It is claimed that this will restore the patent-protected marketing life of a pesticide to what it would have been if the extensive testing to satisfy regulatory authorities was not required. That is not to decry this testing, which everyone agrees is important in producing safer pesticides, but merely assure a protected market life of reasonable length.

Even more worrying is the fact that some countries have reduced or even withdrawn patent protection. This started with India in 1972, was followed by Mexico in 1975 and has been continued in South America (by the Andean Pact Commission).

7.3.4   Summary

We have seen in section 7.3 that the major aim in pesticide development is perfect selectivity of biological action. This is very difficult to achieve, and most of the problems encountered in past pesticide usage are due to lack of selectivity. However, remarkable success is being achieved with modern pesticides, e.g. for control of wild oats in cereal crops.

The development of a new pesticide is a very long and very expensive process and requires the close collaboration of a wide variety of scientific and commercial personnel. Currently, to obtain one new marketable product a company has to start with 10 000 compounds and carry out a testing programme continuously for ten years at a cost of at least £70 million. These amounts are increasing year by year, partly as a result of public pressure for safer and safer products, and this has resulted in: (i) a reduction in the rate of new pesticides being introduced on to the market; (ii) a fall in the number of companies carrying out basic research work, due to the greater risk involved; and (iii) a reduction in the length of patent life remaining during which the product may have 'protected' sales. This also increases the risk that a new pesticide will not be profitable. One consequence of these results is that it is now not worth while attempting to produce a pesticide unless it will be used on a major crop, because eventual sales in excess of £1 million per annum will be necessary in order to justify the risks, and only major world crops fit this criterion.

Most of the remainder of this chapter will be devoted to some important and representative examples of the major groups of pesticides—fungicides, herbicides and insecticides. Each of these groups will be considered in turn.

<table>
<tr><td>

**7.4
Fungicides**

</td><td>

7.4.1 Introduction

Although over 100 000 species of fungi have been classified, only about 200 are known to cause serious plant diseases. Nevertheless, their effects can be far-reaching, as evidenced by the destruction of the whole of the potato crop by a fungal disease commonly known as 'potato blight' in Ireland in the 1840s. This resulted directly in over one million deaths because potatoes had been the main item in the diet of most of the population. Catastrophes like this are now avoidable with the aid of modern fungicides. In a more encouraging vein, in the 1880s it was the custom in that important wine-growing area of France, Bordeaux, for farmers to paint vines growing near the roadside with a mixture of copper sulphate and lime (Bordeaux mixture) in order to discourage theft of the grapes. At this time, although vines were being destroyed by the fungal disease downy mildew, Millardet observed that the vines near the roadside were unaffected and he soon carried out experiments that confirmed the effectiveness of Bordeaux mixture against vine mildew.

Even today fungal disease is still a serious problem and current crop losses due to it have been put at in excess of £60 000 million per annum.

Fungi do not contain chlorophyll and must therefore obtain their food or energy from other sources. Differing sources enable two main types of fungi to be distinguished, (a) saprophytic and (b) parasitic fungi. Saprophytic fungi live on decaying matter and are largely responsible for the breakdown of animal and plant remains in the soil. They thus contribute to soil fertility and some are also actually cultivated as food, e.g. mushrooms and truffles. In contrast, parasitic fungi grow at the expense of living plants or animals and it is this type which is primarily the cause of fungal disease in plants. In many ways, new fungicides are more difficult to develop than insecticides, because the close relationship between the fungus and its host plant means that high selectivity of action is required.

Tables 7.1 and 7.2 have already demonstrated the scale of fungicide sales and perhaps also suggested that, in contrast to insecticides, they are more widely used in temperate than tropical climates.

Western Europe is evidently the dominant market, whereas North America is another major user of pesticides in general, as has already been highlighted.

Fungicides traditionally consisted of inorganic compounds of sulphur, copper and mercury. However, problems with phytotoxicity have led to the development of organic fungicides which also have increased fungicidal activity. Before considering examples of these we must be aware that fungicides are divided into two groups based on the way they are used and act.

</td></tr>
</table>

These are (i) contact (or surface) fungicides and (ii) systemic fungicides. Contact fungicides are applied to the surface of the plant, preferably prior to fungal attack. Their effectiveness on plants already affected by disease is clearly limited. The protective layer of the fungicide must be maintained and several reapplications may be necessary. All the traditional fungicides plus some of the modern organics such as dithiocarbamates belong to this type, which still dominates the market, and may account for about 80% of all fungicide sales.

In the developed world systemic fungicides are becoming increasingly important. These actually enter the plant and are translocated throughout it by the vascular system. Clearly they will be much more effective than contact fungicides against spores which have already entered the plant. Well-known examples are Benomyl, Carboxin and Dimethirimol.

7.4.2   Inorganic fungicides

These are largely compounds of sulphur, copper and mercury. Elemental sulphur is probably the oldest effective fungicide known and was recommended as early as the beginning of the nineteenth century to control mildew of fruit trees. The use of Bordeaux mixture on vines has already been mentioned. Other simple copper compounds (e.g. copper sulphate, cuprous oxide) have also been used as fungicides. Problems with phytotoxicity have led to their replacement by the more effective modern fungicides.

Organomercury compounds such as phenylmercuric acetate were widely used as seed dressings but their high toxicity led to the deaths of birds and other animals which ate the treated seed. This, coupled with the generally strong public feeling against heavy metals such as lead and mercury in the environment, has led to their replacement.

7.4.3   Dithiocarbamates

These were the first purely organic fungicides to be developed and they currently dominate the market. The first were introduced in the 1930s. Interestingly, they were originally synthesized as accelerators for rubber vulcanization. They contain two thiocarbamate groups

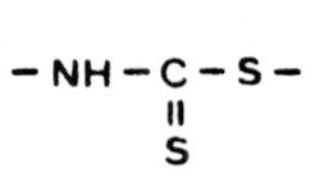

$$-NH-\underset{\underset{S}{\overset{\|}{}}}{C}-S-$$

and are not only very effective fungicides but are also extremely efficient, i.e. they have low dosage rates, are long-acting and have very low phytotoxicity. In addition they are easily and cheaply manufactured, as the following examples show.

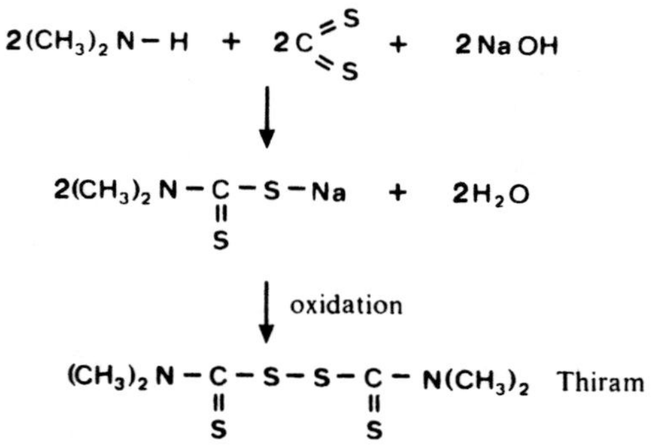

$$2(CH_3)_2 N-H \ + \ 2C\!\!\begin{smallmatrix}\nearrow S\\ \searrow S\end{smallmatrix} \ + \ 2NaOH$$

$$\downarrow$$

$$2(CH_3)_2 N-\underset{\underset{S}{\|}}{C}-S-Na \ + \ 2H_2O$$

$$\downarrow \text{oxidation}$$

$$(CH_3)_2 N-\underset{\underset{S}{\|}}{C}-S-S-\underset{\underset{S}{\|}}{C}-N(CH_3)_2 \quad \text{Thiram}$$

Thiram is used to control grey mould on lettuce and strawberries and as a seed dressing.

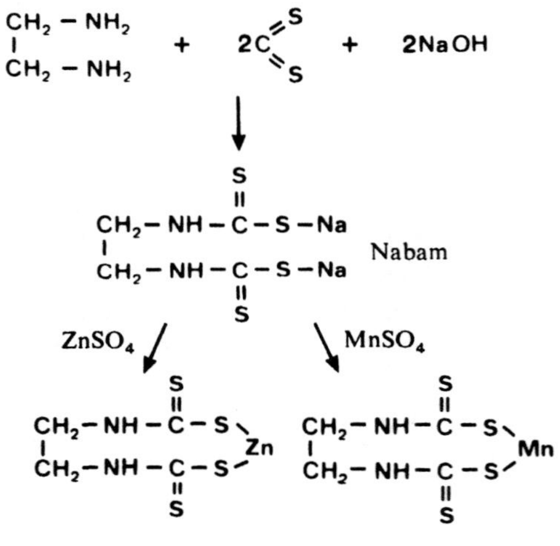

$$\begin{aligned}CH_2 &- NH_2\\ |\\ CH_2 &- NH_2\end{aligned} \ + \ 2C\!\!\begin{smallmatrix}\nearrow S\\ \searrow S\end{smallmatrix} \ + \ 2NaOH$$

$$\downarrow$$

$$\begin{aligned}&\quad\;\overset{S}{\overset{\|}{}}\\ CH_2 &- NH-C-S-Na\\ |\\ CH_2 &- NH-\underset{\underset{S}{\|}}{C}-S-Na\end{aligned} \quad \text{Nabam}$$

$$\overset{ZnSO_4}{\swarrow} \qquad\qquad \overset{MnSO_4}{\searrow}$$

$\begin{aligned}&\quad\;\overset{S}{\overset{\|}{}}\\ CH_2 &- NH-C-S\\ |\qquad\qquad Zn\\ CH_2 &- NH-\underset{\underset{S}{\|}}{C}-S\end{aligned}$	$\begin{aligned}&\quad\;\overset{S}{\overset{\|}{}}\\ CH_2 &- NH-C-S\\ |\qquad\qquad Mn\\ CH_2 &- NH-\underset{\underset{S}{\|}}{C}-S\end{aligned}$
Zineb	Maneb

These compounds are used to control downy mildews, potato blight and tomato blight, and are noted for their very low mammalian toxicities.

7.4.4 Sulphenimides

Many compounds containing the $\geq N-S-CCl_3$ grouping have been shown to have broad-spectrum potent fungicidal activity. Captan is the best-known example and was introduced in the early 1950s as a seed dressing and for the control of apple and pear scab and botrytis. Note its easy and cheap synthesis which commences with a Diels–Alder reaction.

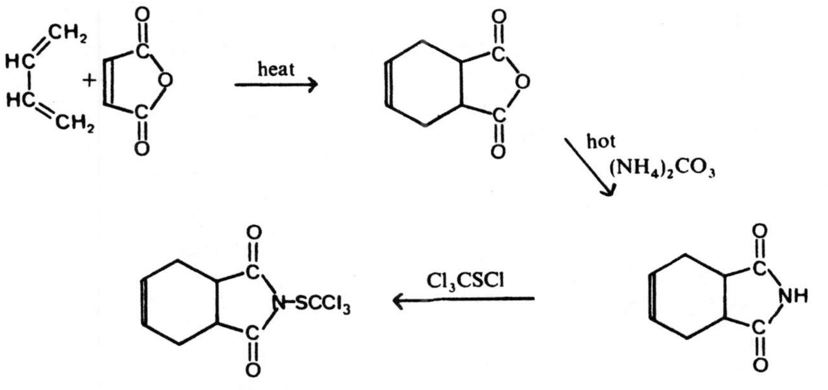

Captan

Two analogues of Captan—Folpet and Difoltan—were introduced in the early 1960s. They are also used as foliar fungicides and are more effective against potato blight than Captan.

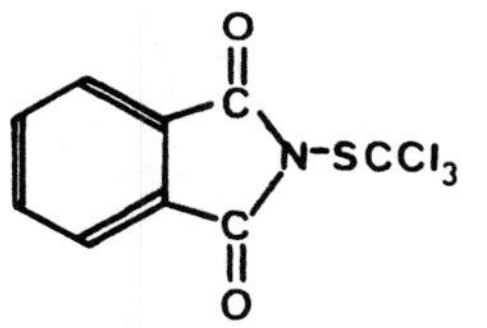

Folpet

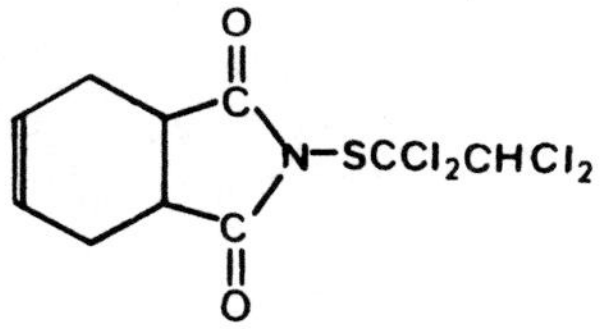

Difoltan

Like the dithiocarbamates they function by inhibiting enzyme activity in the fungi.

7.4.5   Systemic fungicides

This type was only introduced commercially in the late 1960s. Since these compounds move within plants into the plant tissues they have obvious advantages over the contact or surface types that have been discussed so far. Nearly all of them are heterocyclic compounds. As examples we will consider (i) benzimidazoles, (ii) pyrimidines, (iii) 1,4-oxathiins, (iv) morpholines, (v) imidazoles and triazoles and (vi) acylalanines. The first four groups are now well established but the last two represent emerging groups whose importance is increasing fairly rapidly, so much so that the imidazoles and triazoles alone have captured around 25% of the systemic fungicide market.

7.4.5.1   *Benzimidazoles.*   The best-known examples, which were introduced in 1967/8, are Benomyl and Thiabendazole. Both have been widely accepted as broad-spectrum systemic fungicides active against botrytis, powdery mildews and apple scab, and as a seed dressing against common bunt of wheat.

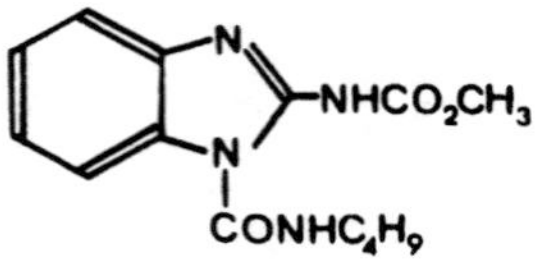

Benomyl 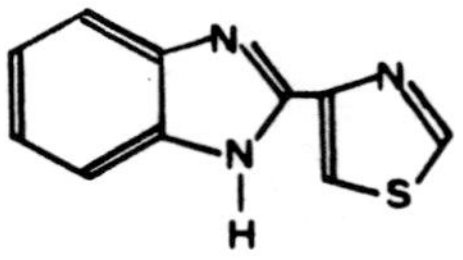   Thiabendazole

Benomyl can be synthesized as follows:

$$NCNH_2 + ClCO_2CH_3 \xrightarrow{-HCl} NCNHCO_2CH_3$$

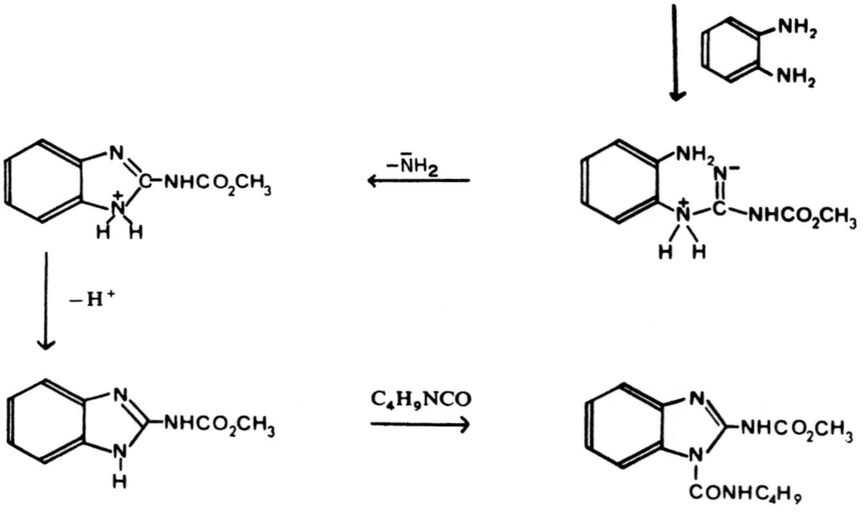

They act by interfering with DNA synthesis in the fungi.

7.4.5.2 Pyrimidines.

Dimethirimol shows outstanding systemic activity by root application against powdery mildews of melon, cucumber, etc. A single application to the soil gives control for up to eight weeks, in contrast to surface fungicides which must be applied every 10 days and still give inferior control.

Dimethirimol may be made by the following condensation reaction. Its analogue Ethirimol has an —NHC_2H_5 group at position 2 in place of the —$N(CH_3)_2$ group. It is very effective as a seed dressing against barley powdery mildew.

Both appear to act by enzyme inhibition.

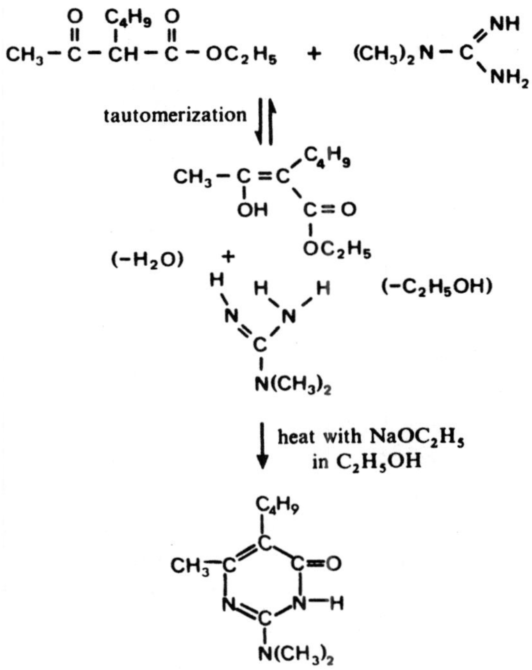

Dimethirimol

7.4.5.3 1,4-Oxathiins.

Carboxin and Oxycarboxin, introduced in 1966, were the first commercially successful systemic fungicides. They are selectively toxic to a number of cereal diseases such as smuts and rust. Again the mode of action involves enzyme inhibition.

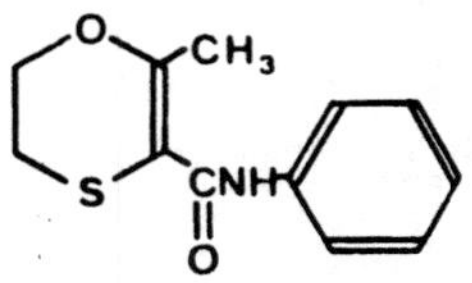

Carboxin

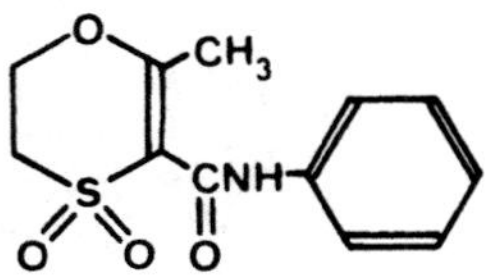

Oxycarboxin

7.4.5.4 Morpholines.

The morpholine derivative Tridemorph is used to control powdery mildew in cereals.

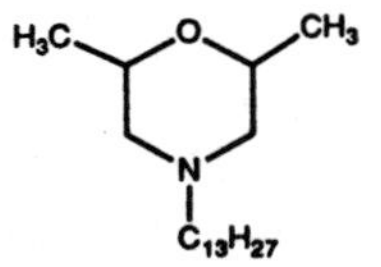

Tridemorph

More recently Shell introduced Dimethomorph which is sold as a 50:50 mixture of the E- and Z-isomers.

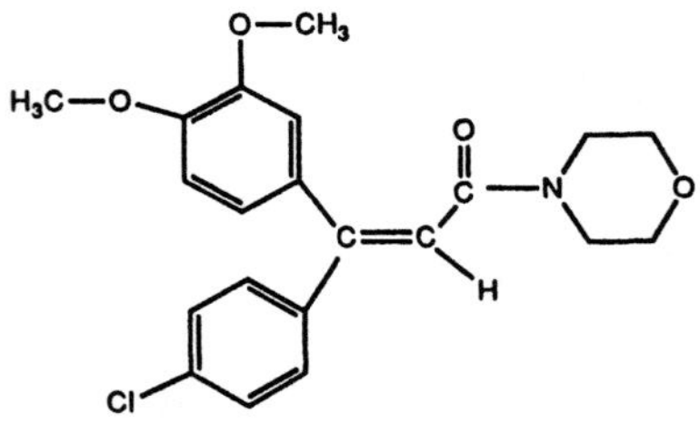

Dimethomorph (Z-isomer)

It controls some important fungal diseases of vines and potato and tomato blight.

7.4.5.5 Imidazoles and triazoles.[4]

These types of fungicides were first described in the 1960s, and by 1982 accounted for about 20% of the systemic fungicide market. Prochloraz is an example of the imidazole type and is a protectant and eradicant fungicide which is effective against a wide range of diseases affecting field crops, and fruit and vegetables at application rates of $400-600 \, \text{g ha}^{-1}$.

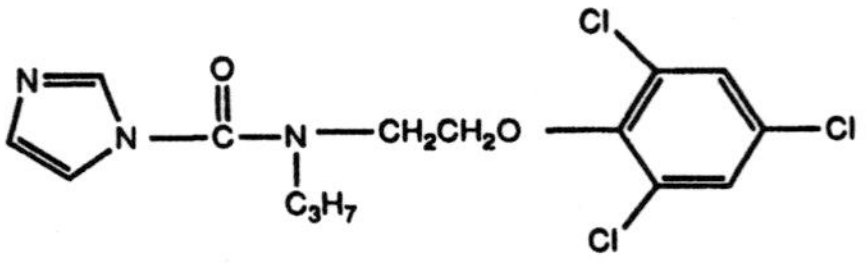

Prochloraz

Triadimefon and Triadimenol are triazoles which are systemic fungicides. The former is a protective and curative against mildews and rusts attacking a variety of crops, whereas the latter is used as a cereal seed treatment for protection against certain fungal spores. Their mode of action is to inhibit

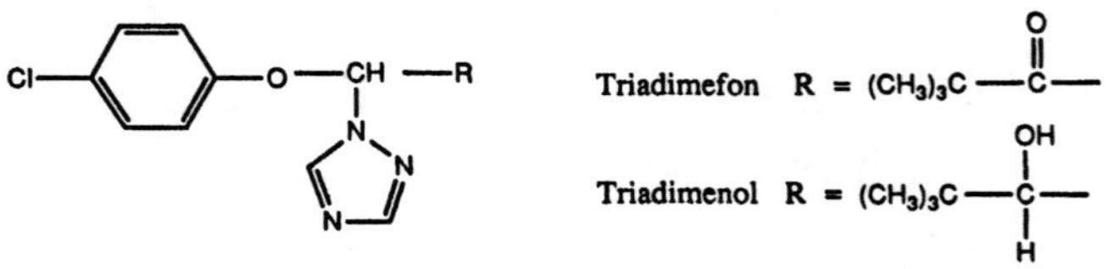

ergosterol biosynthesis by the fungal spores. Ergosterol is the major sterol in many fungi and plays an important part in membrane structure and function. Examples of other heterocyclic fungicides having this mode of action have a pyridine or pyrimidine ring as the heterocyclic grouping.

7.4.5.6 Acylalanines.

These are another important new class of fungicides introduced by Ciba–Geigy in recent years, but already resistance to them has been encountered in some countries and thus they have had to be used in mixtures with other fungicides.

Furalaxyl is a systemic fungicide used for preventive or curative control of diseases caused by soil-borne oomycetes, whereas the related Metalaxyl is used for control of diseases caused by air- and soil-borne members of the *Peronosporales* on a wide variey of crops.

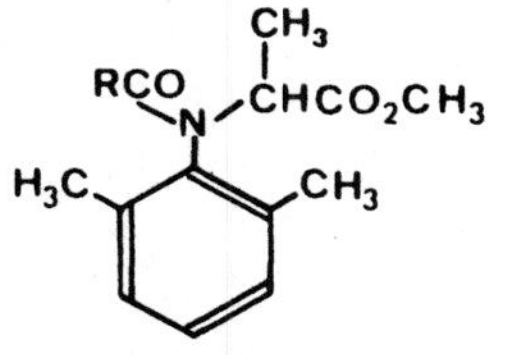

Furalaxyl R =

Metalaxyl R = CH₃OCH₂—

Systemic fungicides are amongst the safest pesticides in current use and they also do not appear to have any adverse environmental effects. Their mode of action (excepting Carboxin and Oxycarboxin) involves interfering with synthetic processes in the fungus. Contrast this with the contact fungicides which act by interfering with energy productioin or transport processes in the fungus.

7.4.5.7 Natural fungicides. Since most plants are clearly resistant to the vast majority of fungal pathogens, it was reasoned that this might be due to the presence of natural antifungal agents in the plant. Indeed some compounds have actually been isolated and their antifungal activity demonstrated *in vitro.* An example is Wyerone, which was isolated from broad-bean seedlings.

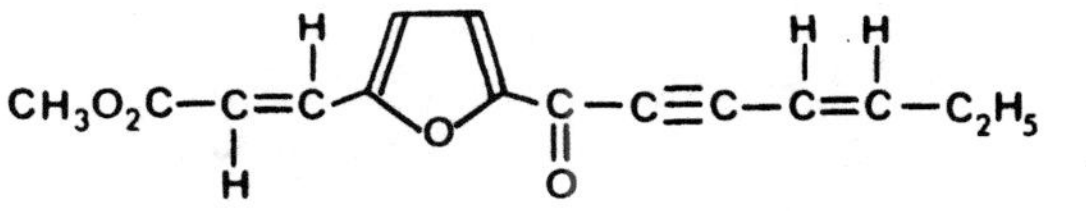

Wyerone

Phytoalexins are antifungal agents that are produced by plants in response to injury or attack, e.g. infection of beans with tobacco necrosis virus leads to the production of phytoalexins. With the development of strains of fungi resistant to systemic fungicides this line of research could have a promising future.

Finally, many antibiotics have been shown to have activity against plant diseases but their only important area of current use seems to be Japan where Blasticidin-S and Kasugamycin are used against rice blast.

7.5.1 Introduction

7.5 Herbicides

Tables 7.1 and 7.2 have already shown clearly that herbicides are the most important single group of pesticides. Although in world-wide terms herbicide sales are somewhat greater than those of insecticides (44% of the total as against 29%), for UK production herbicides (in terms of value) dominate the scene and account for almost 50% of total pesticide production. Use pattern geographically is similar to that of fungicides, with heavy usage in temperate zones such as Europe and North America, and rather small amounts, in comparison, being used in tropical areas like Africa and Asia. Similar comments apply in each area in terms of relative usage of the major classes of pesticide, with herbicides dominating in the former, whereas insecticides are the most important class in the latter tropical areas. It has been estimated that in the UK more than three-quarters of all land under cultivation is regularly treated with herbicides.

Herbicides are used to kill or control weeds, which we may define as any unwelcome form of plant life. Weeds, as we are all aware from our experiences in the garden at home, seem to be ubiquitous, and appear and thrive under all sorts of growing conditions. Not only are weeds themselves a pest—they may compete with the growing crop for water, light and nutrients—but they may also act as hosts for crop pests and diseases and carry these over to the next growing season. The variety of weeds is considerable and all soil contains weed seeds which are ready to germinate when conditions become suitable.

Herbicides can be divided into two types which are (i) total (non-selective) and (ii) selective. Total herbicides, as the name suggests, tend to kill off all vegetation, i.e. both crops and weeds. They therefore tend to be used immediately after a crop has been harvested for clearing the land, or before planting a crop. Some of them, such as bipyridyls, have eliminated the need for ploughing. Clearly they must be used very carefully. Selective herbicides are selective in their action; for example they may kill weeds growing in the midst of a crop without harming the latter. The timing of their application is also very important and will vary from compound to compound. Thus some will be applied before the emergence of crops and others later. We shall be concerned mainly with selective herbicides, which again emphasize the major goal of pesticide researchers—that of selectivity of biological action. It is also necessary for us to be very selective in the groups of herbicides that we can look at. We will therefore concentrate on the following groups: (i) phenoxyalkane carboxylic acids; (ii) substituted ureas; (iii) triazines; (iv) bipyridylium salts; (v) glyphosate; and (vi) sulphonylureas, before finally turning briefly to plant growth regulators.

7.5.2   Phenoxyalkane carboxylic acids

Indolylacetic acid (IAA) was the first plant growth hormone to be discovered (1928). Since it exerts a powerful stimulus to plant growth it was reasoned that application of large amounts of it would cause excessive plant growth, resulting in death of the plant. However, it was found that plants are able to regulate the level of IAA by metabolism and it was therefore not herbicidal. Following from these observations, other plant growth hormones were discovered, e.g. α-naphthylacetic acid which is used to stimulate root growth in plant cuttings. This line of research culminated in the early 1940s with the discovery of the herbicidal activity of the chlorinated phenoxyacetic acids,

2,4-dichlorophenoxyacetic acid (2,4-D) and 2-methyl-4-chlorophenoxyacetic acid (MCPA). These are plant growth stimulants like IAA, but differ from IAA in that they are not metabolized. Hence when applied in higher concentrations they cause lethal, uncontrollable and grossly distorted growth. They were the first hormone weed killers and the first really selective herbicides.

Over 50 years of extensive use testifies to their safety and efficiency, and in the USA consumption of 2,4-D alone is in excess of 32 000 tonnes per annum. It is produced by some six companies and marketed in over 15 000 different formulations. 2,4-D and MCPA possess several advantages:

(i) Low mammalian toxicity.
(ii) Selectivity of action (they kill broad-leaved weeds but do not affect cereal and grass crops, and so great is their selectivity that relatively little skill is needed in their application).
(iii) Effectiveness at low concentrations, e.g. $100\,g\,ha^{-1}$ (approx. 100 ppm) (interestingly, at much lower concentrations they can be used as plant growth stimulants—25 g spread over 100 000 pineapple plants causes them to start flowering, and this is reportedly done in Hawaii to have plants flowering throughout the year for the benefit of tourists).
(iv) Their synthesis is easy and cheap, e.g. the Dow process for synthesis of 2,4-D.

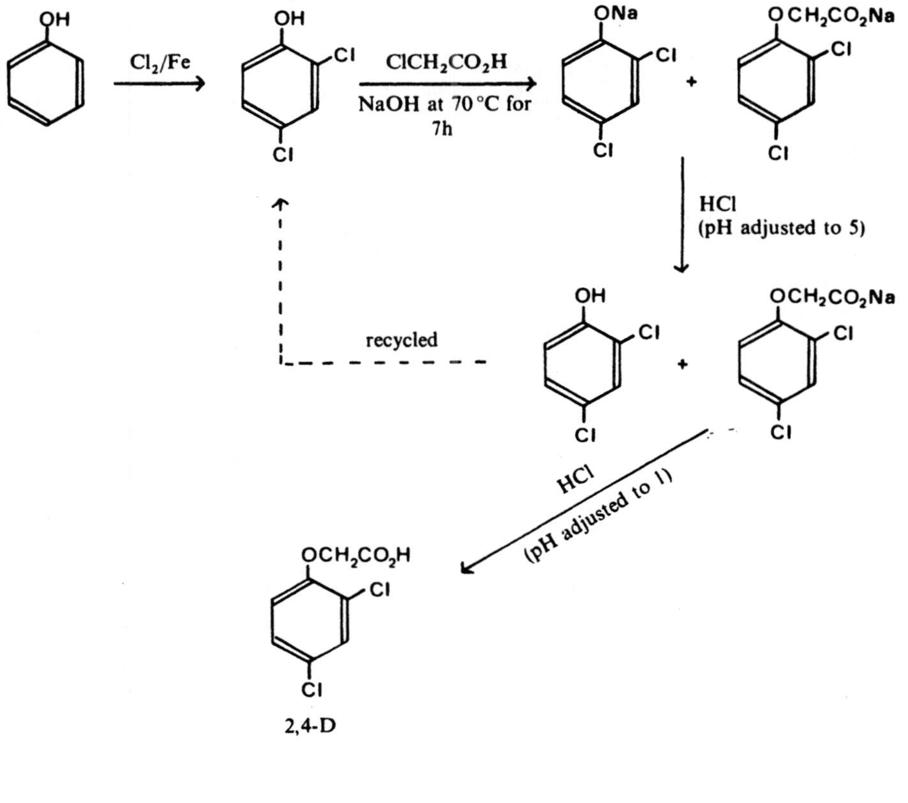

2,4-D

The Dow process is a batch process, and in the second step the solution must be kept alkaline and a 50% molar excess of 2,4-dichlorophenol is used in order to minimize formation of $HOCH_2CO_2H$. The synthesis of MCPA is similar to that of 2,4-D but the starting phenol is *o*-cresol.

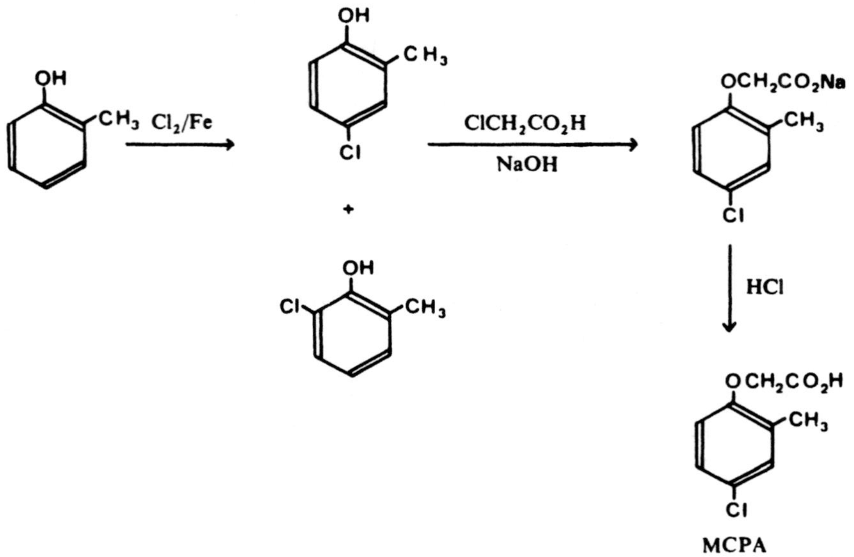

In the USA 2,4-D is very much preferred to MCPA, but the reverse is true in the UK; in the rest of Europe both tend to be used. The main reason for this is historical, and was the availability of the starting phenol when the compounds were introduced in the 1940s—*o*-cresol was fairly readily available in the UK from coal-tar whereas in the USA synthetic phenol was available.

Although the sodium salt of MCPA is fairly soluble and may be used directly in aqueous sprays, 2,4-D and its sodium salt are not very soluble in water. 2,4-D can be made more water-soluble by formation of salts with amines such as triethanolamine. Alternatively, alkyl esters may be made for formulation in soil or oil/water sprays. In the UK MCPA is widely used domestically, either alone or in combined fertilizer–weedkiller formulations for controlling dandelions and daisies in lawns.

Although 2,4-D and MCPA are such valuable aids to the cereal- and grass-grower, they cannot be used on leguminous crops such as clover and lucerne because they kill the crop as well as the broad-leaved weeds! Research by Wain showed that the chemically very similar phenoxybutyric acid analogues of 2,4-D and MCPA, rather surprisingly, controlled the weeds and did not harm clover and lucerne. Further research produced another surprising result—the phenoxybutyric acids are inactive, i.e. non-herbicidal, when applied to the weed. The weed (only) rather foolishly converts them into a lethal herbicidal form. This process is known as *lethal synthesis*. What is the herbicidal compound and how is it formed?

The conversion involves a standard biochemical process known as β-oxidation, shown in the case of MCPB.

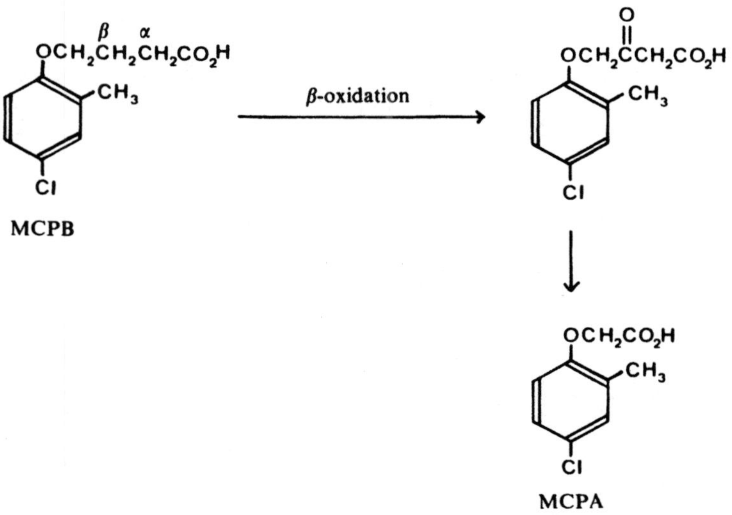

MCPB

β-oxidation

MCPA

The final product is none other than MCPA, which is being formed only within the weed. This β-oxidation pathway suggests that other analogues of MCPA or 2,4-D which also have an odd number of methylene groups could also, by a series of β-oxidations, be degraded into MCPA or 2,4-D and should therefore act as herbicides; analogues having an even number of methylene groups cannot be converted into these products and will not be herbicidal. Synthesis and testing of appropriate compounds confirmed these suggestions.

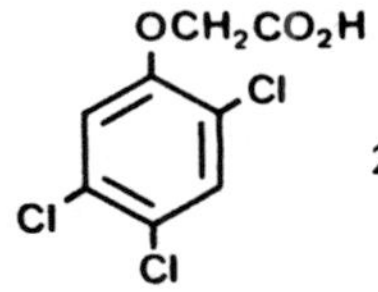

2,4,5-T

Whereas 2,4-D and MCPA have been extensively used for over 50 years without any problems, the chemically similar 2,4,5-T has been surrounded with controversy after it was widely used as a defoliant (combined with 2,4-D as Agent Orange) by the USA during the Vietnam war in the 1960s. A whole book has been devoted to the arguments ranged for and against its use— Alistair Hay's *The Chemical Scythe*.

Let us consider both sides of the argument, firstly its advantages as a herbicide and then its adverse effects. As well as being a defoliant, 2,4,5-T is very good at controlling woody perennials. Thus in the UK it has been widely used to control brambles encroaching on footpaths or railway lines by the Forestry Commission and British Rail respectively. It has low mammalian toxicity and it is easily and cheaply manufactured.

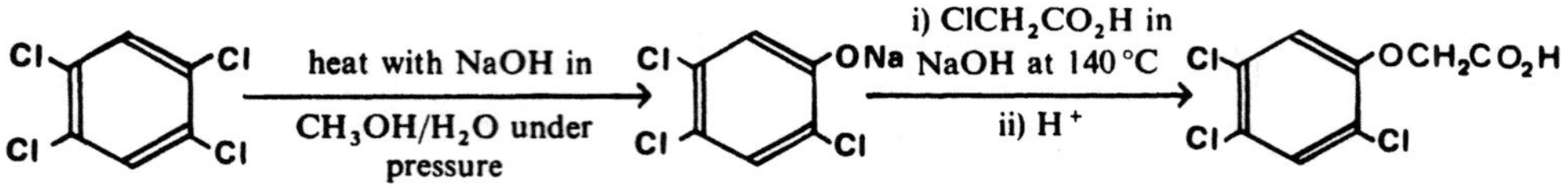

(Note the similarities to the synthesis of 2,4-D and MCPA.) The temperature in the second step requires careful control because if it rises above 160 °C a side reaction between two molecules of sodium, 2,4,5-trichlorophenate occurs.

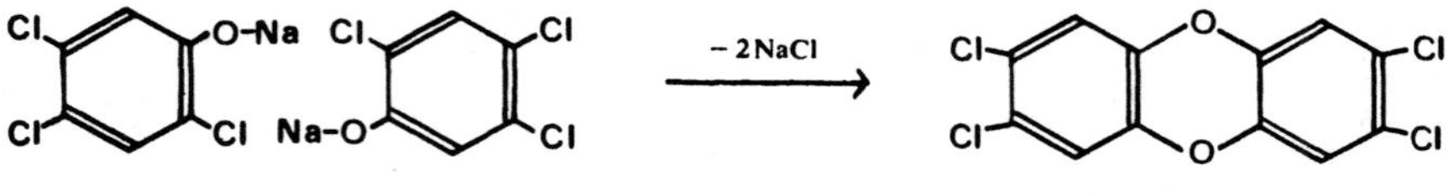

Tetrachlorodioxin
(one isomer)

It is the small amounts of this dioxin impurity in 2,4,5-T which have been responsible for the problems associated with its use. These first became apparent at the time of its use in Vietnam, when it was found that the incidence of malformed babies in villages close to zones that had been sprayed with 2,4,5-T from aircraft was very much greater than would have been expected. Clearly 2,4,5-T could have drifted towards the villages, exposing pregnant women to it, and it therefore became the prime suspect. Studies commissioned at the US National Cancer Institute duly cleared 2,4,5-T itself. However, they did establish that dioxin, present in the 2,4,5-T at concentrations in the 10–30 ppm range, was a very potent teratogen, i.e. it caused pregnant mammals to give birth to malformed young. Although samples of 2,4,5-T containing 30 ppm of dioxin were shown to cause malformations in rat foetuses, those containing only 0.5 ppm did not.

Arguments for and against the banning of the use of 2,4,5-T have consequently hinged around the concentration of the dioxin impurity present. Current manufacturing methods ensure this is kept well below 0.5 ppm, a level which manufacturers claim is safe, although opponents dispute this. Litigation is currently in progress in the USA for compensation for veterans of the Vietnam war who claim that exposure to Agent Orange has been responsible for birth defects in their children and the development in themselves of cancer. Although carcinogenic and teratogenic effects of relatively high concentrations of dioxin in 2,4,5-T can clearly be demonstrated in test animals, extrapolating these results to the lower (often imprecisely known) concentrations to which people may have been exposed, perhaps several years previously, introduces a considerable degree of uncertainty. Much of the evidence against 2,4,5-T has been circumstantial or has been based on statistical surveys on the number of birth defects in an exposed population sample compared with that for an unexposed sample.

Nevertheless, 2,4,5-T has been banned, either completely or for some applications, in certain countries such as West Germany and the USA. In the

UK consideration of the evidence for and against this herbicide by the Pesticides Safety Committee led it to conclude that a ban on its use was not justified. However, continued pressure against its use on behalf of agricultural workers caused use of 2,4,5-T in the UK to drop by about 50% in the mid-1980s. Alternative herbicides such as glyphosate are now available, but they are much more expensive.

Public attention was again drawn to dioxin in July 1976 when an explosion at Givaudan's factory at Seveso, Northern Italy, released a vapour could of 2,4,5-trichlorophenol, containing dioxin, over the surrounding area (see section 3.2.4). The accident was caused because process operatives did not follow the set procedure. As a result many people were evacuated from their homes for up to two years and a number of women elected to have abortions because of the potential danger to their unborn children.

The mode of action of 2,4-D, MCPA and 2,4,5-T is varied, but results in lethal, uncontrollable and grossly distorted growth, therefore affecting respiration, food reserves and cell functions.

7.5.3 Substituted ureas

These were introduced by DuPont in the early 1950s as persistent and total herbicides, typical examples being Monuron and Diuron.

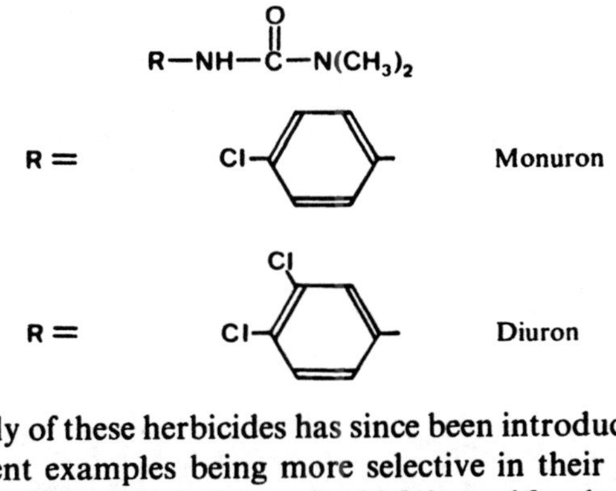

$$R{-}NH{-}\overset{\overset{\displaystyle O}{\|}}{C}{-}N(CH_3)_2$$

R =   Cl—⟨benzene ring⟩—   Monuron

R =   Cl—⟨benzene ring with Cl⟩—   Diuron

A whole family of these herbicides has since been introduced by a variety of companies, recent examples being more selective in their action, e.g. Fluometuron (R = 3-trifluoromethylphenyl) which is used for the control of broad-leaved weeds and grasses in cotton.

Heterocyclic ureas have been introduced recently, including Tebuthiuron (1974) which is a broad-spectrum herbicide for the control of herbaceous and woody plants when applied at the rate of about 7.5 kg ha.$^{-1}$ As expected, at

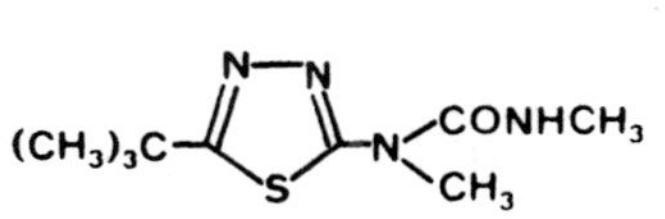

$(CH_3)_3C$—⟨thiadiazole ring⟩—N(CONHCH_3)(CH_3)

Tebuthiuron

higher concentrations it can be used as a total herbicide. The urea herbicides act by interfering with photosynthesis by inhibiting the Hill reaction, thus causing the weed to 'starve'.

7.5.4   Triazines

When included with the urea herbicides of the previous section, the triazines consititute one of the most important and widely used groups of herbicides today. Like the ureas, in high concentrations ($10-20\,kg\,ha^{-1}$) they are total herbicides and are used to clear paths, industrial sites and railway lines, whereas at lower concentrations ($1-3\,kg\,ha^{-1}$) they are used as selective herbicides.

The first examples, reported by Geigy in the early 1950s, were Atrazine and Simazine. They are widely used, particularly the former, for weed control in maize and sugar cane. Atrazine provides a direct contrast with the phenoxy-butyric acids discussed in section 7.5.2. In the latter case the weed carried out a chemical conversion on the applied chemical, thereby producing the selectivity of action. With maize the reverse type of selectivity is present, because the enzymes in the maize are able to detoxify the Atrazine by hydrolysing off the chlorine. The weeds do not possess the appropriate enzymes and therefore succumb to the herbicidal effects.

Atrazine, in terms of tonnage produced, is probably the single most important herbicide at the present time. Again, an easy and cheap synthesis is a key factor.

$$Cl_2 + NaCN \longrightarrow CNCl + NaCl$$

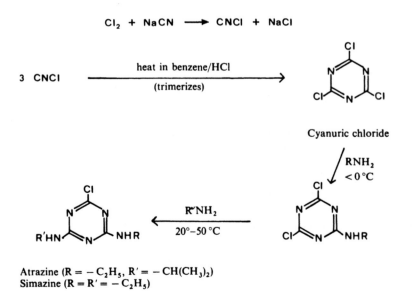

Atrazine ($R = -C_2H_5$, $R' = -CH(CH_3)_2$)
Simazine ($R = R' = -C_2H_5$)

Cyanuric chloride is also an intermediate in the production of melamine plastics. The chlorines are very readily replaced by nucleophiles, hence the low reaction temperature. However, as each successive chlorine is replaced the reactivity drops and it is therefore quite feasible to introduce different substituents in a stepwise manner.

Effective herbicides are still obtained if the 2-chlorine is replaced by a methoxy- or thiomethyl-group. However, substitution by other groups results in loss of herbicidal activity. Variations in the alkylamino side chains extend the range of biological activity and persistence.

The mode of action of the triazines is the same as that of the substituted ureas discussed in section 7.5.3.

7.5.5 Bipyridylium salts

The first examples of this type of herbicide, introduced by ICI in 1958, were Diquat and Paraquat. They are contact herbicides which are phytotoxic, rapidly killing top growth by cell membrane destruction and desiccation of the plant. Uses include preharvesting treatment of crops such as potatoes, soybeans and sugar cane. They are also widely used as total herbicides to clear land prior to planting of crops. With the development of direct-drill sowing of seeds, this has in some cases eliminated the need for ploughing, thereby reducing soil erosion. A major attraction is that Diquat and Paraquat rapidly lose their activity on contact with the soil because they become so strongly adsorbed. Thus the treated area can immediately be resown without danger to the seed. Diquat is also used against aquatic weeds.

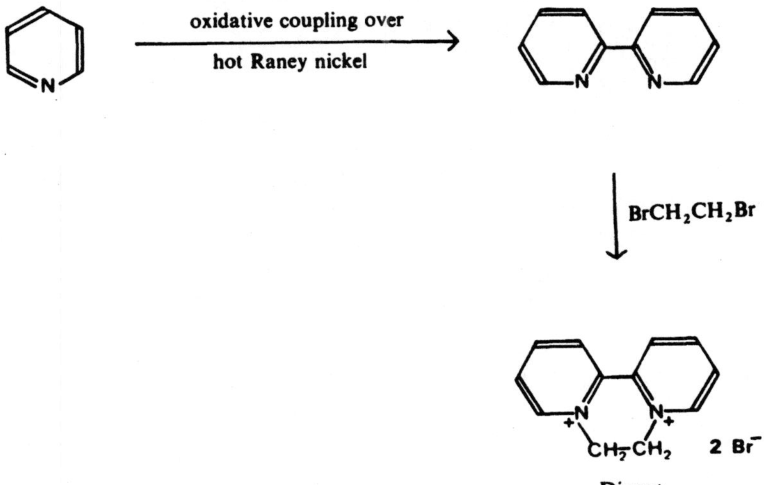

Diquat

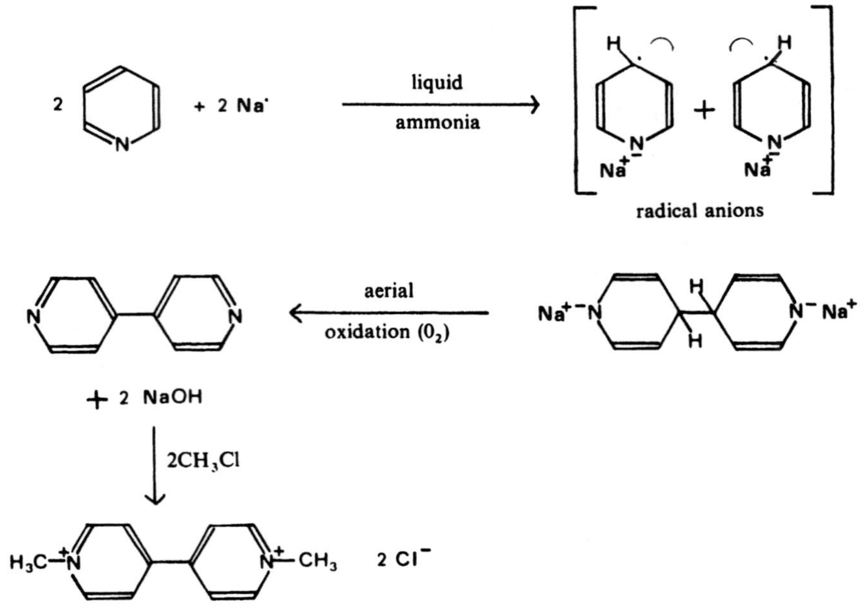

Paraquat

Although both are quaternary salts of bipyridyls, Diquat is a 2,2'-bipyridyl whereas Paraquat is a 4,4'-bipyridyl.

Given their relatively high costs relative to the herbicides discussed so far (this is due to the high cost of pyridine), the rapid expansion in their use is remarkable. Clearly it relates to their novel applications. Their mode of action involves reduction of the cation to a free radical during interference with photosynthesis. Re-oxidation back to the original cation by molecular oxygen then takes place, accompanied by production of hydrogen peroxide which is extremely toxic to plant cells.

Although Paraquat has only a moderate mammalian toxicity, when ingested in large amounts by humans it causes lung damage leading to respiratory failure and death. Unfortunately this chemical has received a quite unfounded notoriety because of several deaths from accidental poisoning after it was placed in unlabelled or incorrectly labelled bottles. When correctly used in its intended agricultural applications it has not caused any serious problems, and has been a very efficient herbicide.

7.5.6 Glyphosate

This recent herbicide was introduced by Mosanto under the trade name Roundup in 1971 and has since become a major revenue earner—so much so that it has been christened the first 'million dollar' pesticide. Structurally it is novel because it bears no resemblance to any other herbicide. It is a non-selective herbicide which is particularly effective against annual and deep-rooted perennial weeds, e.g. couch grass, in which it is translocated into the rhizomes. It is a foliar applied herbicide which can be used at any time of the year and is inactivated by being strongly adsorbed on to soil. In this way it is similar to the bipyridylium salts. However, whereas the latter bind strongly on to the clay minerals in the soil, glyphosate is thought to bind by chelation with metals, such as iron, in the soil.

Conversion to the isopropylamine salt is carried out before it is used as a herbicide. Its mode of action appears to be interference with the biosynthesis of aromatic amino acids such as phenylalanine. This in turn inhibits nucleic acid metabolism and protein synthesis.

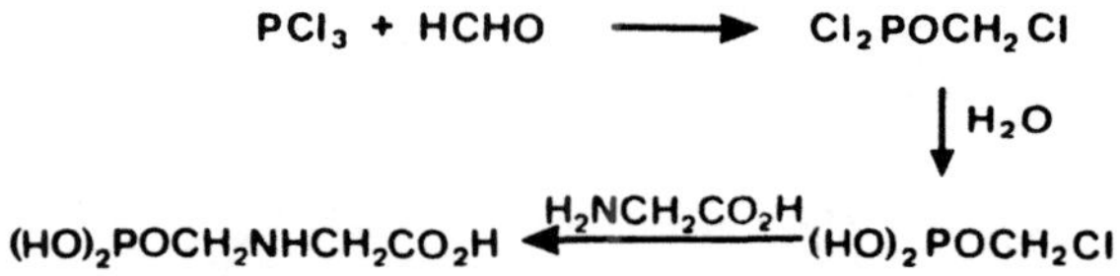

$$PCl_3 + HCHO \longrightarrow Cl_2POCH_2Cl$$

$$\downarrow H_2O$$

$$(HO)_2POCH_2NHCH_2CO_2H \xleftarrow{\ H_2NCH_2CO_2H\ } (HO)_2POCH_2Cl$$

Glyphosate

7.5.7   Sulphonylureas

This group is discussed separately from the substituted ureas because it is more recent (introduced by DuPont at the end of the 1970s), is active at extraordinarily low application rates and has a different mode of action. Chlorsulfuron controls most broad-leaved weeds and some grasses in cereals at application rates of only 10–40 g ha^{-1}, whereas Sulfometuron-methyl is used for the long-term control of Johnson grass without harming Bermuda grass; it is, however, not selective on most agricultural crops. (Note that most current herbicides have application rates of the order of kg ha^{-1}.) They also have a wide margin of safety for mammals.

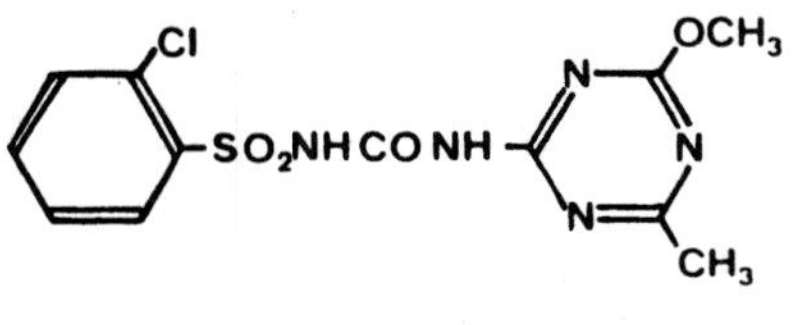

Chlorsulfuron 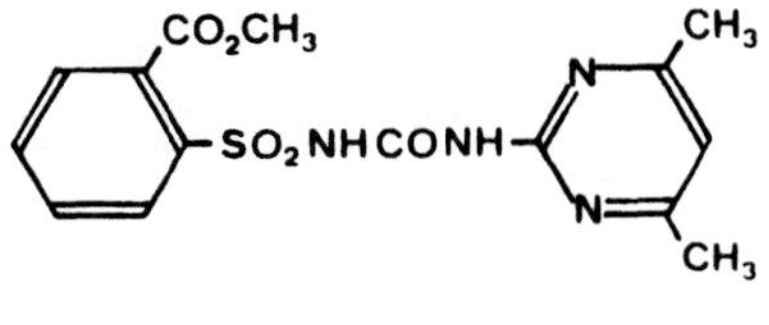   Sulfometuron-methyl

They are prepared from the appropriate sulphonamide.

DuPont have continued to introduce herbicides of this type which are both selective and environmentally acceptable. They include (crop protected): Tribenuron-methyl (cereals), Bensulfuron-methyl (rice), Rimsulfuron (maize)

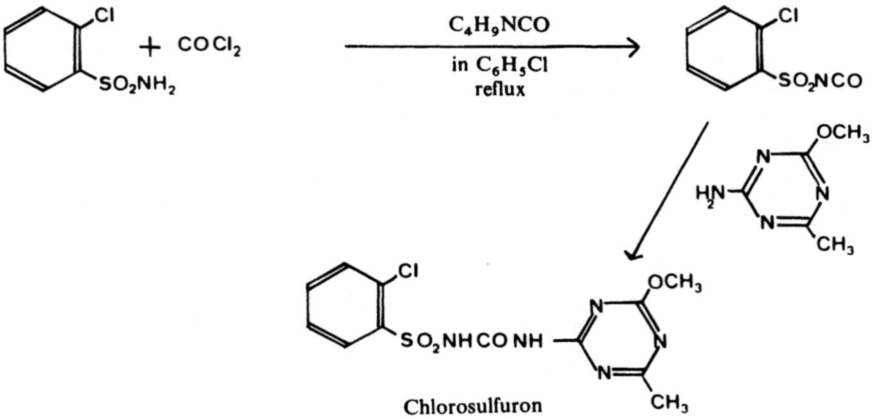

Chlorosulfuron

and Triflusulfuron (sugar beet). These are active at the 10–200 g ha^{-1} level.

Their mode of action is to interfere with some process which is essential for cell division in the plant.

7.5.8 Other herbicides

Other significant groups of herbicides include polysubstituted benzenes, e.g. Trifluralin; carbamates, e.g. Asulam; acetamides, e.g. Propachlor; and the triazoles, e.g. Amitrole.

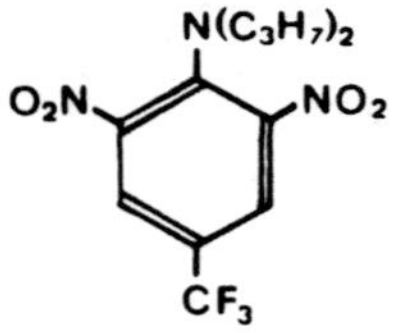

Trifluralin

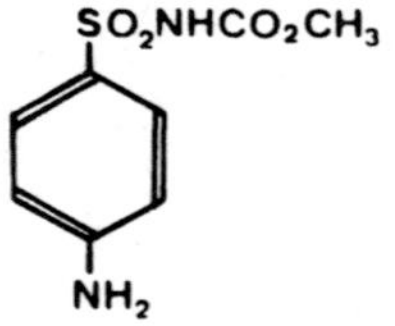

Asulam

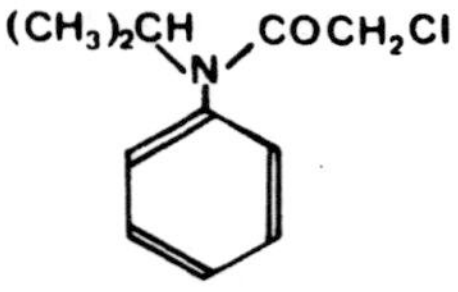

Propachlor

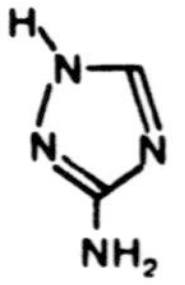

Amitrole

Several cyclohexanedione herbicides have been introduced to the market during the 1980s. They have the general structure

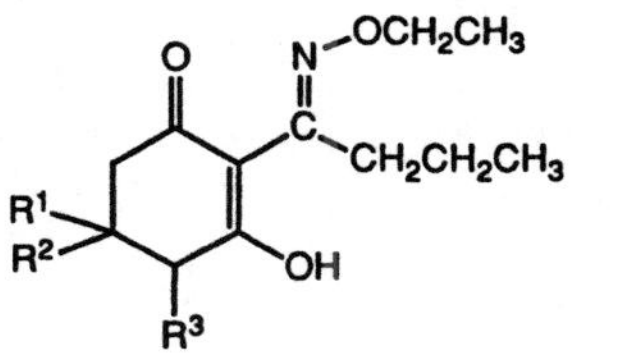

Examples are

Alloxydim where $R^1 = R^2 = CH_3$ and $R^3 = CO_2CH_3$

Sethoxydim where $R^1 = CH_2CHSCH_2CH_3$ and $R^2 = R^3 = H$
CH_3

They are used to control grasses and other perennial weeds growing among cotton, sugar beet, potatoes, rape seed, etc. Their mode of action is to inhibit fatty acid synthesis and prevent cell division.

Fusilade is a herbicide which selectively kills grass in broad-leaved crops. It therefore has exactly the opposite selectivity to 2,4-D.

It is considered here as an illustration of a growing trend among biologically active compounds like drugs and pesticides—that of marketing just the active enantiomer or stereoisomer rather than the racemate or mixture of stereoisomers. The reasons for this should be perfectly obvious, and as was noted in section 3.1, the thalidomide tragedy would not have happened if only the active enantiomer (rather than the racemic mixture) had been marketed. Clearly the big attraction with plant protection agents is that only half as much chemical is applied compared with the racemic mixture. Therefore there is less chance of adverse environmental effects occurring.

Let us look at Fusilade and consider its synthesis to produce the single herbicidally active isomer—the R-enantiomer. It has the structure

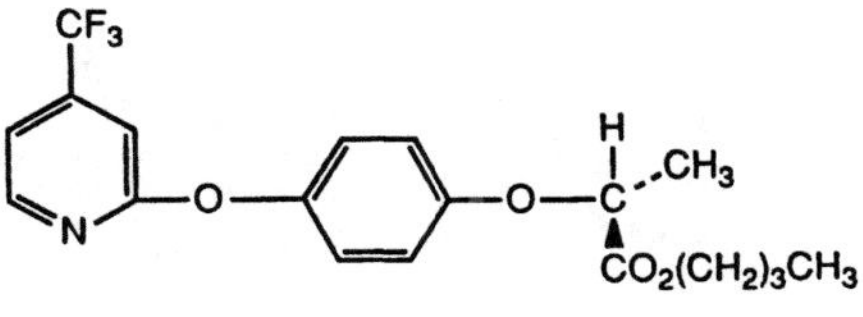

Fusilade (R - enantiomer)

This has only the one asymmetric carbon atom and can therefore exist in two enantiomeric forms. Hence to produce the required R-enantiomer we must either carry out an ordinary chemical synthesis but resort to optical resolution via, say, diastereoisomer formation at some stage, or carry out an asymmetric synthesis. Clearly, the extra costs incurred, compared with just producing the racemate, must be compensated for in a higher selling price!

The approach actually used is one of asymmetric synthesis. Note that the asymmetric centre is incorporated late in the synthesis in order to (a) minimize possible losses of optically active material and (b) reduce possibilities of racemization occurring. The synthesis is shown below. Interestingly the required S-enantiomer of the 2-chloropropanoic acid is obtained from its racemate by getting certain bacteria to destroy the unwanted R-enantiomer only!

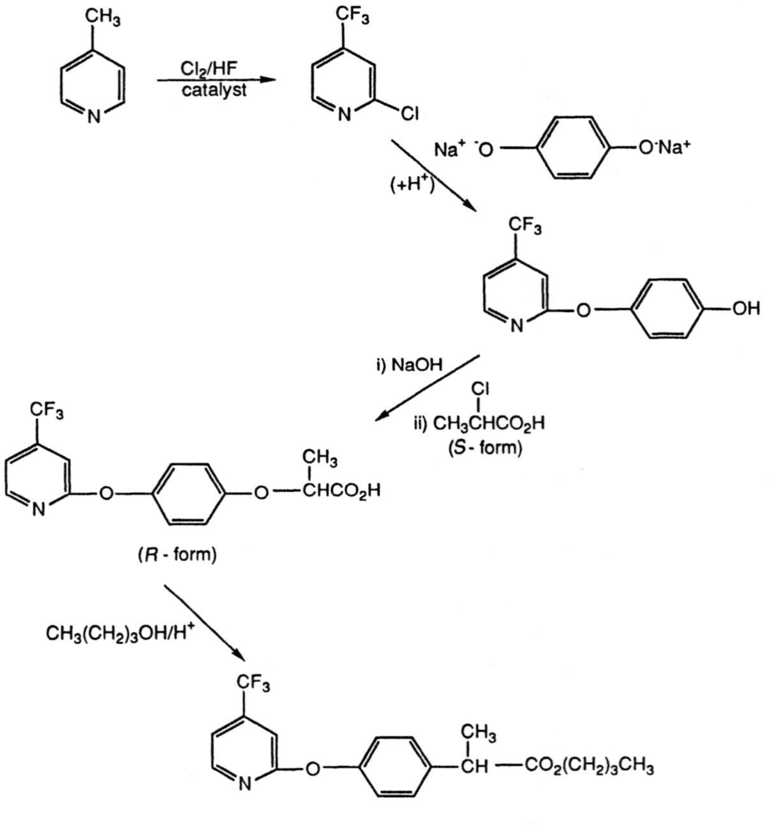

R-Fusilade

Fusilade is a relatively simple compound to make in an optically or stereo pure form since it has only one chiral centre and therefore two stereoisomers.

A much greater challenge has been overcome with the synthetic pyrethrin insecticide Karate, which has 16 possible stereoisomers of the basic structure, since it has one chiral centre, one C=C bond and a disubstituted cyclopropane ring. Here stereoselective synthesis produces a mixture of only four stereoisomers. Fractional crystallization separates out the two enantiomers which are marketed as a racemic mixture.

7.5.9   Plant growth regulators (plant hormones)

Some examples of plant growth regulators, IAA, 2,4-D, MCPA and 2,4,5-T (all plant growth stimulants) have already been discussed.

Plant hormones control the remarkable physiological changes which plants undergo, e.g. flowering, blossom and fruit formation, shedding of leaves, etc. Some naturally-occurring compounds are used to control some of these aspects. A good example is the gibberellins, which are a group of plant hormones (over 50 structurally similar compounds which have a complex pentacyclic structure) that stimulate cell elongation and division. They are used to induce germination in barley during brewing.

Many synthetic compounds are available, and one of the earliest examples was the use of ethylene to promote ripening of fruit. Compounds are now available which when sprayed on to crops slowly decompose to give ethylene. An example is Ethrel which is used in orchards, where it is sprayed on to trees and the ethylene it produces promotes ripening and loosening of the fruit.

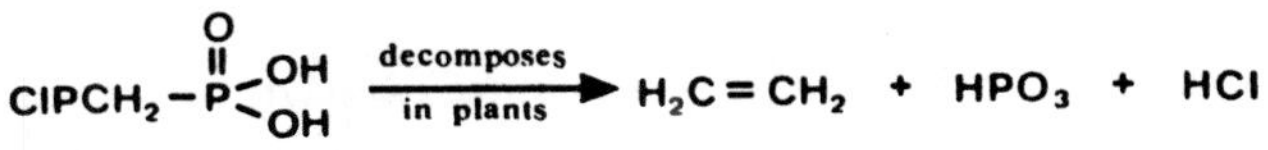

$$ClPCH_2-\overset{\overset{\displaystyle O}{\|}}{P}\!\!<\!\!\overset{OH}{OH} \quad \xrightarrow[\text{in plants}]{\text{decomposes}} \quad H_2C=CH_2 \;+\; HPO_3 \;+\; HCl$$

Ethrel

Interestingly, 2,4-D at a concentration of 12 ppm is used to prevent fruit drop and increase fruit size in citrus trees, e.g. grapefruit, lemon and orange.

Cytokinins, which are adenine derivatives, are used to prolong the storage life of green vegetables, cut flowers and mushrooms. A well-known example is Zeatin.

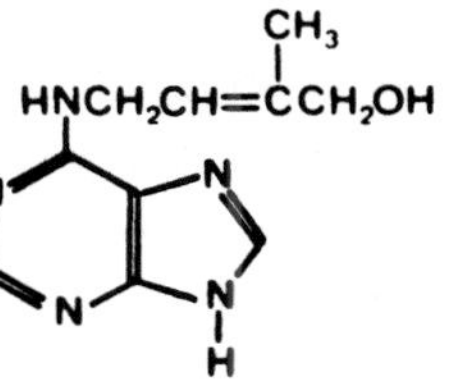

Zeatin

Equally, compounds are available which retard plant growth. Again both naturally occurring and a variety of synthetic compounds exist. Examples of the naturally occurring group are gallic acid (3,4,5-trihydroxybenzoic acid) and cinnamic acid, both of which prevent sprouting in stored onions and potatoes.

Amongst the synthetic growth retardants Chlormequat Chloride (2-chloro-ethyltrimethylammonium chloride), introduced by Cyanamid in 1959, has several uses, including reducing the height of cereals to minimize wind damage and facilitate mechanical harvesting.

Diaminozide is used to control the growth of fruit trees, and the shape and height of ornamental shrubs and flowers. One application is to stunt the growth of chrysanthemums to make them suitable as indoor pot plants.

$$CH_2CONHN(CH_3)_2$$
$$|$$
$$CH_2CO_2H$$

Diaminozide

Ancymidol is effective as a plant growth regulant on a variety of flowering plants, e.g. tulips and dahlias. Its growth-inhibition effect can be reversed by application of gibberellic acid.

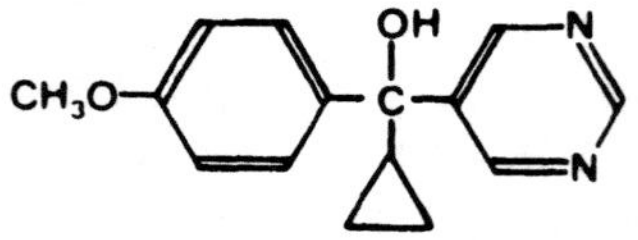

Ancymidol

Several triazole and pyrimidine ergosterol biosynthesis inhibitors (see section 7.4.5.5) also inhibit gibberellin synthesis, and therefore act as plant growth retardants, e.g. Paclobutrazol.

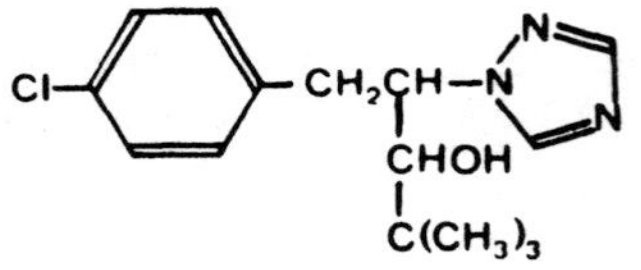

Paclobutrazol

7.6 Insecticides

7.6.1 Introduction

Insecticides are the second most important group of pesticides on a world-wide basis and are of prime importance in tropical countries, where their use (measured in monetary terms) may exceed that of all other groups of pesticides put together.

The vast majority of insects are not regarded as pests, and indeed some are beneficial to man—bees produce honey and aid pollination, and ladybirds eat aphid pests (greenfly). This is fortunate for us, because there are more than 700 000 species of insect, but only about 10 000 of these are regarded as pests. Insects can be pests in either of two ways. Firstly, in the same sense as trouble-some fungal disease or weeds, they may consume or destroy growing crops, thereby reducing our yield of food. (It is worth noting in passing here that almost 30% of all food grown in the world is consumed by insects. This very high figure is perhaps more readily accepted when it is realized that a very large swarm of just one insect, the locust, can consume 3 000 tonnes of green crops in a single day.) It is this aspect of insect pests that we are directly interested in. The second way in which insects are regarded as pests is as transmitters of some very unpleasant (tropical) diseases. Example are malaria (mosquito), typhus (body louse), and sleeping sickness (tsetse fly). Clearly development of chemicals to control these insects and prevent them spreading disease can do much to relieve human suffering and, as we shall see, save many millions of lives.

It is convenient to divide insecticides into two categories, (a) natural and (b) synthetic. We shall study each of these groups in turn but it is worth pointing out at this stage that in terms of quantities used the synthetics completely dominate the market. An interesting, fairly recent development has been the production of improved synthetically-modified naturally occurring insecticides, the best-known example being the pyrethrins. Finally, there are several alternatives to the use of chemicals for insect control which are being actively studied and developed. They include use of pheromones and growth-regulating hormones and will be considered briefly towards the end of this section.

7.6.2   Naturally occurring insecticides

These are compounds obtained from plant sources and they have all been used for at least 100 years, although their use has been declining as more efficient synthetic analogues or alternatives have been introduced. We will consider the three most important examples which are, in ascending order of scale of use, nicotine, derris and pyrethrum.

7.6.2.1   Nicotine.

This is well known as an ingredient of tobacco and is the cause of addiction to tobacco smoking. Tobacco was introduced to the UK in 1585 by Sir Walter Raleigh, and before 1 700 water extracts of tobacco were being used to kill aphids on garden plants. Today nicotine is extracted from tobacco (which may contain up to 8% of it) by either steam distillation or solvent extraction. Being an organic base it is sold mainly as its sulphate and used to control aphids and lice, mainly in the Far East. It is highly toxic to all mammals and the fatal dose for a human being is about 40 mg. This highlights the dangers of smoking tobacco, since a cigarette may contain 14 mg and a cigar 70 mg of nicotine. Why is smoking not (immediately) fatal? Fortunately the body rapidly metabolizes nicotine and therefore keeps its concentration below the danger level. Organophosphate insecticides have now largely replaced nicotine.

CH₃ Nicotine

7.6.2.2   Derris.   Derris powder is obtained from the roots of *Derris elliptica* in Malaysia and *Lonchocarpus* in South America, where it has been used by the natives for several centuries to paralyse fish, causing them to surface for easy collection. It has for many years been used as an ideal garden insecticide since it has low toxicity to mammals, is harmless to plants, but is highly toxic to many insects, particularly caterpillars, and does not leave residues on vegetables. It has also been used in cattle and sheep dip formulations. The active ingredients of derris are a group of compounds called rotenoids, the most important being rotenone, a pentacyclic oxygen heterocyclic compound.

7.6.2.3   Pyrethrum.   Pyrethrum is solvent extracted from the flowers of *Chrysanthemum cinerariaefolium* and is the most important naturally-occurring insecticide. This variety of chrysanthemum is grown mainly in Kenya but also in Ecuador. Pyrethrum owes its importance to its very rapid knockdown action against a variety of flying insects, e.g. houseflies and mosquitoes and its low mammalian toxicity. The active ingredients are a series of structurally similar compounds called pyrethrins, which have the general formula shown below.

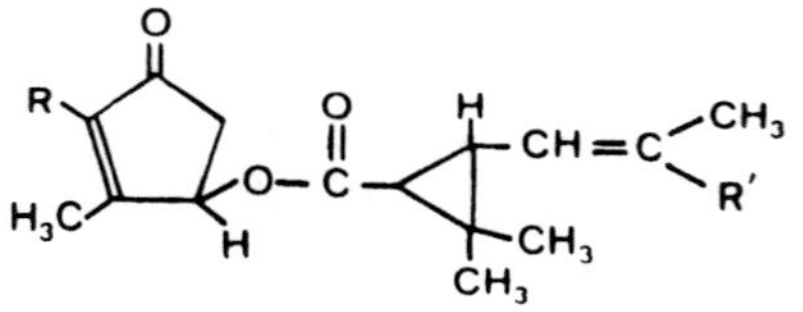

For example, pyrethrin I has $R = CH_2 = CH - CH = CH - CH_2 -$ and $R' = CH_3$. The natural pyrethrins suffer from the disadvantage that they rapidly lose their activity on exposure to light and oxygen. A great deal of research work over many years has therefore been devoted to producing synthetic analogues having the desired stability and activity, and this has resulted in the commercialization of a number of pyrethroids. Early examples (1967) were Resmethrin and its stereoisomer Bioresmethrin. These are respectively 20 and 50 times more effective in knockdown sprays than natural pyrethrins, and although more stable, they are also still fairly rapidly decomposed by air and sunlight.

Agriculturally useful synthetic pyrethroids have only been developed since 1972. They have exceptional activity (approx. 0.1 kg ha^{-1}) as broad-spectrum insecticides and are photostable, lasting for up to seven days on foliage. Permethrin is a well-known example.

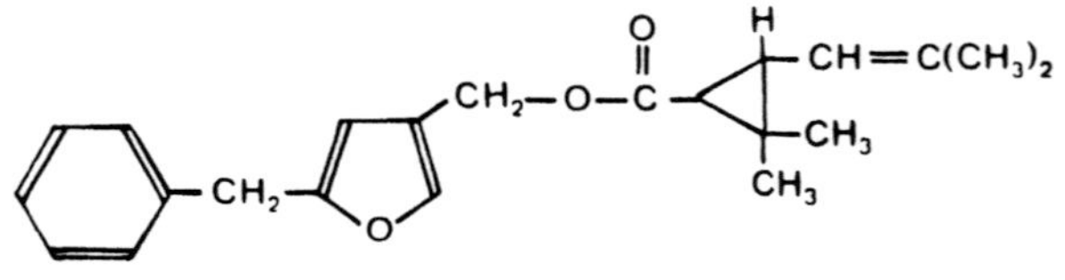

Resmethrin

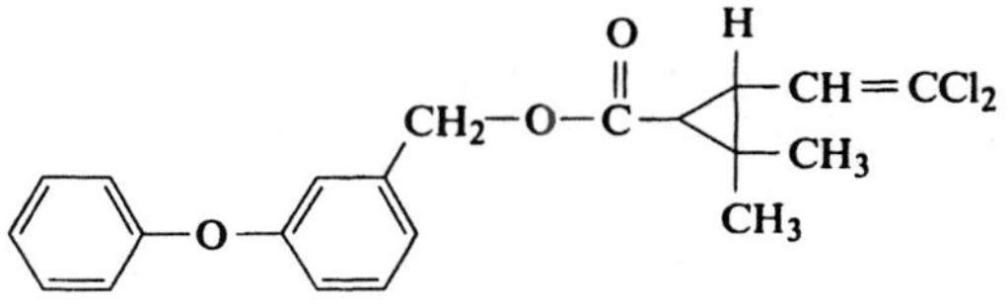

Permethrin

More recent examples still have substantially increased activity (0.01 to 0.05 kg ha^{-1}), are photostable, and provide long residual effectiveness in the field. Cypermethrin is an illustration.

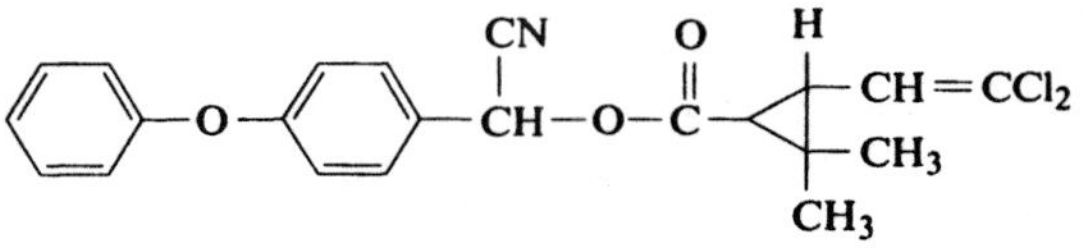

Cypermethrin

Tefluthrin (introduced at the end of the 1980s) is the first pyrethroid soil insecticide and is active against a variety of soil insect pests.

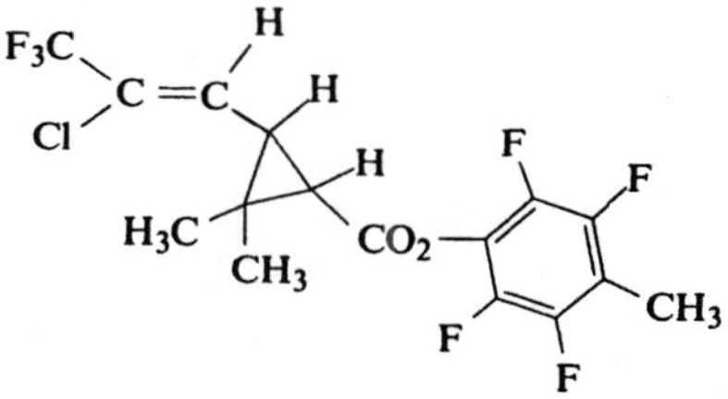

Tefluthrin

Fenvalerate is a highly active contact insecticide which is very useful because it controls some strains which are resistant to organochlorine, organophosphate and carbamate insecticides. It has the structure shown below.

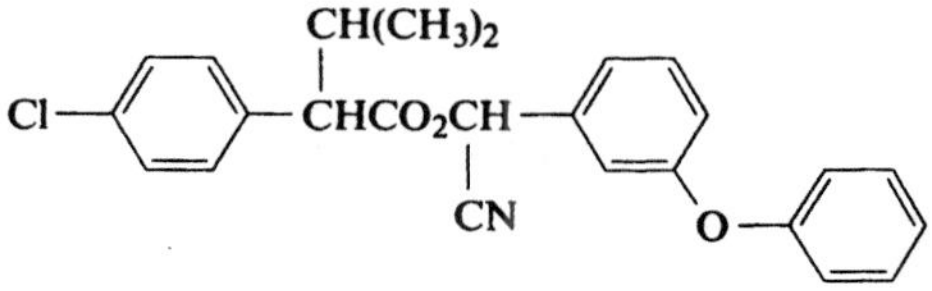

Fenvalerate

7.6.3   Synthetic insecticides

As indicated previously, these dominate the market and their importance in tropical climates is stressed again. Tables 7.1 and 7.2 demonstrate their commercial importance.

Pre-1940 synthetic insecticides, like the fungicides in use at that time, were largely inorganic materials—arsenicals, sulphur and sodium fluoride (and alumino-and silico-fluorides). By today's standards these would be quite unacceptable because of their high mammalian toxicity. Since 1940 development of new synthetic insecticides has mirrored the evolution of the pesticides industry as a whole, with interest focused almost entirely on organic molecules (except for 'mixed' compounds like the organophosphates) and the discovery of an extremely diverse range of chemical structures which show the required biological activity. The following important groups may be indentified: organochlorines, organophoshates, carbamates, dinitrophenols, bridged biphenyls, formamidines and microbiologically-synthesized insecticides.

7.6.3.1 Organochlorines Historically these are the most important and controversial insecticides, and include the most widely known pesticide of all, DDT.

DDT (dichlorodiphenyltrichloroethane) was first synthesized in 1873, but it was not until 1939 that Paul Muller of Geigy discovered its insecticidal activity, a discovery that brought him the Nobel Prize for medicine in 1948. It was a remarkable molecule, extremely effective against a very wide variety of insect pests ranging from houseflies, body lice and mosquitoes to Colorado beetles, pink bollworms and gypsy moths. For the important lessons which it has taught us it is worth studying in detail, and since any effect chemical is never 100% advantageous we must consider the advantages and disadvantages, since the balance between the two determines the utility of any chemical. Note that its most beneficial role has been in human health rather than agriculture, although its use in the latter has been widespread.

The advantages of DDT are (i) high insecticidal activity against a wide variety of pests (see above), (ii) little or no mammalian toxicity, (iii) a very stable molecule, and (iv) simple and very cheap manufacture. In fact DDT manufacture involves almost the perfect organic 'synthesis'—heating the reactants together in a single step, pouring the reaction mixture into water and filtering off the crude DDT, which can be used without further purification in

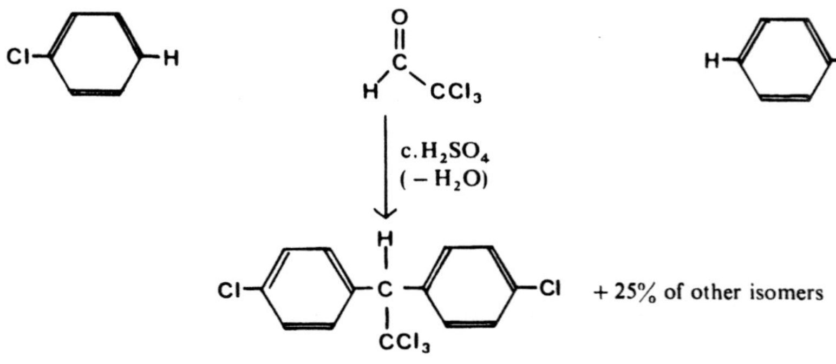

DDT

many applications. So cheap was this synthesis that in 1970 DDT was selling for only 18.5 US cents per pound weight.

Its great chemical stability is on the face of it very desirable, since this means the compound will remain unchanged and continue to exercise its insecticidal properties. Re-application will not be necessary and therefore the cost of treatment will be kept very low. Ironically, therefore, as we will see shortly, it was this very property that led to DDT's downfall.

DDT first came to the general public's attention in 1943 during World War II, when it was used by the US Army in Naples to arrest, for the first time in history, a typhus epidemic. This one disease killed more than 2·5 million Russians during the same war.

Once it had been recognized that DDT killed the mosquitoes which spread malaria, a world-wide malaria eradication programme was quickly put into operation. This programme alone was at one stage consuming 60 000 tonnes of DDT per annum. In its early days it was extremely successful. For example in India alone the annual number of deaths dropped from 500 000 in 1960 to only 1 000 by the early 1970s, and world-wide the annual number of deaths attributed to malaria has been reduced from over 6 million to less than 2 million because of the success of DDT and other insecticides. The scale of DDT's success in these and agricultural applications may be gauged by production statistics. In its heyday (late 1950s) over 100 000 tonnes per annum were being manufactured, although by the early 1970s the figure had fallen to around 20 000 tonnes per annum. (The reasons for this are discussed in the next subsection.) Since its introduction in 1939 a total of more than 2 million tonnes of DDT have been used in agriculture and human health—a remarkable amount for a pesticide.

Rachel Carson in her book *Silent Spring* (1962) was the first person to draw attention to the possible adverse environmental effects of pesticides. By the mid to late 1960s concern was being expressed about the plight of wildlife, particularly birds, in areas where DDT had been and was still being used extensively. Birds, for example, were producing eggs with very thin shells which broke before hatching was due, and deaths occurred amongst birds of prey such as sparrowhawks and peregrine falcons. Investigation showed that some of the latter had a very high concentration of DDT in their bodies—up to several thousand ppm.

These problems are a consequence of two properties of the DDT molecule—firstly, its solubility characteristics (virtually insoluble in water, 2×10^{-4} ppm, but very soluble, 10^5 ppm, in lipids and organic solvents like hexane); secondly, and as indicated earlier rather ironically, its great chemical stability. It is biodegraded only very slowly in the environment and its main degradation product DDE (formed by loss of HCl from the trichloroethane part of the DDT molecule) is even more stable. In contrast, DDT is metabolized slowly in humans to DDA (formed by hydrolysis of the $-CCl_3$ group to $-CO_2H$). Thus it is referred to as a persistent insecticide.

In the environment, therefore, the DDT is slowly leached into rivers and streams. There it enters and passes up the food chain, concentrating in fatty tissues as it does so. The following figures quantifying this refer to the Lake Michigan area of North America, an important fruit-growing area in which DDT was sprayed very liberally. Note that the numbers are directly comparable because they are expressed in ppm of body weight.

	DDT concentration (ppm of body weight)
Bottom muds ⎱ of lake	0.014
Water ⎰	0.00002
Amphipods (eaten by fish)	0.410
Trout	6
Herring gulls	99
Peregrine falcons	5 000

Progressing up the food chain as far as the trout, the DDT will have no appreciable effect, but as far as the birds, particularly the peregrine falcons, are concerned serious problems—even death—will result.

Because of these problems, and also the development of DDT-resistant strains of insects, DDT was banned in the United States in 1973 and has since been banned in other countries in the developed world. Alternative insecticides are available, such as organophosphates, but they cost at least three times as much as DDT. The developed world has the luxury of choosing to use these, as it can pay more for the food which they help protect and for a pleasant environment. However, in the Third World the nations' resources are so small that even DDT can barely be afforded. It is therefore still used on a large scale in these countries.

Any judgement on DDT must balance the great benefits which it has brought humankind against the adverse environmental effects which it has caused. It must be based on scientific facts and not emotion. Clearly, on balance it has been of great value to the world; nevertheless, we should now use alternative more environmentally acceptable insecticides where possible.

What lessons has DDT taught us? Firstly, the theoretical attractiveness of great chemical stability (i.e. persistence coupled with broad-spectrum activity) should be viewed with caution. Secondly, when coupled with high lipophilicity it means that the pesticide may be an environmental danger. Finally, investigations into the mode of action of DDT illustrate the difficulty of unravelling the complexities of chemical processes in living systems. These investigations have been going on for 40 years and many thousands of research papers have been published; we know that DDT interferes with the insect's nervous system, preventing the transmission of nerve impulses, and does not appear to react with any particular enzyme, but full details at the molecular level still elude us. This highlights the difficulties in adopting the attractive, logical, biorational approach to developing new pesticides, where interference with a specific biochemical function is the target. Despite this, many would argue that this is the direction in which pesticide research should increasingly move.

Most of the *cyclodienes* were introduced in the late 1940s and early 1950s. Aldrin and Dieldrin are well known examples and are synthesized commencing with a Diels–Alder reaction. Like DDT they are persistent insecticides which are also very lipophilic, but they are much more toxic. They have been widely used as soil insecticides and seed dressing and are extremely good for the control of termites. Dieldrin proved excellent for the control of ectoparasites e.g. lice on sheep and cattle ticks. Their properties, similar to DDT, caused the same sort of problems, made worse by their higher mammalian toxicity, and in the 1950s they were responsible for the deaths of a significant number of seed-eating birds. These adverse environmental effects plus development of insect resistance have led to complete banning of the cyclodienes or severe curtailment of their use in much of the developed world.

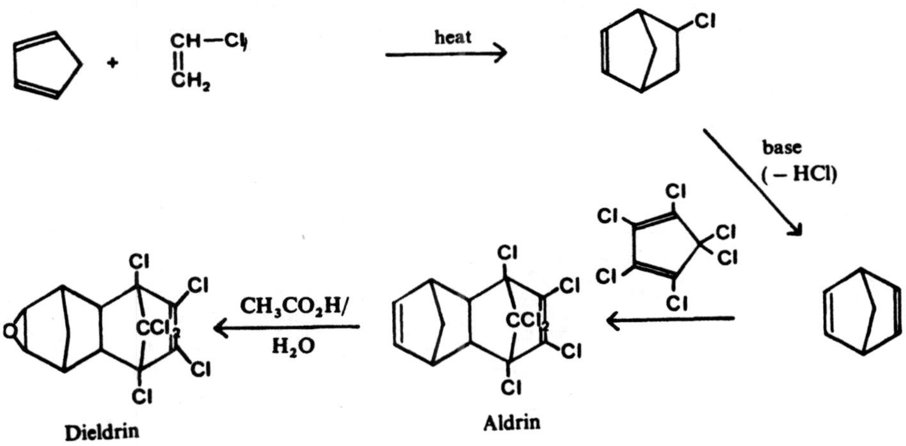

Dieldrin   Aldrin

The more expensive organophosphates and carbamates have taken their place. As might be expected, Aldrin and Dieldrin have a similar mode of action to that of DDT.

γ-BHC (Lindane) still tends to be incorrectly called BHC (benzene hexachloride) since it is actually a hexachlorocyclohexane. Its preparation involves an unusual reaction of benzene—that of *addition* of chlorine.

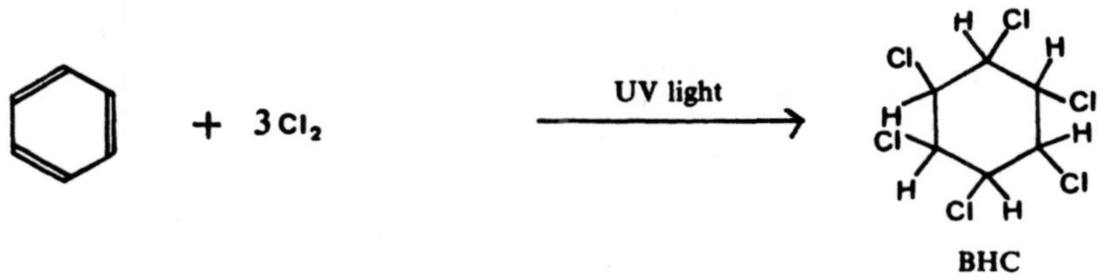

BHC

The BHC obtained consists of a mixture of several stereoisomers and only one of these actually possesses insecticidal properties. This is designated the

γ-isomer and is easily remembered because when the cyclohexane ring is drawn in its correct chair form all six chlorine atoms occupy equatorial positions. If necessary, purification to concentrate the γ-isomer can be carried out, and a product which is $> 99\%$ γ-BHC is marketed under the trade name Lindane.

BHC was yet another important pesticide to be discovered in the early 1940s. It was first synthesized in 1825 (compare 1875 for DDT). It has broadly similar properties to DDT, is an effective seed dressing for protection against soil insect attack, and has also been used as a fumigant. One big advantage over DDT is the fact that it is not persistent and therefore does not create environmental problems. It has therefore not been restricted in the same way as other organochlorine insecticides.

7.6.3.2 Organophosphates. Organophosphates have now largely replaced organochlorine insecticides, but the first insecticidal organophosphates originated from studies by Schrader at I. G. Farben in Germany before and during World War II when work was in progress to develop nerve gases. As a result many of the early compounds which showed promising insecticidal activity were too toxic to mammals to be acceptable.

World-wide interest in organophosphate insecticides was triggered off by Schrader's synthesis in 1944 of Parathion. Since then many thousands of active

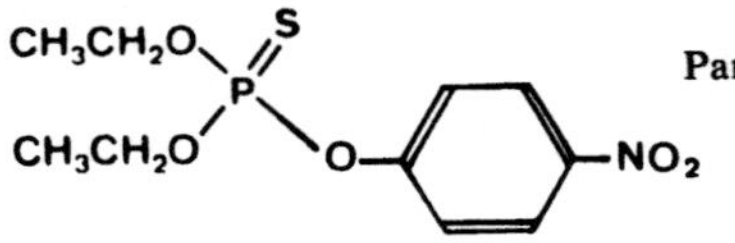

$$CH_3CH_2O\diagdown \underset{CH_3CH_2O\diagup}{P}\overset{\displaystyle S}{\diagdown}O{-}\langle \text{ring} \rangle{-}NO_2 \qquad \text{Parathion}$$

compounds have been synthesized, but not all of these are distinguished by improved selectivity of action, reduced mammalian toxicity or more acceptable levels of persistence. Most of these compounds fit the general structural formula.

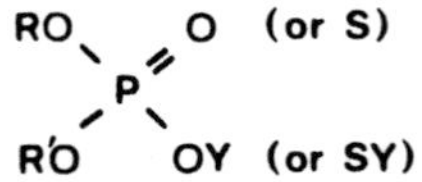

$$RO\diagdown \underset{R'O\diagup}{P}\overset{\displaystyle O \text{ (or S)}}{\diagdown}OY \text{ (or SY)}$$

The substituents R and R' are commonly $-CH_3$ or $-CH_2CH_3$ but Y can be any of a wide variety of alkyl, aryl, mixed or heterocyclic groups which in turn may or may not be substituted.

Malathion, introduced in 1950 by Cyanamid, was a major step forward with its very much lower mammalian toxicity.

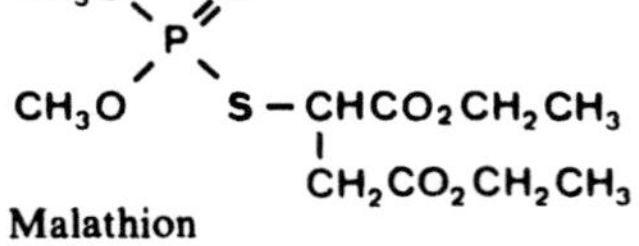

$$CH_3O\diagdown \underset{CH_3O\diagup}{P}\overset{\displaystyle S}{\diagdown}S-CHCO_2CH_2CH_3$$
$$\quad | $$
$$\quad CH_2CO_2CH_2CH_3$$

Malathion

It has been widely used to control an extensive range of insect pests, e.g. aphids and red spider mites, both in agriculture and around the home, and is made as follows:

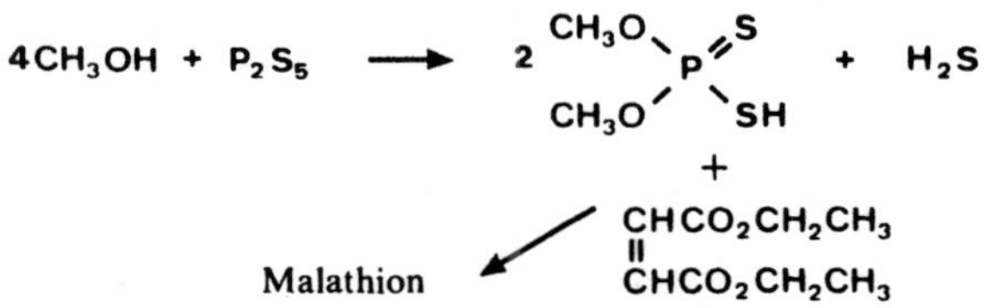

$$4CH_3OH + P_2S_5 \longrightarrow 2\ \begin{matrix} CH_3O \\ \\ CH_3O \end{matrix}\!\! P \!\!\begin{matrix} S \\ \\ SH \end{matrix} + H_2S$$

$$+$$

$$\begin{matrix} CHCO_2CH_2CH_3 \\ \| \\ CHCO_2CH_2CH_3 \end{matrix}$$

Malathion

In practice this is a one-stage process with the dimethyl maleate being added once the dithoic acid is formed.

Dichlorovos (Vapona), introduced in 1951 by Shell, is more volatile than most organophosphate insecticides and is impregnated on strips for use in houses. It is slowly released over a matter of months, killing flying insects. Although it has fairly high toxicity it is rapidly detoxified in mammals

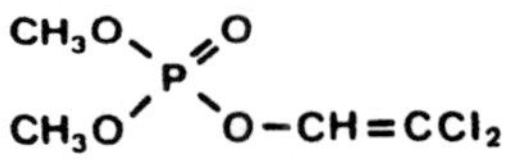

$$\begin{matrix} CH_3O \\ \\ CH_3O \end{matrix}\!\! P \!\!\begin{matrix} O \\ \\ O-CH=CCl_2 \end{matrix}$$

Vapona

Turning to examples where the group Y is a heterocyclic system, as we have already seen earlier in this chapter, and will see also in the pharmaceutical field, the pyrimidine and triazine ring systems are prominent in many biologically-active molecules and the same is true in the organophosphate field. Diazinon, introduced by Geigy in 1951, is a non-systemic insecticide with fairly low mammalian toxicity, good residual activity and a wide spectrum of activity, e.g. aphids, carrot flies, spider mites.

Pirimiphos-methyl, introduced by ICI (1970) and containing a pyrimidine ring, will control insects which affect stored crops, e.g beetles and moths, and also those which are a health hazard, e.g. fleas, lice and cockroaches. It has very low mammalian toxicity and acts very quickly by both fumigant and contact action.

Menazon, also introduced by ICI (1961), is a much more selective systemic insecticide which specifically controls aphids. It is environmentally very acceptable because it does not harm beneficial insects like bees and ladybirds and is very safe because of its low mammalian toxicity. In common with many organophosphates where Y in our general structure is a heterocyclic group, it is made by reacting the appropriate phosphorus acid salt with a chloro-heterocyclic group.

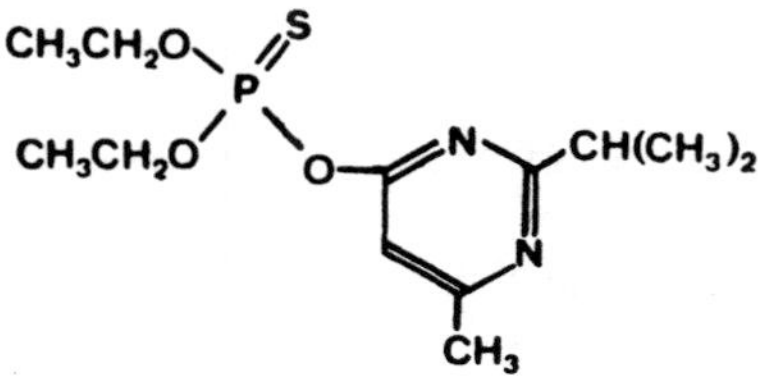

Diazinon

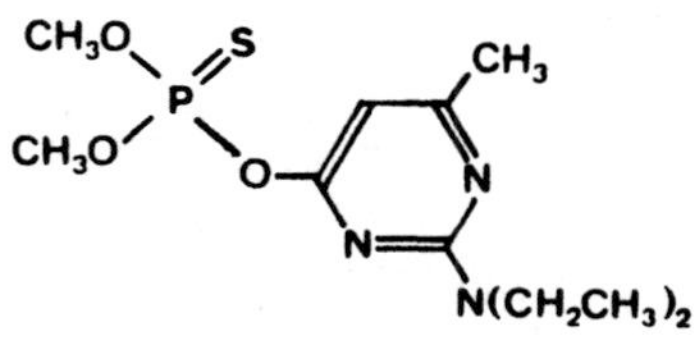

Pirimiphos-methyl

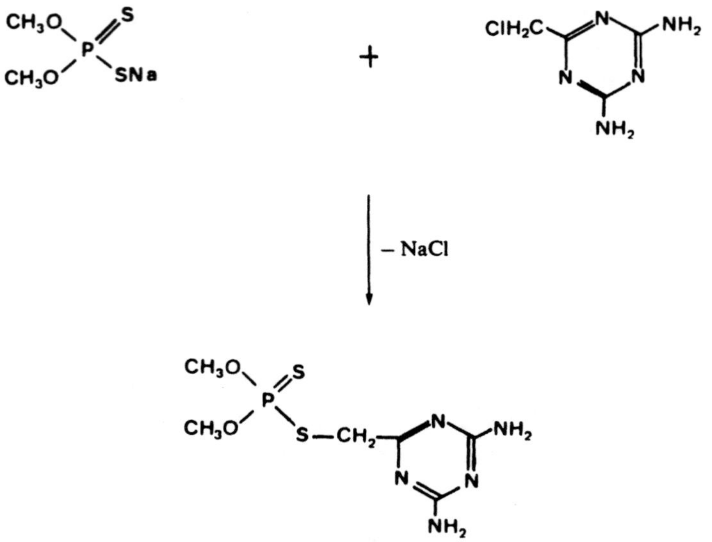

Menazon

Organophosphates act by inhibiting the action of important enzymes of the insect's nervous system, particularly the cholinesterases which maintain the organization and transmission of nerve impulses. The magnitude of their insecticidal activity is roughly proportional to their ability to phosphorylate (and hence tie up) acetylcholinesterase. We may formulate this in simple chemical terms as follows:

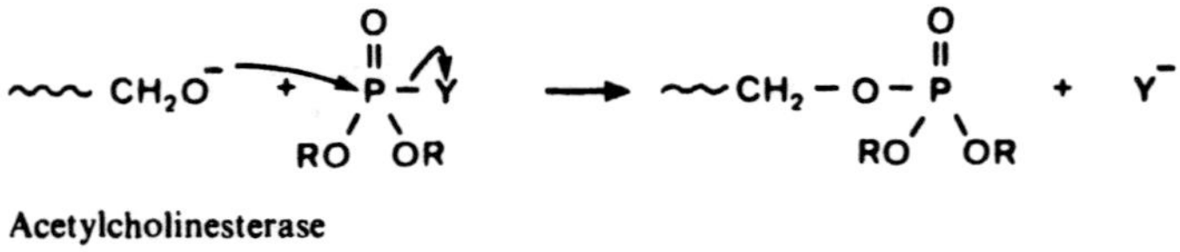

Acetylcholinesterase

They differ from the organochlorines in that they owe their activity to only part of the molecule—the toxophore—rather than the whole molecule, and they are biodegradable by hydrolysis of the phosphate ester toxophore. It is finally worth emphasizing again that these advantages over the organochlorines have a price—organophosphates are more expensive by a factor of at least three.

7.6.3.3 Carbamates. The third important group of synthetic insecticides is the carbamates. Like the organophosphates they act by cholinesterase inhibition; again the process may be formulated in simple chemical terms.

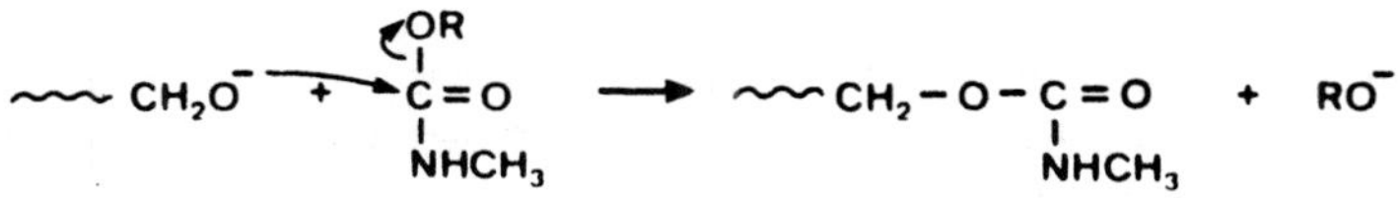

$$\sim CH_2O^- \ + \ \overset{\displaystyle OR}{\underset{\displaystyle NHCH_3}{C=O}} \ \longrightarrow \ \sim CH_2-O-\underset{\displaystyle NHCH_3}{C}=O \ + \ RO^-$$

Acetylcholinesterase

However, whereas the Y leaving group in the organophosphates could be varied quite widely with relatively little change in the insecticidal activity, here the nature of the OR leaving group is critical.

Carbaryl, introduced by Union Carbide in 1956, was the first successful carbamate and has one of the widest spectra of activity of any insecticide, being used on fruit, vegetables and most other crops. Its other big advantage is its low mammalian toxicity. Its use has been greater than that of all the other carbamates combined.

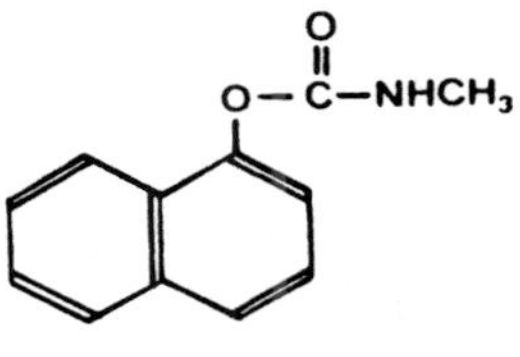

Carbaryl

Carbofuran (1967) is a systemic heterocyclic carbamate which has a wide range of insecticidal, acaricidal and nematicidal activities. It is therefore active against a wide variety of soil and foliar pests.

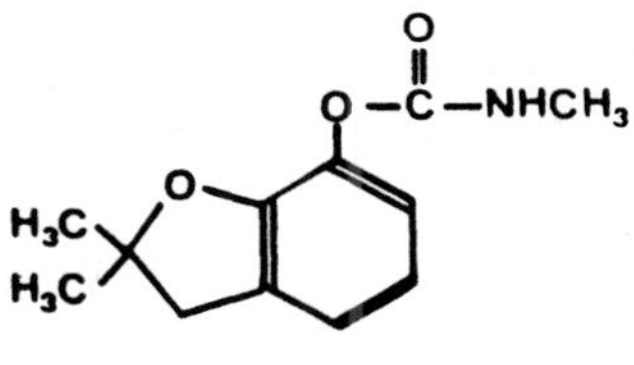

Carbofuran

7.6.3.4 *Other insecticides.*   Relatively few other groups of insecticides have been developed. Therefore mention is made only of the dinitrophenols, bridged biphenyls and the formamidines. An example of the latter is Chlordimefon which is toxic by virtue of its effect on the insect's nervous system.

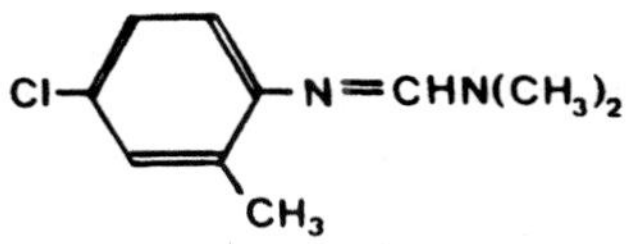

Chlordimefon

7.6.3.5   Microbiologically synthesized insecticides.   Production of pesticides by culturing micro-organisms (fermentation) is an area of great interest to the agrochemicals industry, and the fungicides Blasticidin-S and Kasugamycin (p. 237) are obtained in this way.

The avermectins are a group of macrocyclic molecules (actually pentacyclic lactones) isolated from a fermentation broth of the soil organism *Streptomyces avermitilis*, and consist of eight major components with high anthelminthic, acaricidal and insecticidal activity.[5] They are very potent compounds: application rates as low as $0.05\,kg\,ha^{-1}$ are effective against some pest species.

Although they interfere with the insect's nervous system their mode of action is different to that of organophosphates and carbamates and is novel. It involves inhibition of α-aminobutyric acid (GABA)-mediated inhibitory potentials as well as the excitatory postsynaptic potentials at the neuromuscular junction. This stops GABA-mediated nerves from regulating peripheral muscles.

Avermectins offer some grounds for optimism since they are effective against insects resistant to other insecticides, and their differing mode of action mentioned above may prevent the development of cross-resistance (see also section 7.7).

7.6.4   Alternative methods of insect control

There is a great deal of interest in several alternatives to traditional chemical insecticides, and much research and development work is being carried out on these by major agrochemical companies. Examples which we will briefly consider are (i) release of sterilized males, (ii) biological predators, (iii) microbial insecticides, (iv) pheromones and (v) growth-regulating hormones.

7.6.4.1   Release of sterilized males.   In this method very large numbers of, say, the male of the pest insect species are bred and are then sterilized by exposure to X- or γ-radiation. They are then introduced into the problem area in such numbers that there are far more of them than there are natural, fertile males, so that mating is much more likely to involve a sterile insect. As a result no offspring will be produced. Thus in time the population will decline and insects will no longer be a pest.

Although the drawbacks are obvious—the vast numbers needed and the slowness of the processes—the method has clear advantages in that it is highly specific and will not create any adverse environmental effects. It was extremely successful in the eradication of the screw-worm from Curaçao. This, however, was a rather special situation involving an island. A similar campaign in the south-western USA involving release of 4 billion sterilized male screw-worms was only partly successful.

An alternative approach is to use chemosterilants, i.e. chemicals which sterilize directly the natural population of insects. One such compound is Apholate.

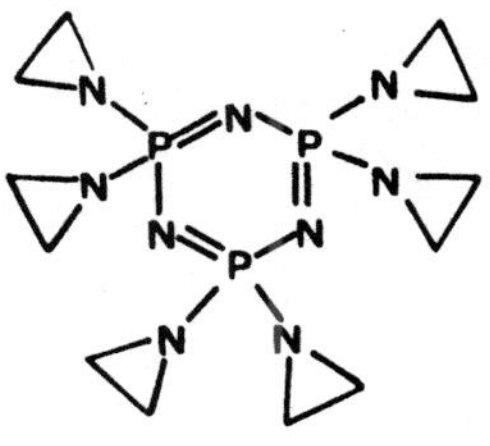

Apholate

7.6.4.2 Biological predators.   This method consists of releasing a biological predator to control the insect pest. In the UK its only successes have been in enclosed environments such as glasshouses. However, the technique was used very successfully nearly 100 years ago in California when the Australian ladybird beetle was introduced as a predator for cottony cushion scale which was ravaging citrus trees.

One example of where this approach went drastically wrong was the introduction of the cane toad into Northern Australia to control greyback beetles, which were destroying sugar cane crops.[6] They were introduced from Honolulu in 1935 and being free of natural predators begin to breed prodigously. They ignored the greyback beetles in favour of easier pickings of small prey. Unfortunately, when eaten or attacked by larger animals, such as snakes and birds, they secrete a powerful venom from their shoulder glands which rapidly kills the animal. They have spread south from Northern Queensland down beyond Brisbane and even cats and dogs have been killed by them. There is serious concern for the threat that they pose to indigenous creatures and scientists are trying to identify a suitable virus with which to kill the toads.

7.6.4.3 Microbial insecticides.   This is a much more recent approach to insect control. The microbe can be either bacterial, viral or fungal, and may be regarded as a 'living insecticide'. *Bacillus thuringiensis* is already being marketed as a larvicide for blackflies and mosquitoes and to control caterpillars. In 1983 Lisansky produced a virus which was aerially sprayed on to 4000 hectares of Scots pine to control an infestation of sawfly. His company, named Microbial Resources Limited, offers two bacterial and two fungal pesticides which have been cleared for use in the UK. One of the latter—*Verticillum lecanii*—devours aphids and whitefly. Several of the multinational chemical giants are also marketing products, e.g. Abbott Laboratories, Sandoz and Solvay et Cie. The cost of this method is about 10 times that of using DDT—but of course it is much more specific and utilizes naturally-occurring organisms.

7.6.4.4   Pheromones.   The term pheromone was originally used for insect sex attractants but its usage has now expanded to include any behaviour-affecting chemicals released by living organisms. It is a fascinating area of study, but our interest will be confined to insect sex attractants. These are chemicals released in minute amounts (microgram quantities) by one sex of the insect and detected anywhere up to a few miles away by another insect of the same species of opposite sex.

Chemists have isolated, identified and then synthesized the compound which is then placed in sticky traps to lure the attracted insects to their deaths. The attraction of this approach is that it is very selective (different species of insect each produce their own different chemical(s)) and will not create any environmental problems. Unfortunately the cost is several times that of using conventional chemical insecticides.

A number of pheromones are commercially available, Zoecon Corporation being a prominent source. In the USA the number of pheromones on the market increased from 9 in 1978 to 90 in 1982. An example is Dispalure, the sex attractant of the gypsy moth.

$$CH_3CH(CH_2)_4 - \overset{O}{\overset{/\ \backslash}{CH - CH}} - (CH_2)_9 CH_3$$
$$\underset{CH_3}{|}$$

Dispalure

Their use as pest control agents has not always been successful. In Scandinavia the technique was used to try to control moths that were affecting trees; a vast number of moths were trapped, but the population level had hardly fallen at all—those trapped had been replaced by others coming into the controlled area from outside. Biologists, however, use this technique to monitor insect population densities.

7.6.4.5   Growth-regulating (juvenile) hormones.   These are identical with or very similar to chemicals produced by insects which regulate the beginning and end of various stages of their growth, such as moulting and metamorphosis. Insect growth regulators (IGRs) alter the growth and development of insects.

Great success has been achieved with juvenile hormones in the laboratory but in the field it is much more difficult to judge the precise timing of application of the IGR to achieve maximum efficiency. The best-known commercial product is Methoprene (Zoecon Corporation, 1975).

$$CH_3O - \overset{CH_3}{\underset{CH_3}{\overset{|}{C}}} - (CH_2)_3 \overset{CH_3}{\overset{|}{CH}}CH_2\overset{H}{\overset{|}{C}} = C - \overset{CH_3}{\overset{|}{C}} = C - CO_2CH(CH_3)_2$$

Methoprene

It controls mosquito larvae by preventing their development beyond the pupal stage, when they die. It is also used to control fleas on domestic pets.

They are also prevented from reaching the adult stage and therefore cannot breed.

Dimilin or Difubenzuron (Thompson-Hayward Chemical Co.) is registered by the EPA for use against gypsy moth caterpillars in forests and boll weevil in cotton. It is also awaiting approval for use against a variety of other insect pests. It is active against mosquitoes at as low a concentration as $1.5\,g\,ha^{-1}$. Structurally it is a substituted urea.

Difubenzuron

Dimilin acts against the larval stage of most insects by blocking the synthesis of chitin. This is a vital part of the hard outer covering of insects known as the exoskeleton, and lack of it renders the insects much more susceptible to attack by predators.

7.7 Pesticide resistance

Application of a pesticide virtually never achieves 100% mortality of the pest; those individuals that survive tend to have an inbuilt resistance to the pesticide. They then breed and their offspring develop a greater resistance to that pesticide. This can only be partially overcome by applying even greater concentrations of the pesticide. Even so, in time—usually a few to several years—the pest will become totally resistant to that particular pesticide and control can only be regained by trying an alternative chemical. The problem of resistance now applies to some members of all three major groups of pesticides, although it has been most apparent and. widest-ranging with insecticides.

The development of resistant strains of insects has, rather ironically, been aided by the great success of the early insecticides like DDT. Their high toxicity meant that the insects susceptible to them were rapidly killed, and because of the compound's broad spectrum of activity the predators of the insect pest were often also eliminated. The few surviving insects, with their slight genetic difference conferring resistance, were therefore able to multiply rapidly, and resistance became a major problem. Even worse, generally speaking, they tend to develop resistance to a second insecticide (i.e. cross-resistance) much more rapidly than a non-resistant strain of the same insect. An example of this occurred in the malaria eradication programme in Ceylon (now Sri Lanka), where DDT was very effective in the early 1960s in controlling the disease. As the mosquitoes which spread the disease developed resistance, organophosphates like malathion were introduced in the early 1970s in place of DDT. Within a few years mosquitoes were discovered which were becoming resistant to malathion.

Although examples of resistance to fungicides have only been noted more recently, many examples are now known. For example, although Dimethirimol was only introduced in 1968 cases of resistance were already reported in 1969. Even though its use was quickly discontinued, widespread resistance to it was still found in 1971. Interestingly, it seems that resistance develops rapidly towards systemic fungicides but not towards surface or contact fungicides. This may be a consequence of the former interfering with a specific enzyme reaction whereas the latter affect more general processes.

In contrast to insecticides and fungicides, resistance to herbicides is much rarer. One example is the development of resistance of a common annual weed (*Erechtipes hieracifolia*) in sugar beet to 2,4-D.

This relatively rapid onset of resistance to many pesticides is another powerful argument for somehow streamlining and reducing the time-scale for the discovery and development of new pesticides. It also provides an incentive to use the integrated pest-management approach.

7.8 Integrated pest management

The general approach to crop protection has changed somewhat in recent years. The tendency is now towards controlling the pest population to an acceptable level rather than trying to wipe it out completely, because of the danger of 'knock-on' effects when the balance of nature is disturbed. Integrated pest control is also now the norm. In this strategy a combination of biological and chemical methods of pest control allied with good agricultural practice are used, instead of only chemical pesticides. This should not only be more effective but is also less likely to lead to pesticide resistance.

The British Agrochemicals Association and its counterparts in Germany and Spain are involved in an independent project entitled Linking Environment and Farming (LEAF). Its objective is to develop and promote the concept of integrated crop management, combining the best of traditional practices and of modern technology as the way forward in producing a reliable supply of high-quality food economically, whilst maintaining and enhancing the environment.

7.9 The future

The agrochemicals industry clearly has a secure future because of the ever-increasing need to produce more food for the world's growing population. Although the potential for alternative methods to chemical pesticides (e.g. pheromones, biological predators, growth-regulating hormones for pest control) is considerable, much remains to be done before they can be considered as viable alternatives. It is clearly recognized that for the foreseeable future the major weapon in our armoury will continue to be chemical pest-control agents. Some of these chemicals will also be the front line in the battle against the insects responsible for the spread of virulent tropical diseases such as malaria, sleeping sickness and elephantiasis, which cause so much suffering.

The great public interest in the safety of chemicals and their effect on the environment has been discussed in section 7.3. There is no disputing the improvements in safety which have taken place if one compares the modern pesticides with the very early ones, such as those containing arsenic and mercury, and it would seem that, at least in the developed world, the lessons of the environmental problems of the organochlorine insecticides are being heeded. Nevertheless there is still scope for further improvement, as the controversy surrounding 2,4,5-T indicates. However, the effect which public pressure has already had must be recognized and considered. As section 7.3.2 showed, public concern for environmental effects has been a major cause of a marked decline in the rate of introduction of new pesticides, and this has been accompanied by reduction in the success rate, increase in the time-scale and a marked increase in the cost of discovering a new pesticide. We are now arriving at the stage where a balance must be reached between the amount of testing carried out and the safety of the compound. The number of tests laid down must soon level off instead of constantly being increased. Insecticides, in particular, present compelling reasons for this; in several cases seven years of development before marketing have been rewarded with the detection of insect resistance after only two to three years of use. It is worth restating that it is in the interest of the pesticide producer as well as that of the public for the products to be safe and effective, since the vast expense of developing the product is not justified if problems in use mean its withdrawal from the market after only a few years.

Remarkable selectivity of action has now been achieved in some cases, such as the herbicides controlling the pest wild oats in cereal crops, and insecticides like Menazon attacking aphids but not bees and ladybirds. The increasing attention to selectiveness in activity also promises to reduce problems of resistance and cross-resistance which have been encountered with broad-spectrum pesticides like DDT and the organophosphates.

The marketing of only the bio-active enantiomer (or stereoisomer) to minimize the possibility of adverse environmental effects should continue to grow (see Fusilade, section 7.5.8).

All commercial pesticides so far developed have been discovered by using one of two approaches to producing compounds for testing—firstly, a purely random approach whereby any chemical either readily available or generated through any type of research is screened for biological activity, and secondly analogue synthesis whereby compounds similar to known pesticides are synthesized. Pointers are gained by scanning academic research literature and patents. Although this is an improvement on the first method, its success rate is still low, being less than one in every 10 000 compounds tested. A newer approach—which we may term the biorational approach—is gaining increasing support, although its first commercial success is still awaited.[7] In this, an attempt is made to design a compound which will interfere with a specific biochemical process which is essential to the pest's survival. Clearly this is extremely difficult to achieve, but the method represents a logical scientific approach, and as we gain a more detailed understanding of the mode of action of existing pesticides at the molecular level (and hence of structure–activity relationships) progress will ensue with the biorational approach. Its great advantage should be in selectivity of action and in increasing the success rate.

Remarkable strides have been made in increasing the potency as well as the selectivity of pesticides. Thus typical application rates for herbicides have fallen over the years from several $kg\,ha^{-1}$ to $0.1\,kg\,ha^{-1}$ (2,4-D) down to $0.01\,kg\,ha^{-1}$ (sulphonylureas). In the insecticide field, application rates for the synthetic pyrethrins have fallen from $0.1\,kg\,ha^{-1}$ (Permethrin) to $0.01\,kg\,ha^{-1}$ (Cypermethrin). Despite these order-of-magnitude changes we have by no means reached the limits of potency, since Corbett et al.[8] have calculated (having made several assumptions) that for a post-emergence wild-oat herbicide only about $2.5 \times 10^{-9}\,kg\,ha^{-1}$ of the active ingredient is required to effect control. The difficulty lies in getting it to the point where it can start working, by more effective spraying.

Although biotechnology, and specifically genetic engineering, of plants has really only become significant in the last ten years, major advances in the subject are now being made. Early work was directed to producing higher-yielding, more disease-resistant strains of crops. Notable successes have been achieved with rice. This ability to transfer genes from one species to another has markedly improved the precision and speed with which new plants with beneficial traits are being developed.

Attention is also being given to controlling ripening rates and flavour. For example, early in 1993 the American company Calgene was awaiting Food and Drug Administration (FDA) approval to market genetically engineered tomatoes which not only look good but taste delicious. Normally tomato growers can either pick tomatoes before they are fully ripe (and therefore more resistant to bruising) or pick them fully ripe, which is excellent for flavour but, because they are soft, they are very easily damaged. A gene has been inserted which prevents the tomato producing the enzyme which causes softening.

Over 100 genetically engineered crops are undergoing field trials and include pest-resistant cotton and drought-resistant cereals. More information on these topics can be found in chapter 9.

The future of the agrochemicals industry may therefore be viewed with confidence, with a continuing high level of demand for its products, although maintaining the remarkably high growth rates of the last few decades may prove more difficult. Emphasis continues to move from complete annihilation of a pest to its control at an acceptable level by an integrated pest control programme. Environmental acceptability is a must.

References

1. British Agrochemicals Association. *Annual Report and Handbook*, 1992.
2. For a typical procedure see U.S. Patent 3 707 556 (Dec. 26, 1972), Stauffer Chemical Co.
3. 'Chemicals for Crop Protection and Pest Control,' M.B. Green, G.S. Hartley and T.F. West, Pergamon, 1977, 10.
4. 'Recent developments in the chemistry of azole fungicides,' P.A. Worthington, in *Proc. Br. Crop Prot. Conf. Pests Dis.*, 1984, 3, 955.

5. 'Avermectins: their chemistry and pesticidal activity,' R.A. Dybas and A.St.J. Green, *Proc. Br. Crop Prot. Conf. Pests Dis.*, 1984, 3, 947.
6. 'The toads are coming,' Paul Raffaele, Readers Digest (UK edition), 1992, p. 118.
7. As an example see 'The Biochemical Mode of Action of Pesticides,' 2nd edn, J.R. Corbett, K. Wright and A.C. Baillie, Academic Press, 1984.
8. *Ibid.*, pp. 343–4.

'The Biochemical Mode of Action of Pesticides,' 2nd edn, J.R. Corbett, K. Wright and A.C. Baillie, Academic Press, 1984.
'Chemistry and Agriculture,' April 1992, Chemical Society Special Publication, 1979.
'Chemicals for Crop Protection and Pest Control,' M.B. Green, G.S. Hartley and T.F. West, Pergamon, 1977.
'The Scientific Principles of Crop Protection,' H. Martin and D. Woodcock, 7th edn, Crane-Russak, 1983.
'Pesticides: Preparation and Mode of Action,' R.J. Cremlyn, Wiley, 1978.
'The Chemical Industry,' eds D. Sharp and T.F. West, Ellis Horwood, 1982, chapters 36–38.
'The Chemist in Industry 2: Human Health and Plant Protection,' E.S. Stern, J.F. Cavalla and D. Price Jones, Oxford University Press, 1974.
'Pesticides: Theory and Application,' G.W. Ware, Freeman, 1978.

Bibliography

8 The pharmaceutical industry

Craig Thornber

The pharmaceutical industry today is very largely a product of the post-war period. Few of the medicines used now were available before the war and many were not introduced until after 1960. A very limited range of products was available in the 1930s and natural products such as morphine, digitalis and quinine were important. Only a few synthetic compounds, such as aspirin and the barbiturates, were available. The war itself provided a great stimulus to developments for two reasons. Firstly, the strong German position in the pre-war pharmaceutical field meant that the UK and the USA had to find alternative sources of medicines with the outbreak of hostilities. Secondly, the need to treat wounded servicemen and to protect them from tropical infectious diseases in Africa, Asia and the Pacific islands led to a burst of research on infectious diseases. The first sulphonamide antibacterial agents had been known in the 1930s and the possibilities of synthetic drugs were increasingly appreciated. Research on penicillin and antimalarials led to new products which were developed quickly and produced on a large scale.

Those firms in a position to take part in this wartime work were able to grow and gain experience; this helped them to adapt to new challenges in peacetime. The success of the new antibacterial agents led to industrial growth and to the establishment of a research-based industry. New companies have entered the field since the war, supported by parent companies whose major interests were in fine chemicals, oil, food, dyestuffs and perfumery.

The industry has diversified from its earlier focusing on infectious diseases to look for therapies in areas such as heart disease, gastric ulcers, mental disorders, fertility regulation, arthritis, allergy and cancer.

There was a period of rapid growth in the number of products and in the world pharmaceutical market in the 1960s, which attracted new companies to enter the area. In the 1980s and early 1990s, there have been important developments; Japanese companies, which started relatively late compared to their Western counterparts, have started to become significant.

Throughout the developed world, there is an increasing proportion of the population over the age of 60. This is a consequence of longevity and, in some places, because of lower birth rates in recent decades. Patients over 60 are the main users of health-care systems. Moreover, some of the new therapies that have been developed are costly. As a consequence, there is a tendency for total health care costs to rise as a proportion of the gross national product. The efforts of governments to restrain this growth by a number of mechanisms, including price control, present limits to the growth of health-care providers, including the pharmaceutical industry. At the same time, companies face higher costs for research and development in the attempt to produce improved medicines and higher levels of competition as the market for pharmaceuticals becomes more international.

The result of these pressures has been for companies to seek mergers and strategic collaborations so that risk can be spread among a larger portfolio of products. Major new companies have emerged in the last few years by this means, such as Bristol-Myers Squibb, SmithKline Beecham and Rhone-Poulenc Rorer.

The third recent development is the growth of the biotechnology sector. A large number of small companies have been set up, particularly in the USA, often by small groups of scientists, supported by venture capital. So far, few have made trading profits. Some biotechnology companies seek collaborations with established pharmaceutical companies because of their need to gain access to capital, and to expertise in distribution and marketing; some will become absorbed by their larger partners. Others, with early successes, may emerge in due course as major pharmaceutical companies. By the end of 1991, it was estimated that a total of 132 biotechnology-based drugs and vaccines were in development in the USA, an increase of 63% since 1988. More than 20 biotechnology-derived therapeutic agents are on the market now. Those with the highest sales in 1991 were Epogen and Neupogen, both from the American company, Amgen.

The pharmaceutical industry has many features that distinguish it from other parts of the chemical industry. In addition to research, manufacturing and sales, we need to consider the interfaces that the industry has with medicine, law and government. The industry is very largely international in nature. This arises from the international nature of science and medicine and the ease with which relatively low tonnages of modern medicines can be transported. In many countries the pharmaceutical industry has a relationship with its customers that is unique in manufacturing. The industry provides drugs, the physician decides when to use them, the patient is the consumer and the bill is paid by private or national insurance.

As a consequence, the industry does not have a simple direct relationship with a customer but a complex interaction with the medical profession, the public and state health bureaux. Our personal involvement as patients, taxpayers and voters gives us an emotional, financial and political stake in health-care provision.

It is not possible to discuss the industry on a purely national basis. All the major companies have extensive export business and all the drug-using countries draw their supplies from several nations. Consequently, this chapter will set out to describe the world-wide pharmaceutical business, drawing in, where appropriate, the British perspective. Some of the processes of research and development will be described in general terms and some special features of the pharmaceutical market will be highlighted. Against this background, the costs of research and price of drugs can be discussed firstly as they relate to the companies involved and secondly as they relate to the overall economy of drug-producing nations. Finally, we shall look ahead at likely trends within the industry.

8.2
The world pharmaceutical market

8.2.1   Introduction

The pharmaceutical market falls into two segments. Ethical products are those that are not promoted to the general public; most of them are available only on prescription. Over-the-counter (OTC) products are freely available. The latter are generally well-established treatments for minor conditions. The distinction between these two categories is very desirable and is quite clear in countries that have a well-developed health-care system. For the purpose of this chapter, I shall consider the pharmaceutical market as comprising both prescription medicines and OTC preparations for human use. On a geographical basis it is necessary to concentrate upon Western Europe, Japan and the USA. This is because Eastern Europe, the countries of the former Soviet Union and China have made only a small contribution to pharmaceutical research, relying in the main upon copying Western drugs. In Eastern Europe, Hungary, which has several pharmaceutical companies with long traditions, is the best-placed of the former Soviet satellite states to develop a modern pharmaceutical industry.

In 1991, the total market for pharmaceutical products, excluding Eastern Europe, the former Soviet Union and China, was estimated to be £105 340 million, with £87 960 million of this for ethicals and the remainder for OTC products. Such an estimate, and comparisons for different regions and years, are complicated by constant currency fluctuations. The sales figures for the major markets are shown in Table 8.1. These data show that the top six countries account for about 70% of the world market. Their identity is not surprising since they are the most populous, wealthy and industrially advanced nations. Many of the smaller Western European nations have a *per capita* expenditure on pharmaceuticals similar to that of their larger neighbours.

8.2.2.   Companies

In Table 8.2 we examine the production rather than the consumption of pharmaceutical products by looking at the top 15 corporations. Seven of them are American owned, two British, two German, three Swiss and one a French and American partnership. The nations from which the top 30 companies are

Table 8.1 The world pharmaceutical market by country in 1991

Market	Value, £ billion	% of world market
USA	26.8	29
Japan	17.4	19
Germany	7.2	8
Italy	6.8	7
France	6.5	7
UK	3.2	3
Spain	2.5	3
Canada	2.2	2
Brazil	1.4	2
Mexico	1.4	2
Others	18.1	18

Source: *Scrip*, Review of 1992, January 1993, PJB Publications Ltd.

Table 8.2 The top 15 pharmaceutical companies world-wide in 1991

Rank	Company	Country	Sales, £ billion
1	Glaxo	UK	4.094
2	Merck and Co	USA	4.081
3	Bristol-Myers Squibb	USA	3.337
4	Hoechst	Germany	3.067
5	Ciba–Geigy	Switzerland	2.605
6	Sandoz	Switzerland	2.508
7	SmithKline Beecham	UK/USA	2.468
8	Bayer	Germany	2.434
9	Roche	Switzerland	2.327
10	Eli Lilly	USA	2.277
11	American Home Products	USA	2.270
12	Rhone-Poulenc Rorer	France/USA	2.160
13	Johnson and Johnson	USA	2.144
14	Pfizer	USA	2.129
15	Abbott	USA	1.984

Pound conversions based on the average rate for the year of £1 = \$1.77. Modified from data in *Scrip*, Review of 1992, January 1993, PJB Publications Ltd.

drawn, USA, Switzerland, Germany, Japan, UK and France, reflect not only drug consumption but strengths in chemistry, manufacturing and medicine. Japanese strength in pharmaceuticals has emerged more recently than that of the other producers. In 1989, Japan had a pharmaceuticals trade deficit of £823 million. Most of the European and American companies have many overseas subsidiaries for marketing and some have research bases in more than one territory. For example, the top three Swiss companies have major research-based operations in the USA and several American and continental companies undertake research in the UK. By contrast, Japanese companies are not well-established in Europe and the USA and foreign companies are not well established in Japan for reasons of language and culture, and because Japanese law required for many years that subsidiaries of overseas companies operating in Japan should be jointly owned with Japanese enterprises. Now, however, a number of Western companies do have subsidiaries in Japan and the Japanese producers have been able to license products to Western companies for development and marketing in Europe and North America.

The top 15 companies, with sales ranging from £2 000 million to £4 000 million in 1991, together account for about 38% of world pharmaceutical sales. The remainder is supplied by a very large number of smaller companies. Currently the leading companies each have about 3.7% of world sales and the thirtieth just over 1%. It can be appreciated from these figures that many companies, which are far from being in the top 30, will still be very substantial businesses.

In addition to Glaxo and SmithKline Beecham, which are mentioned in Table 8.2, there are four other British research-based companies. These are Zeneca (ICI), Wellcome, Boots and Fisons. The Glaxo group includes Allen and Hanbury and has research facilities at Ware in Hertfordshire, at Greenford in Middlesex, at Stevenage and in the USA. Beecham had wide ranging interests such as food and toiletries in addition to pharmaceuticals. Since the merger with SmithKline there has been a programme of divestment of business outside the pharmaceutical field. SmithKline Beecham has research facilities at Great Burgh, Brockam Park, Harlow and Welwyn in the UK and in Philadelphia. Bencard is a SmithKline Beecham subsidiary. Zeneca is involved in pharmaceuticals, agrochemicals and speciality chemicals; the pharmaceutical business is concerned exclusively with human medicines, together with the subsidiary, Stuart Pharmaceuticals. It has research facilities in Cheshire, Reims in France and in Wilmington, Delaware, USA. Wellcome has interests mainly in human medicine. It has research facilities in Beckenham, Kent, and a research institute in the USA at Bethesda, Maryland. Calmic is a Wellcome subsidiary. Boots has extensive interests in pharmaceutical wholesaling and in retailing through a nation-wide chain of pharmacies. Its research laboratories are located in Nottingham. Crookes Products is a Boots subsidiary. Fisons was formerly involved in animal health and fertilisers as well as pharmaceuticals but has been restructured to focus on pharmaceuticals. Fisons' research laboratories are in Loughborough. Weddel is a Fisons subsidiary. There are two smaller research-based British companies that have strengths in biotechnology; these are British Biotechnology and Amersham International.

Several overseas pharmaceuticals companies have subsidiaries in the UK. Those that conduct research in Britain include the Swiss companies, Ciba-Geigy, Sandoz and Roche, the American companies, Merck, Pfizer, Eli Lilly and Wyeth and the French companies, Roussel and Servier.

8.2.3   *Products*

The 25 leading branded pharmaceutical products, by value of sales in 1991, are listed in Table 8.3. The use of medicines could be calculated in terms of value, the number of prescriptions or the number of treatment days. Each method has its disadvantage because of variations both in the size of prescriptions and in the costs of different treatments.

The naming of drugs requires some explanation. Each single active ingredient has a full chemical name. Such names are invariably long and have value only to organic chemists. For general use in pharmacology and medicine, it is convenient to have a shorter name. Such a name is sometimes called a generic name or free name. Two specific groups of such names are International Nonproprietary Names (INN) and United States Approved Names ((USAN), which are recommended by the World Health Organisation (WHO) and by the USAN Council respectively.

In addition, each product has a trade name or brand name. Such names are trademarks and usually are the property of the manufacturer. If the same drug is sold by different companies in the same country it will be sold under different

Table 8.3   Top 25 drugs world-wide by sales in 1991

Brand name*	Indication	Company**	Sales, £ million
Zantac	Peptic ulcers	Glaxo	1700
Adalat/Procardia	Angina	Bayer/Pfizer	1100
Renitec	Hypertension	Merck and Co	990
Capoten	Hypertension	Bristol-Myers Squibb	890
Ceclor	Antibiotic	Lilly	700
Cardizem	Angina	Tanabe/Marion Merrell-Dow	700
Tenormin	Hypertension	ZENECA Pharmaceuticals	660
Voltaren	Arthritis	Ciba–Geigy	650
Ventolin	Asthma	Glaxo	650
Tagamet	Peptic ulcers	SmithKline Beecham	620
Mevacor	Hyplipidaemic	Merck and Co	620
Gaster/Pepcid	Peptic ulcers	Yamanouchi/Merck and Co	600
Naprosyn	Arthritis	Syntex	530
Augmentin	Antibiotic	SmithKline Beecham	500
Isoptin/Calan	Hypertension	Knoll/Searle	500
Prozac	Fluoxetine	Lilly	500
Zovirax	Anti-viral	Wellcome	500
Ciprobay	Antibiotic	Bayer	500
Iopamiron/Isovue	Contrast agent	Braccho/Schering AG/Daiichi	475
Losec	Peptic ulcer	Astra/Merck and Co	450
Rocephin	Antibiotic	Roche	440
Seldane	Hay fever	Marion Merrell Dow	430
Xanax	Tranquilliser	Upjohn	400
Becotide	Asthma	Glaxo	400
Sandimmun	Immunosuppressant	Sandoz	375

* Brand names of the drugs are trade marks.
** Originating company named first.
Modified from data in *Scrip*, Yearbook for 1992, PJB Publication Ltd.

brand names. For example Astra and Ciba–Geigy both sell metoprolol but under the trade marks Betaloc and Lopressor. Whereas the generic name applies to the active ingredient, the brand name refers to the formulated product containing the active compound or compounds and other materials, known as excipients, needed to produce a physically and chemically stable dosage form with the desired drug-release characteristics. Proprietary medicines are brand-name products available OTC, which are promoted directly to the public such as the numerous brands of aspirin.

The products listed are likely to encompass a variety of different dosage forms. A drug may be available in tablets of various sizes to enable the physician to titrate the dose for the individual patient. Some drugs may be available as tablets, as injectable forms or in creams for topical application. A compound may be a constituent of a product containing more than one active ingredient. The top 25 products sold from about £375 million to £1 700 million per annum in 1991, and together accounted for about £15 505 million, which was about 14% of the total world market. This is in marked contrast to the share of the world market held by the top 15 corporations of approximately 38%. The remaining 86% of the world market is made up of thousands of products with sales below £375 million a year. The major products are inevitably well-established treatments which have been launched in many countries and are likely to have been sold in the major markets for at least four or five years. Such drugs are likely to have been discovered anything from 10 to 20 years ago. Some drugs have more than one major manufacturer or distributor because of licensing arrangements. When the patent on a drug has expired, the drug can be manufactured by other companies, but Table 8.3 indicates the branded products originating from the inventing company and its main licensees.

The leading products in individual countries vary considerably because of variations in the incidence of disease, divergences in medical practice and the particular strengths of international and local companies in the market. Moreover, a drug is unlikely to be launched in all markets simultaneously and may not be launched at all in some for a variety of medical and commercial reasons. As a result the list of the leading 25 products world-wide in any one year represents an average rather than an accurate picture of any one country. It will be influenced more by the usage in the USA than by that of any other country.

If we were to consider the product usage by prescription number rather than by total value, a somewhat different picture would emerge. This is because a number of very well-established drugs which have not yet been superseded, continue to be used. They are available at lower prices than newly launched products because the patents on them have expired and several companies compete with the originator in the market place. For example, in the UK, minor analgesics (painkillers) such as aspirin, paracetamol and codeine preparations feature among the top 20 most prescribed drugs.

In 1991 the top 25 products by value in the UK achieved sales of from about

£15 million to £130 million. In total, they represented about one-quarter of the total pharmaceutical market in the country. The total prescriptions in the UK for 1990 were 450 million with the leading 25 drugs achieving from 3.2 million to 11.5 million prescriptions, with a total of about 27% of all prescriptions.

8.3.1 General considerations

8.3
The initiation of a project in the pharmaceutical industry

Let us consider the process whereby a drug is produced and launched into the market. If we imagine a situation in which a new pharmaceutical company is formed, we can trace the whole cascade of decisions and processes which lead to a marketable product. In the first instance, a group of people, possibly an existing chemical company, bankers or financiers, have some capital which they wish to invest to produce an attractive return. On the basis of examining the performance of several industries they may decide to enter the pharmaceutical business. In reaching this decision they will need to examine a number of important questions. How much capital needs to be invested to produce a viable business? How long will it take before the business makes a profit? What are the risks? Is suitably skilled manpower available? The most important single question is whether the likely return on capital looks attractive relative to other investments given the risks that have to be taken. These kinds of questions have to be addressed by the potential investors such as investment analysts in merchant banks, pension funds and insurance companies or the board of directors of an existing company.

There are several kinds of pharmaceutical business area; some companies specialise in one branch, some operate across the whole spectrum. Each branch of the business needs to be examined separately from the financial viewpoint. Some pharmaceutical products are sold OTC without a prescription. They are commonly for minor ailments and are sufficiently established as safe and simple to use that no medical guidance beyond instructions on the packet is needed. The active ingredients of such pharmaceuticals have usually been widely used for many years and are not the subject of current patents. The products are commonly creams and ointments for minor skin conditions, cough syrups, mild analgesics, vitamins and indigestion mixtures. To the general public, in the pharmacist's shop, these products may not seem to be clearly differentiated from toiletries, cosmetics and food supplements. Nevertheless, they do form a discrete group of products designed to treat medical conditions. The OTC business could be entered by a new company without doing research, by using well-established, patent-free compositions of long-established safety. The business would be mainly manufacturing and marketing through retailers. The products would have many competitors and profit margins would be likely to be small. Advertising to the general public would be of importance and the brand name and image could be very significant. In this respect the business could be compared with cosmetics, toiletries and food.

By contrast, ethical pharmaceuticals are those which are available only on prescription. These fall into two major classes. Products that are the subject of a current valid patent can be made only by the patent holders or, with their permission, by a licensee who pays the patentees a royalty. These products are sold mainly by the major research-based companies. When all the patent property on a product has expired in any country, other manufacturers are free to make and sell the product, subject to permission being gained from that country's drug regulation authorities. This kind of pharmaceutical product is called a generic drug. Such a product may be identical in composition to that produced by the originator. The active ingredient will have the same full chemical name and free name but it cannot be sold under the same brand name as the original, because that will be a trade mark. Similarly, the colour, shape and size of the product or package may have to be different so as not to infringe trade marks and registered designs.

It can be seen that the generic pharmaceutical business involves little research since it relies on the research of others and concentrates only on successful products whose patents have expired. As with OTC drugs, manufacturing and marketing are the main elements of the business. Competition may be strong from other generics manufacturers. Since little research has been undertaken, the manufacturer can afford to sell the product more cheaply than the originating company, which has to pay for the research on successful and unsuccessful projects. However, the generic drug company, like the research-based company, will be subject to high standards of quality control, will need to satisfy drug regulatory requirements and will be dependent upon sales arising from prescriptions. Consequently it can advertise only to the medical profession and not to the general public.

The research-based pharmaceutical company is concerned with the generation of totally new compounds known as new chemical entities (NCEs). The research is expensive and lengthy and the chance of success on any single project is low. Once a likely compound has been found, the company will need to undertake lengthy studies in animals and then in patients to prove to the medical profession and to government drug regulatory agencies that the compound is well tolerated and effective. This position could be reached eight years after the discovery of the new compound and say 12 to 15 years after initiation of the research programme. Consequently, this type of business is characterised by high research investment over a prolonged period and a long-delayed return on investment. The risks of failure are high, but equally, for the successful product, the rewards can be high if a significantly improved therapy is devised and sold in many countries during the period of patent protection.

In entering the pharmaceutical business, our hypothetical company will need to assess the opportunities and risks posed by each of the three kinds of operation—OTC, generics and new chemical entities (NCEs). For a British research-based pharmaceutical company, the cost of research and development of an NCE to the point where it could be sold in the UK is such that the money could not be recouped by selling only in the home market, which is only 3% of the total world market for pharmaceuticals. Consequently, whereas an OTC or generic pharmaceutical company could have a purely national business, a British research-based company is certain to want to take full advantage of the development of a successful product by selling it world-wide through its own subsidiary companies or through licensing arrangements. Such an operation inevitably involves an appreciation of the pharmaceutical business in each of the major territories. What special national features does each country have in relation to the manufacture and sale of pharmaceutical products? Does the country have a patent system or allow registered trade marks and designs? What are the requirements of the various national drug agencies? Will company law allow the formation of foreign-owned subsidiaries? Can the company take profits back to its own home country? Are there export or import impediments? How are drugs paid for in the country in question? Are prices fixed by governments and, if so, on what basis?

From this brief survey, it is evident that the decision made by a company about entering any business area will be dependent upon many financial, legal and commercial considerations. It is equally evident that entry to the research-based pharmaceutical business can be undertaken only by a large company with very substantial resources that is prepared to take risks, wait a long time for profits, and market its products in many countries. Since it is this sector of the pharmaceutical industry that generates new products and that is involved in the widest range of activities, it is this sector that we will follow in the subsequent discussion.

8.3.2   The research portfolio

Let us consider that our hypothetical pharmaceutical company has decided to enter the NCE sector of the business. According to its resources it may feel able to undertake research on a single disease or upon many. How can this be decided? Since the pharmaceutical market is an international one, any company producing an NCE is in direct competition with the pharmaceutical companies of the major drug-producing countries—the UK, Germany, Switzerland, France, Japan and the USA. It is not a coincidence but indeed a prerequisite that these nations all have a well-developed economy, a high standard of living, a traditional strength in science and a strong patent system. Thus, in any research project, it is necessary to appreciate that the competition will be international and the project must be resourced with a quantity and quality of people and equipment such that it has the chance of beating the competitors in either the novelty of the new treatment or its speed of introduction to the market. While these considerations provide a pressure towards establishing a few large projects, the need to spread the risks associated with any one project produces pressures to increase the number of research areas. In deciding the number and balance of projects within the research portfolio it is necessary to take into account the different risks associated with each kind of research.

In considering the attraction of any single disease area as a subject for research, there are medical, scientific and commercial considerations. Firstly, it is necessary to identify a significant medical need. What are the current treatments for the disease and in what way are they inadequate? Medical opinion from general practitioners and consultants will be sought. What is the overall size and structure of the market? The market may be fragmented with many companies and products or may be dominated by just a few drugs. The products may be mainly old and well-established and consequently cheaper because of the introduction of generic versions, or may be new products with patent protection, which are relatively more expensive. The special circumstances pertaining in each of the major countries all need to be considered. It is possible that a new product with really significant advantages will expand the market by drawing in new patients who have been hitherto untreated. In addition, with a drug that is a major breakthrough, a company will have a better chance of negotiating a good price for the product.

Using this knowledge it is possible to make arguments about the likely commercial success of various new products which have a range of advantages over current therapy and to make an assessment of the sales that could be achieved. Clearly by the time a new project produces a marketable drug, several years will have elapsed and other companies will have introduced new products affecting the structure of the market. Consequently, it is necessary to make some kind of estimate of the scale and quality of other companies' efforts in the area. It is useless to propose a new programme unless attractive research ideas are available with the quantity and quality of people and equipment needed to pursue the study. The overall commercial attraction of the area has to be assessed by comparing the sales potential with the feasibility of reaching the medical improvements sought. Because of the expense of drug research and development, it will be the case that for rare diseases the market is just too small for a company to have any chance of recovering its costs before the patents on the product expire. Thus, market forces result in the situation where research is concentrated on the most common diseases in the developed nations.

If the company is in the position to select a group of research programmes from a large number of options, it may wish to achieve some balance in the overall portfolio by looking at some additional selection criteria. Some research programmes are intrinsically more risky than others. In 'breakthrough' research, the company seeks a totally novel form of therapy. It may be that there is no current treatment for the disease in question or one which is quite inadequate. The objective is to find a new approach and to find a drug that utilises this approach. This type of programme involves the highest risks and the highest rewards. On the other hand, it may be that a concept of treatment has been established already but current drugs acting by the desired mechanism have use-limiting side effects. In this case, the objective is to find a new chemical compound with a more selective action. The risks involved are less than for breakthrough research because the validity of the general approach has been already demonstrated. However, the competition is more severe because other research groups will also be attracted by the lower risk. Consequently, the rewards will have to be shared. So called 'me-too' drugs arise when a company produces a compound that has very similar properties to one already known and which offers in preliminary clinical trials little medical advantage to the physician or patient. While it is easy and popular to deprecate this kind of product it should be pointed out that a product can only be labelled a 'me-too' drug once it has been well studied in the clinic. For example, the change in incidence of some uncommon but serious side effect from, say, 50 in 1000 to 10 in 1000 may be regarded as a significant improvement but it cannot be demonstrated until the new drug has been carefully evaluated in several thousand patients. Thus, a product could ultimately be considered to be a significant improvement or merely a 'me-too' only after extensive clinical testing. By this time, the company involved has already spent most of the money necessary for the research and development and will be anxious to seek some return for this investment.

Moreover, it often happens that sequential modifications to the chemical structure of a drug can produce not only agents of similar mode of action with improved properties but also drugs with completely new and important actions. Thus, the sulphonamide bacteriostatics led directly to the sulphonamide diuretics and then to sulphonamide antidiabetic drugs by a process of continual molecular modification. It does not make commercial sense to seek 'me-too' products since a drug with no advantages over other therapies cannot expect to achieve a large share of the market.

The decisions about balancing the research portfolio between breakthrough and modification will need to be made at the research director level in consultation with both medical and commercial directors, who can together make some estimate of the attraction of the target products and the quality of the research effort available. Research targeted toward different diseases has different constraints in relation to toxicity testing. For example, an antibiotic will be dosed for perhaps 10 days in what is possibly a very serious illness. A high incidence of minor side effects might be acceptable to the patient under such circumstances. Moreover, the short period of treatment reduces the possibility for the drug to have some slow, deleterious effect on the body. Similarly, in the treatment of a life-threatening condition, for which there is no alternative therapy, such as cancer, a higher incidence of more serious side-effects may be acceptable. On the other hand, patients with high blood pressure, who do not feel unwell, may be prescribed an antihypertensive drug on a long-term basis as a prophylactic against heart attacks and strokes. They may find even the most minor side-effect unacceptable. In addition, continuous treatment, over perhaps tens of years imposes very high safety standards upon the drug. In balancing the research portfolio it will be important to consider how these factors influence the probability that a candidate drug selected for development will subsequently become a successful product.

8.3.3   The research phase

Let us imagine now that our hypothetical enterprise has reached a series of decisions about its research objectives. The medical, commercial and research considerations have led to the establishment of a number of research projects. The balance between short- and long-term programmes and high- and low-risk ventures will have been achieved by a process of informed judgement rather than precise calculation. For any particular disease, the therapeutic objectives may be achieved by more than one biological mechanism. For example, high blood pressure is treated by several types of drug. Some act predominantly on the blood vessels, others on the heart, the kidney or the nervous system. Blood pressure is controlled by several inter-related mechanisms and consequently there are several strategies for medical intervention. The judgement concerning which strategy to adopt will be made mainly by biologists and physicians after considering the various advantages and disadvantages of each approach.

Suppose that the decision has been made to seek a drug that causes relaxation of blood vessels. The contraction and relaxation of the vessels is controlled by several circulating hormones and by messenger substances released from nerves. An understanding of the physiology and biochemistry of these control mechanisms may allow a decision to be made on the preferred form of intervention. If the biochemical mechanisms are understood, the medicinal chemist will be in a position to contribute to the decision by speculating on the feasibility of producing novel substances that will selectively mimic or antagonise some natural chemical substance in the body or which will prevent its formation or destruction.

If a particular biochemical approach to the disease can be conceived then three questions have to be addressed by medicinal chemists and biologists. Which compounds should be made? How will they be synthesised? How will they be tested? The first question falls within the province of medicinal chemistry. This discipline is concerned with the study of the relationship between the structure and biological activity of drugs, which, in the vast majority of cases, will be organic compounds. Medicinal chemistry is a multidisciplinary and interdisciplinary subject. Information contributing to the design of novel organic compounds may be drawn from theoretical chemistry, conformational analysis, physical organic chemistry, biochemistry and pharmacology, and from analogies with previously elucidated relationship between structure and activity. The second question falls within the province of synthetic organic chemistry. In the UK, medicinal chemists are recruited very largely as synthetic organic chemists. In the course of a few years, they become increasingly familiar with the factors involved in drug design and become medicinal chemists. They design novel drug candidates and then devise and execute syntheses to produce them. The third question, relating to the testing of compounds, falls within the provinces of biochemistry and pharmacology. Physiology is the study of the working of the body under natural conditions, whereas pharmacology is concerned with the effect of foreign substances of the body.

It can be perceived readily that biochemists and pharmacologists face an enormous problem in deciding how candidate drugs should be tested. The final target is a drug that works in human disease. It would be unethical to test novel compounds directly in patients. To find a suitable drug it may be necessary to synthesise hundreds of compounds and it would be absurdly expensive to subject all of these to safety testing prior to investigation in humans. A system has to be devised whereby a large number of compounds can be looked at quickly and cheaply to select a group worthy of further investigation in more sophisticated tests. This process may be repeated several times, employing increasingly rigorous selection filters, to give, at length, a short-list of compounds that meet all the criteria necessary to be worthy of exhaustive safety evaluation and subsequent evaluation in humans.

Very commonly, the first tests in a screening cascade will be biochemical ones performed *in vitro* (i.e. in the test-tube) involving pure enzymes, cell cultures or cell membranes. These are commonly cheap, capable of testing many compounds and able to give reproducible results. Alternatively, isolated tissues can be used, such as a strip of blood vessel mounted in a small bath of oxygenated buffer solution. These *in vitro* tests are possible when a drug with a well-defined biochemical or pharmacological mechanism of action is sought. Because the body is so complex, with numerous interactions between its various parts, the effects of a drug on the whole body cannot be predicted from its effects on isolated pieces of tissue. For example, the body may metabolise the drug, producing new compounds, each with its own action. In addition, the body will excrete the drug in the urine or bile. Consequently, at some stage, it will be essential to test the candidate drug in a whole conscious animal to confirm that it can be absorbed by mouth, has a useful duration of action and produces the desired effect without undue side-effects either at the therapeutic does or at multiples of that dose. Two technical problems arise at this stage. While all mammals share similar biochemistry and physiological mechanisms, they each have their own special features, which have developed to suit their particular needs. While rabbits are herbivores, dogs are carnivores and rats omnivores, so it is to be expected that they will have some differences in their digestive system. No laboratory animal can be taken as fully representative of man. Secondly, the drug is needed to treat patients, not normal healthy people. Consequently, it is desirable to use tests which show a biological effect upon a disease process such as an infection or a disorder such as diabetes or high blood pressure. Laboratory animals do not always suffer from the same diseases as man. The biologist may have to develop a laboratory model that measures some process analogous to the human disease. For example, it is not possible to consider that rats suffer from anxiety, depression or schizophrenia as experienced by people. A rat placed in a brightly lit open space will usually seek a hiding place in a dark corner. The speed of this response can be influenced by drugs that in man are known to alleviate anxiety. Using this observation an 'anxiety model' can be developed for the testing of anxiolytic drugs.

At an early stage in the research programme, it will be necessary to define all the properties that any compound must have before it can be considered seriously as a potential drug. It is the objective of the testing cascade to identify a short-list of compounds, based largely on their biological activity. The key features will be effectiveness, selectivity of action, a route of administration appropriate for the disease and a suitable duration of action. Selectivity of action is never absolute. What will be important is the ratio between the dose of drug producing the desired effect and the dose producing some side-effects. The ratio sought will depend upon the severity of the side-effect. The potency of the drug, that is the weight of drug which has to be dosed to achieve the desired effect, is important for convenience of dosing and for manufacturing costs but is not more important than selectivity of action. In addition to these biological criteria, there will be some chemical ones. It is necessary that the compound can be protected by a sound patent, and that it is stable in a pure form and as a formulated product so as to have an adequate shelf-life. The compound must be capable of large-scale manufacture at a cost which is commensurate with the value of the final product. If a compound is found that passes these selection criteria, or if there are strong indications that it will pass them, then such a compound may be selected as a development candidate, to go forward for full rigorous evaluation preparatory to human studies and eventual registration.

Thus in the research phase, one perceives three classes of risk. Is the biological hypothesis, on which the research is based, sound? Are the tests used predictive of the desired therapeutic effect? Can the chemists devise compounds to meet the selection criteria in an acceptable time?

8.3.4   The development phase

The process of drug development encompasses all those activities necessary to take a compound that is a drug candidate from research and to produce an effective medicine for patients. The major drug-producing and -consuming nations have government agencies for the approval of new medicines. In the UK the Medicines Act of 1968 provided for the licensing of all medicinal products and for controls over their labelling, supply, advertising and sale. The Act established a licensing authority (the Minister of Health), an advisory body known as the Medicines Commission, and the Committee on Safety of Medicines (CSM). The CSM has the role of reviewing applications for product licences and for clinical trial certificates. It operates, like the Medicines Commission, under the aegis of the Department of Health.

In the USA the regulatory authority is the Food and Drug Administration (FDA). These agencies are administered by public servants but call extensively on independent outside experts in the field of medicine. To varying degrees in different countries, such agencies lay down detailed regulations about the way in which new drugs will be tested for safety in animals and ultimately for safety and effectiveness in man. A British company wishing to sell its product in the USA or Japan, for example, will need to satisfy the regulations of that country. These regulations now extend to the way studies are carried out, known as 'Good Laboratory Practice' for animal studies and 'Good Clinical Practice' for studies in man. In addition 'Good Manufacturing Practice' covers the production process.

Regulatory agencies are concerned with the results of the work reported and the way in which the results were obtained. They examine closely the full details of all experiments and carry out spot checks on laboratories providing the data. As a result, the drug development process is geared to demonstrating safety and efficacy to the satisfaction of the company's own scientific, medical and ethical requirements and then demonstrating this to the external regulatory agencies. In addition, the full profile of the drug's action in patients, in a wide range of circumstances, must be evaluated to define and establish its use, and its advantages over earlier therapies. During the course of the work, it often happens that the company discovers some major flaw in the compound that will prohibit its use, or some minor drawback that makes the compound less attractive. In either case, the compound's development is terminated. The safety studies carried out are not solely a consequence of drug regulation. They were carried out to varying degrees before the current legislation was enacted and it is easy to see why this should be so. The scientists and physicians within the industry wish to make good medicines that will benefit patients. If they do this, they will enhance their professional reputations and bring themselves and their companies just rewards for their diligence and invention. If ineffective or unsafe drugs were to be sold, it would soon become widely known. Physicians in the community would not continue to prescribe ineffective drugs and post-marketing surveillance would uncover adverse effects. Such a situation would lead both to litigation for alleged drug-induced injuries and to adverse publicity in the media, seriously affecting a company's standing in the eyes of the medical profession and general public alike. Consequently, as with any other serious industrial activity, it makes good science and good business to ensure that the drugs produced are safe and effective and that the medical profession is provided with all the information necessary to use them wisely. Despite the best efforts of companies and regulators, drugs will occasionally be cleared for general use and subsequently be found to have an incidence of side-effects not predicted by all the previous trials. Such observations may require that the drug be limited in its use or withdrawn completely. As with motoring accidents, despite improvements in cars and roads, MOT tests and breathalysers, accidents continue to happen because there are almost infinite possibilities for unforeseen combinations of unusual circumstances.

8.3.4.1   Pre-clinical safety evaluation.   There will be some differences in the development process for drugs intended for very short-term treatments, such as antibiotics, and those for chronic use, such as antihypertensive agents. The complexity of drug development is such that it cannot be fully described in an article of this length, but using the more demanding case of a drug for chronic use, we can sketch out the main processes of drug development, firstly in relation to safety testing. It is the objective of safety testing to find out what adverse effects could occur if a compound is dosed for a sufficiently long period at a sufficiently high dose. It is insufficient to say that at a particular dose the compound was very well tolerated. This is because safety testing will be carried out in tests lasting a few months or at the most one or two years, whereas continuous dosing to patients for some drugs may last for 20 years. The safety test accelerates the onset of adverse reactions by using much higher doses than those used to treat the disease. Relative to the animals' normal life-span, the period of testing is long.

No compound is completely safe. Common salt is an essential component of our diet but drinking sea water leads to death through dehydration. There is a range of doses of salt that are safe and necessary but at higher doses adverse effects ensue. It cannot be assumed that a compound from a 'natural' source is any safer than one that is 'synthetic'. Bacteria, plants, fungi, insects and animals can produce very potent poisons and venoms and some natural compounds have very marked pharmacological actions. We are constantly exposed to thousands of natural chemicals in food other than the nutrient essential amino acids, sugars, fats, vitamins and minerals. Plants produce an enormous number of secondary metabolites, such as alkaloids and terpenes, which contribute to the flavours and aromas we value in food. It is from traditional use rather than scientific investigation that we regard some plants as edible and others as poisonous. The plant kingdom does not fall neatly into two such contrasting categories. In practice, we can recognise easily plants causing acute discomfort or death but the longer-term effects of the vast majority of non-nutrient plant products are unknown.

Food preparation processes such as salting, roasting and smoking cause chemical modifications within the food to produce unnatural compounds. The flavour of roasted coffee beans has been extensively studied and now 700 separate chemical constituents have been identified. Flavourings of this nature are accepted by usage but it is inconceivable that they would all survive the rigours of modern safety testing. Because of the wide range of substances to which we are exposed, it would be difficult to trace the origin of a medical condition to a single substance in food, but even so, the possible connection between coffee consumption and cancer of the pancreas has been studied as has the nature of potentially carcinogenic polycyclic hydrocarbons that preserve smoked food from bacterial degradation.

Long-term safety tests must be performed in two species, only one of which is a rodent. The amount of drug needed to produce a pharmacological action is usually measured in mg per kg body weight. If a drug has an effective dose of say 1 mg kg^{-1}, then it will be necessary to undertake studies using much higher dose levels in the safety evaluation programmes. It is necessary to find out the maximum dose of drug that is completely safe and to ascertain what adverse effects could be expected if a sufficiently high dose is administered. Consequently, after pilot studies, usually three dose levels will be selected, perhaps at 10, 25 and 50 mg kg^{-1}. The maximum dose that can subsequently be given to humans will depend on the maximum dose that was completely safe in both species. If the top dose is too high, the animals become too sick for the studies to continue and consequently the effects of long-term dosing, which may be different to those of short-term dosing, will not be discovered.

In the case of a successful drug candidate, no adverse effects will be found at the lower two dose levels, both of which are at significant multiples of the normal therapeutic dose. Initially the trials will involve 1 month of dosing. If an adequate safety margin is found, a 6-month study will follow in the same two species.

Because of the spontaneous occurrence of abnormalities in animals, it is necessary to ensure that the groups of animals dosed with drug can be compared with a second group of animals that receive no drug and that the numbers of animals involved will give a statistically valid result. Unlike physical sciences such as chemistry, biological science is imprecise because of the large number of variables in the studies. Statistical procedures are necessary for the full evaluation of the results. The course of the studies is monitored by daily inspection of the physical condition of the animals and by the measurement of body weight. In addition, the measurement of drug concentrations in the plasma and biochemical measurements on blood are undertaken. At the end of the study, microscopic and biochemical investigations of all the tissues are undertaken in post-mortem examinations. In some cases, the nature of the pharmacology of the drug will cause difficulties in the interpretation of results. For example, dosing of a drug at ten or twenty times the normal therapeutic dose may produce a pronounced pharmacological effect that interferes with the normal activity of the animal. An antihypertensive agent at high doses may cause a very large fall in blood pressure, possibly accompanied by a compensatory rise in heart rate. Such a phenomenon is a consequence of exaggerated pharmacology rather than direct organ toxicity.

Two other types of study are likely to be required for drugs intended for long-term use in a broad range of patients. Two-year trials in mice and rats are undertaken to look for oncogenicity, i.e. the capacity to produce cancer. In addition, it is important to investigate the effects of a new drug on the development of the foetus and on offspring when born. The drug will be dosed to male and female rats to observe any effects on reproduction and upon the health of the offspring both at birth and when they, in turn, come to breed.

Permission to commence very limited studies in patients could be sought following the successful completion of the 1-month safety evaluation study and, for more extensive trials, after the 6-month study. In the UK, this would involve presentation to the Medicines Control Agency, the operational arm of the committee for the Safety of Medicines, of all the laboratory results generated, to seek a Clinical Trial Exemption Certificate. The corresponding permission in the USA is called a Notice of Claimed Investigation Exemption for a New Drug or IND. At this stage, the long-term oncogenicity and fertility trials will be probably only at an early stage as will any 12-month safety studies required by the FDA. These studies will need to be completed before permission can be sought for more extensive trials.

8.3.4.2 Process development.   Several other important activities will be undertaken in parallel with the safety evaluation studies. Firstly, the synthesis of the drug must be optimised and scaled up to produce tens of kilograms of material. This often requires considerable research into the synthesis of the compound. The original research department route is likely to have been chosen for speed and convenience when working on a small scale, rather than for economy and practicability on a large scale. An important consideration in the development of a manufacturing route will be safety in relation to fire hazards and the exposure of plant operators to noxious reagents. Drug synthesis takes place on a relatively small scale compared to synthesis in most of the chemical industry. Furthermore, it is not possible to know whether a drug will be a success until well after it has been launched. As a consequence, a new specialised plant is not built during the development phase. Indeed, the synthesis is performed by batch processes in general-purpose plant and this may continue for the whole life-time of the drug. The nature of the plant may impose restrictions on the type of reactions that can be performed and consequently a totally new synthesis to the compound of interest may have to be devised which avoids, for example, very low temperatures or pyrophoric reagents. The synthetic route employed will govern which impurities may be found in the final product. There may be no more than 2% impurity in total, with no single contaminant present at a level greater than 0.5%. Any change to the manufacturing process may alter the impurity pattern and cause serious problems with registration. The material which is eventually sold should be similar to that used for all the development work in animals and humans on which product registration is based. The work of the process development chemist requires close collaboration with analytical chemists.

Processes used for the manufacture of drugs are commonly different to the laboratory processes described in the chemical and patent literature, not only because new syntheses are developed but because of the special features discovered during scale-up. Much of this detail is confidential because of competitive pressures. A high-volume chemical manufacturer guards processes by a mixture of patents, confidential know-how and the high capital investment required for entry to the technology. By contrast, in the pharmaceutical area, capital investment in the synthesis of bulk drug is low compared to R & D investment. Once a product patent expires and generic drug manufacturers start to compete, the nature of the market changes, from a monopoly-supplied product based on high technology in research and development, to a competitive situation in which manufacturing costs play a very large role.

8.3.4.3 *Pharmacokinetics and metabolism.*

8.3.4.3 Pharmacokinetics and metabolism. The metabolism of the drug in the species used for safety evaluation will be studied to determine how the drug is degraded. It may be necessary to initiate biological studies on major metabolites. Also, the route and speed of excretion will be studied using radioactively labelled material. The development of very sensitive analytical methods for the drug is essential at an early stage so that blood levels can be measured at the therapeutic dose as well as at the higher doses used in the safety trials. The kinetics of drug absorption, distribution to the various tissues and subsequent elimination will also be studied so as to allow the proper design of biological, safety evaluation and subsequently clinical studies and facilitate the interpretation of results. The phenomena studied will vary between laboratory species and between them and humans, but the animal work will indicate a range of possibilities for humans and enable clinical investigators to focus on key features of the drug's action. For example, it may be that drug absorption and delivery to the target organ is much more rapid and complete in one species than in another or the duration of action varies from one species to another. This will have implications for the frequency of administration and the size of dose required to obtain a constant level of biological effect.

8.3.4.4 Formulation development. A suitable formulation of the active compound must be developed to produce optimal absorption and duration of effect. Pharmacy is the study of the presentation of active principles to patients. It is necessary to test that the bulk drug is stable to manufacture, purification, drying and storage. Subsequently, it will be milled to produce a material of defined particle size. This will be mixed with bland materials known as excipients to facilitate its handling and stability when it is pressed into tablets or dissolved to produce solutions for injection. Whether the drug is used in tablets, in capsules or in solution, it must be stable to the formulation process and subsequently to storage during transit and on the pharmacist's shelf under the conditions of temperature and humidity encountered around the world. The product's stability will be studied for all these circumstances, and storage conditions and shelf-life defined. The tablet or capsule must be designed to disintegrate on ingestion and the crystalline form and particle size optimised for dissolution. In addition to the tablet receiving protective coatings to aid stability and facilitate swallowing, it may be necessary to guard the active ingredient from gastric acid by using a coating that will dissolve only at higher pH in the intestine. Moreover, the drug may be formulated deliberately to ensure slow dissolution, so as to extend its duration of action. There are many routes for delivering drugs other than by mouth, such as by absorption through the skin or mucous membranes, by intramuscular or intravenous injection or by subcutaneous depot. The formulation requirements differ according to the physical properties of the active agent, the route of administration and the types of excipient that are acceptable.

8.3.4.5   Analytical chemistry.   Throughout the production and formulation process there is extensive analytical work. Every starting material, reagent and excipient, including the tap water, is analysed to establish its quality and suitability. Each intermediate is analysed for purity and to confirm its structure. The bulk pure drug is examined in great detail to determine the nature and percentage of each impurity. As soon as bulk drug has been mixed with excipients and pressed into tablets, samples are taken, dissolved and analysed again to ensure that the total weight of drug per tablet and its purity fall within the required range. The requirements of chemical analysis in quality control are such that there may be as many analytical chemists involved in the manufacturing plant as there are synthetic chemists in the drug discovery laboratories.

8.3.4.6   Clinical evaluation.   The programme of clinical trials consists of four distinct phases. In Phase I, the primary objective is to confirm that the drug can be safely administered to human subjects and to provide early information on its absorption, distribution, metabolism and duration of action, so as to facilitate the design of subsequent studies.

Commonly, the first stage of safety evaluation in animals will be a 1-month study. If a suitable safety margin is demonstrated, this will allow a maximum of 10 days dosing to humans in the UK, and this will commonly be in healthy volunteers. The drug will be dosed initially at a small fraction of the anticipated effective dose and parameters such as heart rate, blood pressure, blood and urine chemistry will be monitored, together with the subjective feelings of the volunteer. Following each dose, the volunteer will be observed closely for several hours and subsequently at regular intervals for a few days. As each stage is successfully completed, the dose level is raised until the projected effective dose in humans is reached, or the maximum dose level permitted by the safety studies. If the results allow a decision to be made to continue with development, then the next stage will be to ascertain that the drug has the required pharmacology in humans.

It will frequently be the case that the drug has no effect on normal volunteers, but only on patients with the disorder in question. Repeated dosing over several days or weeks may be required to establish the quantity and quality of effect, and safety trials in animals over periods of six or twelve months will be needed to underpin longer studies in patients.

In Phase II, the therapeutic action of the drug is studied. It is necessary to evaluate the effectiveness of the treatment in patients and the incidence of side-effects and adverse reactions, and to define the optimum dosing schedule. Further information on the drug's distribution in the body and its metabolism will be sought.

When the drug is first administered to patients it will be to a very small group. They will be in a closely controlled environment, such as a hospital, under the care of consultant physicians working in collaboration with physicians from industry. The patients will have given their consent to the treatment and the whole study will have to be approved by the regulatory authority and the hospital ethical committee before work commences. The first trials are 'open trials', where the physician and patient know that a new medicine is being dosed. If the work is successful, then, as confidence and experience in the use of the new drug develops, it is important to investigate it under strictly scientific conditions. At this stage, two groups of patients, matched for age and sex, and subjected to rigorous diagnosis are selected and asked to participate in the study. One group receives the new drug and the other either an established drug or a placebo, i.e. a tablet made to look like the drug but containing no active ingredient. Further comparison can be made by switching the therapies after a few weeks so that every patient is treated by both agents. Neither the patient nor the hospital physician knows which treatment the patient is receiving until the end of the trial, when the code is broken. Such a study is known as double-blind cross-over trial and is designed to eliminate unintentional bias by the patient or the clinical scientist.

In some conditions, where spontaneous improvements or remissions are seen, or where patients respond well to the high level of attention given, even the placebo treatment may appear effective. Consequently, it is desirable to compare the new drug with placebo. In some cases, such a comparison is impossible. It would be unethical to take dangerously ill patients off their current therapy for the purpose of such a trial. Similarly, if there is no current therapy and the new drug has a beneficial effect on the disease, it would be unethical to deny the patient the new drug.

Once suitable efficacy has been demonstrated, Phase III clinical trials can be undertaken to provide information on how the drug will perform when in general use. This will include comparison with other drugs and the identification of subgroups of patients who respond more or less well to the treatment. Interactions with other drugs likely to be used by the patients will be assessed. The studies may be widened to include general practitioners as well as consultant physicians. In addition, trials will be carried out in several countries to confirm safety and effectiveness on different populations with their varied genetic background, diet, life-style and medical practice. Many national regulatory authorities insist that trials be undertaken in their own countries before a new drug is approved for use. Whereas earlier trials will have lasted for a few days or weeks, some of the major proving trials will involve dosing for at least 12 months for a chronic disease such as high blood pressure so that the long-term safety can be demonstrated. Phase IV studies are those which may be undertaken after launch to investigate new uses of the drug and more prolonged treatment.

The entire progress of these trials must be organised and managed by physicians within the company to ensure that all procedures are within the requirements of the regulatory authorities and that the appropriate measurements are made and collated for analysis.

8.3.4.7   Registration.   When all these stages have been successfully negotiated, the documentation describing all the chemical, biological, safety evaluation and clinical data requested by the various regulatory authorities to support sales approval can be assembled. There will be separate and different submissions for each country according to its specific requirements and for any one submission the documentation will stand several feet thick. In the UK, the CSM takes the Product Licence Application and subjects it to examination by a team of independent experts, who may call for further data or clarification. Similarly in the USA, the submission known as an NDA (New Drug Application) will be exhaustively probed. This period of evaluation will take several months and, if further data are requested, may extend into years. The drug's use, if approved, may be limited by the regulatory authorities in some regard, possibly pending the production of more results. The limitations imposed will form part of the instructions to prescribing physicians. For example, a new vasodilator may be developed for use in hypertension. Although there may be good reason to believe that such an agent will have beneficial effects in some other disorder, such as heart failure, permission to use it in this way will require a specific application, as fully supported by clinical data as the initial application for use in hypertension.

8.3.4.8   Production, distribution and promotion.   A successful application to the regulatory authorities will result in the company obtaining, in the UK, a product licence, covering permission to sell the drug, and a manufacturing licence, which is obtained by proving that the company has the facilities to make the drug reproducibly to the required standard.

At this stage, production can go ahead on a large scale to supply the material for sale and the formulated product has to be packaged in a suitable form, possibly in foil or in tins, fully labelled in accordance with the regulations of the country of destination and in the appropriate language. It is likely that several dose strengths and dosage forms will have been produced to allow the treatment of patients of different size, with different severity of disease and for acute and maintenance therapy. The packaging process is a considerable exercise in logistics. Some countries require that all medicines used there be formulated and packaged locally. The company may need to establish a subsidiary for this purpose or enter a collaboration with a local firm that has suitable facilities.

Because of the enormous volume of work associated with seeking approval for launch in a single territory, it is unlikely that a company will be able to launch in several at once. Each set of trials, applications and approvals will have its own time-table. Moreover, a company may not have the resources, or perhaps even the commercial confidence in a totally new treatment, to risk a simultaneous world-wide launch. A slower step-by-step approach with careful observation of a steadily expanding patient population might be prudent with a truly revolutionary therapy, so that further development can be guided by the experience gained.

At the time of launch, it is essential to inform all the physicians likely to use the product about its properties and use. Until this point, only a small percentage of the medical profession will have used it, as part of the trials, but their results will have been published in some of the main medical journals, where papers are submitted to rigorous scrutiny by referees. The company launching the drug will wish to draw to the attention of prescribers the results obtained by clinical trials physicians.

Because the process of drug discovery and development is so long and expensive and the remaining patent life short, the company will be anxious to promote the use of the drug on a large scale so as to receive a return on its investment. Promotion will be through the medium of scientific conferences, advertisements in medical journals, mailings directly to physicians and visits by company representatives to the surgery. Guidelines covering all aspects of promotion have been agreed by the Association of British Pharmaceutical Industries, the British Medical Association and the Department of Health. Because of the scientific advances being made, it is often the case that new treatments involve concepts unknown at the time physicians left medical school. As a result, a considerable education process is involved in alerting them to potential benefits and possible drawbacks of the new treatment and its comparison with older, more familiar drugs. Thus, promotion is essential for the proper use of the drug for the patient's benefit as well as for achieving sales, which will ultimately pay for the research on the successful drug and all the unsuccessful projects. It is an element in a cyclic process of research, development, sales, investment and new research.

8.3.4.9 *After launch.* The companies that are involved in innovative research continue to have a considerable development programme on a drug after initial launch. Major clinical trials will continue to be organised to compare the drug with other treatments over longer time-scales and in a wider range of circumstances. The company will still be engaged upon studies necessary for launch in overseas territories. In the light of early experience, it may prove worth while to bring out new tablet strengths to improve flexibility of dosing or to provide new formulations more suitable for children, such as syrups. Longer-acting formulations and combination products may be required. The latter arise where a patient is normally treated by two drug types. Many hypertensive patients will be treated with a diuretic and a beta-adrenergic blocker or an alternative antihypertensive drug. In such cases, a single tablet containing both ingredients may simplify therapy and avoid confusion. In addition, the drug may be investigated for new or wider uses based upon observations made in earlier studies. Each new development of this nature will have to be submitted for regulatory approval.

The company will be monitoring and collecting all papers on the new drug from scientific and medical journals around the world. Some papers will be by physicians who have collaborated in trials with the company, but after launch there will be an increasing number of studies reported from other sources, including trials initiated by other companies comparing the product with their own, still newer, product. The company will be monitoring both efficacy and side-effects and will be the principal source of information on the drug. Numerous enquiries will be made by physicians seeking information about the drug and about its use in unusual circumstances. It is part of the company's job to back its product with this wealth of data and medical expertise, and this activity will continue as long as the product is sold. A company without a research base, which enters the market once patents have expired, does not usually have this resource available, and although this allows the generic company to sell the drug more cheaply it does not come with the same level of technical and medical support.

In summary, we can see that the drug development process is exceedingly complex, lengthy and expensive, requiring the highest level of scientific and medical expertise from a range of disciplines. There is no certainty of success, and indeed, only a fraction of all the candidate drugs that enter development emerge as products. There are high risks and high costs. While nobody would argue with the principles of safety testing and government regulation, its expense and, to some extent, the duplication required for launch in different countries, inevitably add to the cost of the final product and reduce the speed with which seriously ill patients obtain the benefits of new medicines.

8.4 Financial and economic considerations

8.4.1 Research costs

We are now in a position to examine the costs of drug development and to look at the impact of the pharmaceutical industry on the health and the economic position of the UK.

The average time for the development of a New Chemical Entity (NCE) marketed in the UK has risen from 3–4 years in the early 1960s to about 10–12 years by the late 1980s, although in the very best circumstances the time from discovery to market for a chronic therapy can be cut to about 8 years. During the same period, the cost of research and development for a new drug has risen from £15 million in 1968 to £60 million by 1978 and by 1992 to over £100 million. If all the costs associated with projects that failed are taken into account, the sum is even greater. The number of NCEs introduced on to the

British market has fallen from between 50 and 60 per annum in the early 1960s to an average of about 20 for the period from 1970 to 1990. Following a sharp fall to under 15 per annum for four years in the mid-1980s, the numbers have recovered steadily, and 24 were launched in 1990.

Technological economics have been considered in Volume 1, chapter 5. We can consider a hypothetical discounted cash-flow curve for a successful pharmaceutical project, as shown in Figure 8.1. In the first phase of R&D, money is spent on seeking a new compound with the desired properties for perhaps five years. Annual expenditure increases sharply once the development phase commences and continues to do so as the later stages are reached with international clinical trials, production and marketing expenses as the launch approaches. Sales in the country of first launch may not turn the curve upwards significantly as development for launch in other territories is at its most expensive stage. As only 3% of the world market is in the UK, for a British company the most significant launch will be in the USA, which provides about 29% and, usually much later, in Japan, which has about 19% of the world market. After international launch, a successful product will commonly have increasing sales for a few years. The price includes not only production, distribution and sales costs but also the R&D costs. Even when the product has been on sale for a few years, the project may still be in overall deficit. Eventually all the costs and the interest on the money used are paid and the project breaks even; the curve crosses the x axis. Despite all the risks and expenditure, and the success of the new medicine, the company, at this point, is

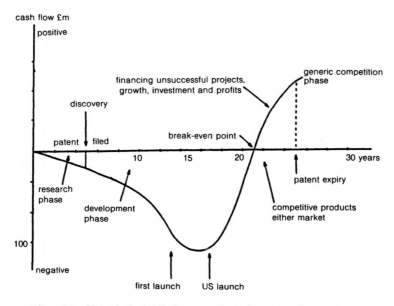

Figure 8.1 Hypothetical cash flow curve for a drug research programme.

no better off than if it had invested its money in a bank. If success continues, the project becomes worth while and money is generated that covers unsuccessful projects, goes into taxes, dividends for shareholders and new capital and revenue investment.

When the patent expires, other manufacturers enter the market, prices fall as the new manufacturers do not have to cover R&D costs, and the originating company loses market share. In addition, by this stage, it is likely that new products have been launched by competitors and that as a consequence sales are already declining in volume. Figure 8.1 illustrates in general terms the time-scales that can apply and the extreme vulnerability of the project at the time of launch when up to £100 million may have been invested.

The patent system is essential to the research-based pharmaceutical industry. In effect, the patent is an agreement between the inventor and the state in which the inventor agrees to reveal the product or process to the public in return for a period of monopoly in which to work the invention to recover costs and make a fair profit. The intention of the system, when founded during the reign of Charles II, and the effect in practice, is to encourage innovation by protecting the rights of inventors. The increased time and expense in bringing a new drug to market are such that companies have to recoup bigger costs in a shorter period of patent life. As a consequence, patent life in Europe was extended to 20 years for cases filed after 1967. The US Patent Term Restoration Act of 1984 allows for extension to the normal 17-year term of 2–5 years according to when the case was first filed and the extent to which product launch was delayed by regulatory procedures.

In the UK, pharmaceutical R&D expenditure has risen from £12 million in 1965, to £30 million in 1970, £251 million in 1980 and £1 082 million in 1990. Since 1965, the rise has been about eightfold in real terms, i.e. after inflation has been taken into account. R&D costs for companies are typically about 15–16% of turnover.

8.4.2   Effects on the UK economy

The gross output of the UK pharmaceutical industry for 1990, including all goods and services and inter-company transactions was £6 753 million. Total exports of pharmaceuticals from Britain were £2 258 million and imports were £1 158 million. On the basis of figures for the first nine months of 1992, the trade surplus for the whole year is estimated to be about £1 270 million.

The pharmaceutical industry is one of Britain's most successful manufacturing sectors. Output rose by 80% in the ten years to 1990, while for manufacturing as a whole it rose by only 23%. The trade surplus of £1 270 million is exceeded only by those from the power-generation machinery and oil products industries, and on the international scene only Switzerland and Germany have higher trade balances in pharmaceutical products. Moreover, the R&D expenditure outstrips that of any other civilian manufacturing sector, overshadowed in total R&D only by electronics and aerospace, which have a large defence-related component. In 1990, the UK pharmaceutical industry employed 87 800 people with about 18 000 in R&D, two-thirds of whom were scientists, technicians and engineers. Total capital expenditure was £570 million.

In 1990, total UK National Health Service (NHS) expenditure on pharmaceuticals amounted to £2 254 million, which is 8.9% of total NHS costs and represents £44 per head of population. Medicines have amounted to between 8 and 9% of NHS expenditure since 1982. The value of OTC drugs sold in 1990 was £862 million. The total number of prescriptions dispensed was 447 million or 7.8 per head. Their average cost, including charges for dispensing, was £4.24.

In the UK, the expenditure per head on pharmaceuticals, including OTC products, in 1989 was £54. This compares with £198 in Japan, £110 in the USA and between £88 and £100 for France, Germany, Switzerland and Italy. The average daily expenditure per person in the UK for 1990 was 12 pence for medicines, 22 pence for newspapers and magazines, 42 pence for tobacco, £1.04 for alcohol, 99 pence for clothing, £2 for food and £2.39 for housing. Public expenditure on defence and education was equivalent to £1.04 and £1.28 respectively. In the UK in 1990, 84% of all prescriptions were dispensed free of charge, of these almost half were for the elderly. In January 1993, the cost of prescriptions, to those patients who do pay, is £3.75 per item.

These statistics help us to see the industry, the cost of research and the price of drugs against the background of NHS expenditure and the economic activity of the nation as a whole. The price of drugs sold to the NHS is controlled indirectly by the Department of Health through the Pharmaceutical Price Regulation Scheme, which began in 1957. It has been revised several times. It requires that companies present details of costs, sales, exports, capital investment and estimates of future sales. Rather than regulate the price of each individual product, the scheme aims to limit the overall return on capital that the company makes on its business with the NHS. The scheme works mainly by negotiation between the companies and the Department of Health, but the Secretary of State for Health can, if necessary, fix prices by order. The scheme limits the percentage of sales that can be spent on promotion and companies can be penalised for exceeding the limits. According to figures produced by the Department of Health, during the 1980s, the rate of return on capital employed for manufacturers' sales to the NHS has declined from about 20% to 6%. This compares with an increase from 15% to 20% for manufacturing industry in general. The rate of return as a percentage of sales to the NHS has fallen from about 13% to 5%.

For companies selling their products overseas, separate price or profit negotiations have to be made with each national pricing authority or with insurance agencies that provide health cover. Each one uses a different formula for the calculation of the price, based upon criteria such as how much R&D the company has undertaken in the country in question, the price obtained in the company's home market and the cost of competitive products. As a consequence, a given drug is likely to be available at different prices in different countries even at launch. Exchange-rate fluctuations over subsequent years could widen the differentials. The market for pharmaceuticals is thus much more regulated than that for consumer products such as cars and household appliances. The requirement for free exchange of goods and services within the European Community has created a situation in which traders can buy drugs in the poorer countries, which have little research-based pharmaceutical industry and where prices are cheaper, and import them into those where prices are higher, often the countries where the originating companies are based. This is known as parallel importing and, pursued to its logical conclusion, would result in a uniform price for drugs across the EC, based upon what the poorest country could afford. This would have a major influence on the ability of companies to fund R&D

**8.5
Health benefits**

It has been shown that health and longevity in the UK and in all developed nations have improved progressively over many decades. This is not solely as a result of improvements in the practice of medicine. Considerable progress was made in the first half of this century before most of the modern medicines became available. General improvements in the standard of living allowed people to improve their nutrition, clothing and housing. Working hours and conditions improved dramatically from those in the early days of the industrial revolution when 12–14 hours a day were worked for six days a week in harsh conditions. Enormous benefits have accrued from public services such as the provision of pure water and efficient sewage treatment. The inspection of farm animals and meat in slaughter-houses, the introduction of hygiene regulations for food preparation and sale, and the availability of refrigeration have had a major impact on reducing the spread of food-borne infections. Industry has provided products to enable us to keep our clothes and homes clean and to improve the standard of personal hygiene. On top of these improvements, which affect mainly infectious diseases and nutrition, we can perceive many direct benefits from pharmaceutical products.

The field of surgery has made dramatic advances in this century, supported by pharmaceuticals. Minor tranquillisers are used to calm the patient before transfer from the ward to the operating theatre. Intravenous induction agents and gaseous anaesthetics are available to allow complicated surgery of several hours' duration. Muscle relaxants are used to facilitate the manipulation of the patient and analgesics are available to control pain both during and after the operation. Antiseptics are crucial for sterilising the operating theatre, and together with antibiotics continue to play a role in protecting the patient from infection during the recovery phase. In the case of organ transplantation, immunosuppressive drugs are used to control tissue rejection. Throughout the operation, the anaesthetist can call on a range of agents to regulate the function of the heart and circulation. Anaesthetics and analgesics also play a significant role in modern dentistry.

A few UK statistics are useful to show the effects of pharmaceutical preparations in other areas of medicine. Nowhere has the effect been so dramatic, in terms of lives saved, than in the field of infectious diseases. Vaccines have played an important role in the decline of infant mortality. Deaths from diphtheria showed little decline between 1931 and 1941 but, in the succeeding ten years, following the introduction of immunisation, a precipitous fall was noted for all age-groups. For those aged 1–14, deaths declined from 304 per million to 2 per million, while from 1971 onwards deaths from diphtheria have been less than one per million for this age-group in the UK. Similar steep declines in the incidence of disease and mortality have been seen for poliomyelitis, whooping cough and tuberculosis as a result of vaccination. In the case of tuberculosis and pneumonia, the declines have been influenced by the availability of antibiotics. Vaccination has eradicated smallpox throughout the world. Overall mortality among those aged 1–14 has declined from 14 900 per million in the age-group in 1870, 9 000 in 1900, 2 700 in 1940, to only 257 per million by 1990. Of these 257, only 17 were caused by infection, 41 by cancer and 79 by accidents. During the same period mortality for children under the age of 12 months fell from 149 per 1 000 live births in 1870, 142 in 1900, 61 in 1940 to 7.9 per thousand births in 1990.

If we look at the population as a whole, of the 20 leading causes of death in 1990 in England and Wales, the only infectious disease was pneumonia, with 53 deaths per 100 000 population, behind ischaemic heart disease (heart attack) with 292, cerebrovascular disease (stroke) with 132, and lung cancer with 68. The combined figure for all types of cancer was 125. These figures show that while infectious diseases have been largely brought under control, diseases of organic dysfunction have come into prominence as longevity increases. These disorders, unlike infections, cannot be cured by a short course of treatment. They are commonly long-lasting conditions involving steady deterioration in the patient's health.

The benefits of modern medicines are not measured solely in terms of death statistics. The relief of symptoms and pain in non-fatal conditions enables patients to lead fuller lives. While the value of health cannot easily be measured in monetary terms the cost of illness does have important economic consequences. A patient in hospital costs the NHS £ 132 per day, during which time the patient is unable to work and may be supported by sick pay. Treatment that allows the patient to return home effects a great saving, and a return to full health allows the patient to work again.

8.6.1   Scientific and technical factors

**8.6
Future trends**

Despite the advances made in the last 40 years, there remain many serious challenges for medicine in the future. The effects of drugs and vaccination on infectious diseases have already been covered. Nevertheless, continual vigilance is required to maintain vaccination and immunisation programmes, to develop new antibiotics to keep ahead of the evolution of drug-resistant strains of micro-organisms and to avoid the conditions where epidemics can spread. There are many diseases caused by infectious agents that cannot be treated with current classes of drugs, and new infections such as Legionnaires' disease and HIV have appeared in recent years. Tropical infectious diseases are still a poorly treated sector. However, it can be predicted with confidence that provided adequate financial resources are applied, research in universities, hospitals and the pharmaceutical industry will make further advances in the treatment of infectious diseases.

For diseases of organic dysfunction, treatments so far have been aimed mainly at the relief of symptoms and reduction of risk factors. For example, it can be shown that the treatment of high blood pressure reduces the incidence of stroke. Conditions associated with ageing, such as arteriosclerosis, dominate the mortality statistics, but what these figures do not show is that as people live to a greater age, large numbers are incapacitated by arthritis or by cognitive disorders such as Alzheimer's disease. These not only reduce the quality of life of the sufferer but impose a heavy burden on health-care facilities.

Enormous progress has been made in biochemistry, physiology and pharmacology in the last ten years. Revolutionary discoveries have been made in molecular biology, immunology and genetic engineering, the like of which chemistry has not seen for a century. Synthetic chemistry and medicinal chemistry have made steady, if not dramatic, progress and they benefit increasingly from the computer revolution in information technology, theoretical calculations, molecular modelling and robotics. These advances lead one to be optimistic that significant progress can be made in moving to a new phase of treatment for degenerative conditions in which the origins of disease are understood and therapy is aimed at halting or reversing progression. For example, studies are in progress to treat genetic disorders by replacing the missing or defective genetic material by gene therapy.

8.6.2 Economic and political factors

While the prognosis for the industry is encouraging in scientific and medical terms, there are a number of serious financial constraints which have already been touched upon in this chapter. On the supply side, R&D costs have risen and indeed some of the new products of biotechnology are among the most expensive therapies available. The pharmaceutical market has become truly international and the leading corporations in Europe, North America and Japan are in direct competition. While this encourages innovation it can lead to a situation where ultimately the available market is too small to allow all the players to cover their costs.

On the demand side, it is clear that, in the developed nations, patients have a high expectation of longevity and a good quality of life. It is the elderly that are the main users of health-care facilities. In 1990, men over 65 and women over

60 each received 22 prescriptions per annum on average, which is about three times the figure for the population as a whole. In the middle, between the suppliers and the users, are the health-care providers such as national or private insurance agencies. They need to keep expenditure in balance with income. In the case of governments, the income will be from various forms of tax, and in the case of private insurers, premiums. Given that the demand for medical care is infinite at its point of use, no expenditure can ever satisfy all the claims made on the health services. In the UK, annual NHS costs are now £28 481 million out of a total public expenditure of £192 000 million. This is about £500 per person, up from £200 in 1980. Over the same ten years, the average cost of a prescription has risen from £1.78 to £4.24 and the daily cost for an in-patient has tripled from £48 to £132.

Eventually, it becomes a matter of national policy to decide what proportion of the gross national product (GNP) will be spent on health as opposed to education, transport, defence and all the other calls on the public purse. In 1990, UK expenditure on health was 5.22% of GNP. These factors provide a limit to growth and the measures to control drug expenditure in Japan, Europe and the USA have already provided a stimulus for reorganisation of the industry. There have been company mergers, strategic alliances and, in some cases, retrenchment with attendant job losses. This process is expected to continue throughout the 1990s, leading, as in the car industry, to a small number of large corporations.

From the point of view of global trade, countries that do not have the expertise and facilities to undertake drug research, and these are the majority, tend to see pharmaceutical trade as a negative balance of payments problem. It is in their short-term interest to reduce prices, to reduce patent life or abolish it altogether so that cheap copies can be made locally without benefit to the inventors. On the other hand, the major developed nations see the pharmaceuticals industry as a successful high-technology industry, well-suited to their scientific, medical and commercial skills, which benefits health and leads to export trade. Such countries are more likely to provide the conditions that will encourage innovation, but they are torn between these economic needs and those of containing costs for their own people.

The pharmaceutical industry concerns us all as both taxpayers and prospective patients. With the costs and time-scales involved in R&D, it is evident that if the industry is damaged by rapid economic changes, it will take a long time to recover and the main losers will be patients.

8.7 Sources of information on drugs

This chapter focuses upon how the pharmaceutical industry operates so that the reader can see drug research and development in the context of the whole business, and see the operation of the industry in the context of health-care provision. It is not possible here to do justice to the chemistry and pharmacology involved in the drug discovery process or to cover the mechanism of action and therapeutic uses of the major classes of drugs. A bibliography is appended and the following paragraphs are intended to guide the reader in selecting more detailed information.

Drug naming is a frequent source of confusion, as compounds have generic names and brand names, as indicated at the beginning of section 8.2.3. Generic names are used in the scientific literature, whereas the brand names are often used by those involved in marketing. The Merck Index[1], now in its eleventh edition, is an encyclopaedia covering the structure and properties of more than 10 000 synthetic and natural substances of biological interest. The USA and the British pharmacopoeias[2,3] provide details on the identification and analysis of drugs approved in their respective countries, the latter having the distinct advantage to the chemist of including structural as well as molecular formulae.

Books on medicinal chemistry come in three varieties. The more traditional approach, exemplified by Foye,[4] structures the material according to the disease area, whereas the books by Krosgaard-Larsen and Bungaard[5] and by Coulson[6] present the material according to mechanistic class, with chapters on enzyme inhibitors, ion-channel regulators and receptor ligands. A third approach, which has appeal to those seeking more information on organic synthesis, is that taken by Lednicer and Mitscher,[7] which covers drug synthesis according to compound type. The major modern work is *Comprehensive Medicinal Chemistry*.[8] Volume 1 covers historical background and socio-economic factors in drug development and health-service provision. Volumes 2 and 3 cover enzyme inhibitors and receptor agonists and antagonists respectively, while Volume 4 addresses physicochemical aspects of drug design. Volume 5 covers the medicinal chemistry considerations in drug kinetics and metabolism while Volume 6 has a drug compendium and an extensive index.

The two-volume *Chronicles of Drug Discovery*[9] contain a total of 23 chapters each describing the discovery of important drugs, written by the inventors. Similarly, *The Role of the Organic Chemist in Drug Research*[10] contains eight chapters on major drug classes by leading workers in the various fields.

The Pharmacological Basis of Therapeutics,[11] now is its eighth edition, describes the mechanism of action of all the main drugs by therapeutic class with chapters, for example, on drugs for the central nervous system, cardiovascular drugs and microbial diseases. Individual agents are described in detail with information on pharmacology, toxicity and clinical use.

Martindale: The Extra Pharmacopoeia,[12] gives information on about 900 drugs including uses, details of dosing and adverse effects. The two-volume *Therapeutic Drugs*[13] gives comprehensive coverage of about 730 drugs with a three- or four-page account of each agent, including the structural formula, pharmacology, toxicology, clinical pharmacology, kinetics, metabolism and therapeutic use.

Drug Discovery and Development[14] contains chapters on safety testing and the clinical evaluation of drug candidates, with case histories on histamine antagonists for the treatment of peptic ulcers, antidepressants and calcium channel blockers for use in cardiovascular medicine written by the inventors. *Multinational Drug Companies*[15] gives a 100-page overview of drug discovery and development before considering corporate organisation and R&D management, and the relationships that pharmaceutical companies have with academic institutes and government agencies.

References and bibliography

1. 'The Merck Index: An Encyclopaedia of Chemicals, Drugs and Biologicals,' 11th edn, Merck and Co., Inc., Rahway, NJ, USA, 1989.
2. 'The United States Pharmacopoeia and National Formulary', USP Convention, Rockville, MD, USA, 1990.
3. 'The British Pharmacopoeia,' HMSO, London, 1988 and supplements.
4. 'Principles of Medicinal Chemistry,' W.O. Foye, 3rd edn, Lea and Febiger, Philadelphia and London, 1989.
5. 'A Textbook of Drug Design and Development,' P. Krogsgaard-Larsen and H. Bungaard (eds), Harwood Academic, Reading, 1991.
6. 'Molecular Mechanisms of Drug Action.' C.J. Coulson, Taylor and Francis, London, 1988.
7. 'The Organic Chemistry of Drug Synthesis,' D. Lednicer and L. Mitscher, John Wiley, New York, Vols I, II and III, 1977, 1980 and 1984 respectively.
8. 'Comprehensive Medicinal Chemistry,' C. Hansch, P. Sammes and J.B. Taylor (eds), Pergamon Press, Oxford, 1990.
9. 'Chronicles of Drug Discovery,' J.S. Bindra and D. Lednicer, John Wiley, New York, Vols I and II, 1982 and 1983 respectively.
10. 'The Role of the Organic Chemist in Drug Research,' S.M. Roberts and C.R. Ganellin (eds), Academic Press, London, 1993.
11. 'Goodman and Gilman's The Pharmacological Basis of Therapeutics,' 8th edn, A. Goodman Gilman, T.W. Rall, A.S. Nies and P. Taylor (eds), Pergamon Press, Oxford, 1990.
12. 'Martindale: The Extra Pharmacopoeia,' 29th edn, J.E.F. Reynolds, The Pharmaceutical Press, London, 1989.
13. 'Therapeutic Drugs,' 2 vols, Sir Colin Dollery (ed.), Churchill Livingstone, London, 1991.
14. 'Drug Discovery and Development,' M. Williams and J.B. Malick (eds), Humana Press, Clifton, New Jersey, 1987.
15. 'Multinational Drug Companies: Issues in Drug Discovery and Development', B. Spilker, Raven Press, New York, 1989.

Statistical data on companies, products and diseases are drawn from:

'Pharma Facts and Figures,' The Association of the British Pharmaceutical Industry, 12, Whitehall, London SWIA 2DY, 1992 which quotes data from the Departments of Health and Employment.
Scrip, World Pharmaceutical News, Year-book for 1992, PJP Publications, 18–20 Hill Rise, Richmond, Surrey TW10 6UA, 1992.
Scrip, World Pharmaceutical News, Review Issue for 1992, January 1993.
Annual Report, Glaxo Co., 1992.

9 Biological catalysis and biotechnology

Mike Turner

9.1
Introduction

In 1980 the Spinks Report[1] defined biotechnology as the 'application of biological organisms, systems or processes to manufacturing industry'. This is a very broad definition. Its logical conclusion, with ourselves included amongst the biological organisms, might lay claim to the sum total of all human activity. This review will take a much narrower definition which focuses on the industrial uses of biological catalysis for the manufacture of chemical products.

The committee chaired by Dr Spinks set out 'to review existing and prospective science and technology relevant to industrial opportunities in biotechnology'. Their review needed to encompass not only those ancient industries like baking, brewing and the manufacture of cheese, but also newer processes such as the manufacture of organic compounds of high and low molecular weight by fermentation or with enzymes; the biological treatment of large bodies of water, whether for its purification or for the leaching of ores; and the development of biological routes to convenient energy sources. Already by 1980 the scale of many of these processes was very large, even by comparison with others which come under the umbrella of industrial chemistry (Table 9.1).[2] The review made clear that there was also a small number of key techniques (Table 9.2) which, given the right encouragement, would have a substantial impact on the future development of this large industry based on biotechnology.

The power of a biological process resides in the catalytic action of the living organism, or, more strictly, of the proteins it contains. The Spinks Report drew together the effect of three decades of research on our ability to manipulate and modify these proteins to fit them for use in manufacture. When the technique of genetic engineering was developed in the second half of the 1970s it provided a breakthrough in this research. With this technique it is possible to intervene with great specificity in the process by which an organism delineates the amino acid sequence of a protein. At best the ability to synthesize a protein can be transferred from one cell to another, even to one in a different organism, thereby crossing boundaries as great as that between a mammal such as *Homo sapiens* and a bacterium such as *Escherichia coli*. Furthermore, in the process the amino acid sequence of the protein can be changed. Biotechnologists are slowly discovering the full power of genetic

Table 9.1 World and UK production of biological and other products (1979) in tonnes per annum. From Dunnill, P. (1981) *Chem. Ind.*, 205

Product	UK	World
Water*	3 700 000 000	n.d.
Crude steel	20 400 000	670 000 000
Milk	15 840 000	458 000 000
Sugar	984 000	90 000 000
Naphtha	4 630 000	81 000 000
Beer	6 590 000	80 130 000
Frozen foods	730 000	20 000 000
Aluminium	550 000	13 500 000
Cheese	216 000	11 000 000
Baker's yeast	100 000	1 750 000
Polyurethanes	83 500	n.d.
Citric acid	15 000	300 000
Benzyl penicillin	2 500	10 000
Human plasma	125	5 000
Cephalosporins	n.d.	900
Vitamin B_{12}	n.d.	6

* From sewage receiving biological treatment.   n.d.—no data.

Table 9.2 Key areas presenting outstanding opportunities for biotechnology

Genetic manipulation
Enzymes and enzyme systems
Monoclonal antibodies and immunoglobulins
Waste treatment (detoxification, by-product utilization)
Plant cell culture and single-cell protein
Production of fuels (ethanol and methane) from biomass

engineering, and the Spinks Report was right to place it at the top of its list of key techniques (Table 9.2).

If the catalytic action of an organism is largely dependent on the proteins it contains, and of a protein on the sequence of its amino acids, then, in principle, the biotechnologist now has the tools to alter that action at will. Biological catalysis is no longer a process which must be taken as it is found, making use only of natural variation, even if that were induced by a process of mutation and selection in the laboratory. The rational design of catalytic proteins, that is, of enzymes, is now possible, and that should allow us to make greater use of the specificity which is inherent in the enzymatic reaction. However, this is unlikely to lead to large-scale biological catalysis displacing chemical technology. It is much more likely to extend the range of possible processes, which is precisely what biotechnology has done in the past.

What has been too frequently forgotten in the excitement of the last few years is that the new techniques are only the latest in a long line of advances from which our modern understanding of biological and chemical catalysis is derived. They stretch back at least to 1833 when Payen and Persoz separated diastase (amylases) from germinating barley (malt)[3] and to the development of the theory of catalysis by Berzelius in 1835. I shall try to place modern advances in the context of older processes, because this may be the only way to show what they will achieve in this field of manufacture. It would be wrong to present the techniques of genetic engineering and large-scale enzymology without their antecedents, which, if they have done nothing else, have made it possible to apply the new techniques to manufacture. Besides, I feel grateful for an echo of sympathy from Payen and Persoz,[3] who, at the end of their 1833 paper, wrote:

> It will seem less remarkable that we have advanced only such a little distance along this new road if it is considered that we have been caught up in a millrace of newborn applications and that we have not thought it right to refuse our collaboration to the manufacturers who have requested it from every direction.

9.2
Microbial synthesis of primary metabolites

9.2.1   Organic solvents

9.2.1.1   Ethanol.   The process in which a yeast ferments a sugar to ethanol has many claims on the interest of a biologist. It is, of course, an old process, and where the product was drunk, it features in the earliest written records. It operates throughout the world, and brewers and vintners have used a variety of ingenious methods to prepare the input sugar. Many of the processes are remarkable for their complexity, and the underlying experimentation which led to the manufacture of alcoholic drinks from barley or rice would represent a considerable achievement for today's biotechnologist, let alone prehistoric people.

The nature of the fermentation process was not clearly understood until, in the 1860s and 1870s, Pasteur's studies helped to lay the foundations of microbiology.[4] When in 1897 Buchner showed that a cell-free extract of the yeast could degrade sucrose, fructose and glucose, but not lactose, to ethanol, this was a watershed in our understanding of biological catalysis. No longer was the process of fermentation dependent on the living cell, and the analysis of the seperate enzymes that were contained in Buchner's 'zymase' was possible.

The process should also interest the industrial chemist. Even at the start of the industrial revolution, brewing (particularly of beer) was a large industry. In 1790 Samuel Whitbread employed a capital of £271 240 and he brewed 30 000 m^3 of beer in 1796. However, it was not easy to transform this process, which makes a dilute, and very impure, aqueous solution of ethanol (say $50 \, g \, l^{-1}$), into one capable of manufacturing large quantities of the pure alcohol. That had to await the invention of efficient distillation equipment.

The alchemists of the Middle East were familiar as early as the first century AD with the technique of distillation which was necessary to separate the ethanol from the water. Indeed, developments of their simple pot stills were used to prepare concentrated spirits with a high ethanol content (say $500 \, g \, l^{-1}$)

well before the Industrial Revolution. Yet the true industrial production of ethanol directly from a crude fermentation broth with its high solids content became possible only after Aeneas Coffey, working in Dublin, developed his multistage continuous still in 1830 (Figure 9.1). The last traces of water in the azeotrope which the still produced were removed with a drying agent.

In 1855, the ethanol manufactured in this process of fermentation and distillation was exempted from excise duty, provided it was first denatured with methanol. Only then was the resulting product (industrial methylated spirit) able to develop as an industrial solvent. Nor was it the last occasion on which a biological process needed the assistance of chemical technology (continuous distillation) and government intervention (exemption from duty) in order to become established.

The essential biochemistry of the process is relatively straightforward, and it features a pathway of carbohydrate metabolism that is common to all organisms. A sugar, glucose, is first degraded to pyruvate (Figure 9.2). The yeast cell derives energy from this process, which it stores in the terminal pyrophosphate group of ATP. It also reduces the catalytic intermediate NAD^+ to NADH. While the ATP is efficiently recycled through its involvement in many synthetic reactions in the cell, the NADH would accumulate under the anaerobic conditions of the fermentation were not a special mechanism available to recycle it. In the yeast cell the enzyme pyruvate decarboxylase degrades the pyruvate to acetaldehyde, and NADH reduces the latter to ethanol in a reaction catalysed by alcohol dehydrogenase. An overall balance of NAD^+ reduction and oxidation is thus maintained.

The sugar is not, however, supplied as pure glucose. A common input is molasses, which is a by-product of cane-sugar refining. It contains 35–40% (w/w) of sucrose, and about 15–20% (w/w) of glucose and fructose. The molasses is diluted to give about $150–200 \, g \, l^{-1}$ total sugar, and the pH is adjusted with mineral acid to between 4 and 5. This solution is then inoculated with yeast, usually *Saccharomyces cerevisiae*. Intially the fermentation is aerobic to allow the yeast to grow vigorously, but when a good biomass is reached the process is run anaerobically. The enzyme invertase catalyses the hydrolysis of the sucrose to glucose and fructose, which are then metabolized to ethanol. After about 36 h at 30 °C the fermentation contains $60–90 \, g \, l^{-1}$ ethanol, which is distilled off.

Unfortunately the yeast cells cannot produce ethanol indefinitely, since it is a toxic product which limits the growth of the yeast cells and at high concentrations may kill them. It is this effect which usually limits the final ethanol concentration to between 60 and $90 \, g \, l^{-1}$. It is possible to increase this final concentration in a variety of ways. Some strains of yeast are better able than others to tolerate ethanol, and if one of these is used it will ferment the sugar to a higher ethanol concentration. The yeast cells will also continue to ferment glucose even though they are unable to grow. A second stage of fermentation in which the non-viable (dead) cells convert more sugar to ethanol could therefore be grafted on to the first stage which used viable cells.

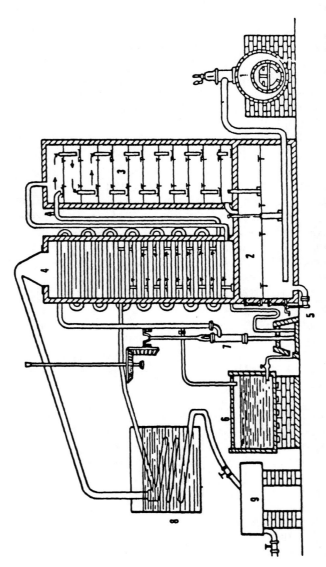

Figure 9.1   The Coffey still for the separation of ethanol from a fermented beer mash: 1, boiler; 2, spent mash chamber; 3, stripping column (analyser); 4, rectifying column; 5, residual mash (stillage) outlet; 6, fermented mash (feed); 7, feed pump; 8, condenser; 9, product tank. From Packowski, G.I. (1978), in *Encyclopedia of Chemical Technology*, 3rd edn, Vol. 3, p. 851.

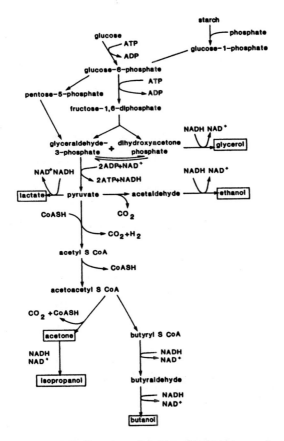

Figure 9.2 Outline of the metabolic pathway by which carbohydrates are degraded to industrial solvents (and lactic acid). The arrows do not necessarily indicate a single reaction step, nor does a single strain or species of an organism catalyse all the reactions. ATP, adenosine-5′-triphosphate; ADP, adenosine-5′-diphosphate; NAD⁺, nicotinamide adenine dinucleotide; NADH, reduced NAD⁺; HSCoA, coenzyme A.

It may also be possible to extract the ethanol continuously from the fermenter, either by solvent extraction or under vacuum at a raised temperature. In the latter case one of several thermotolerant bacteria which grow at 60 °C, and which also metabolize pyruvate to ethanol, could replace the yeast, which prefers a temperature close to 30 °C.[5] Although any of these ideas may be practical, they are usually irrelevant to the main constraints on the process. These are the source of the sugar for the fermentation, and the overall efficiency with which the energy in the sugar is recovered in the ethanol.

The main direct sources of sugar for the fermentation are sugar cane and sugar beet, both of which contain primarily sucrose. Indirect sources of sugar are also used, the principal one being starch from corn (maize), but the starches from potatoes and barley were once important. The starches, which are

Table 9.3 Energy balance for the manufacture of ethanol from sugar cane. From Mouris, E. (1984) *Chem. Ind.*, 435

Energy (GJ) per tonne of sugar cane					
Input		Agricultural yield		Output	
Source	GJ	Source	GJ	Source	GJ
Agriculture	0.855	Bagasse	2.059	Ethanol (0.77 tonne)	1.933
Irrigation	0.619	Sugar	2.509	Surplus bagasse	0.632
Transport	0.012			Trash	0.716
Lime manufacture	0.830				
Storage	0.002				
Total (GJ)	1.571		4.568		3.281
%	100				207

polymers of glucose, are degraded largely to mono- and disaccharides (see section 9.6.1) before the fermentation. The storage polysaccharide inulin, which is a polymer of fructose obtained from Jerusalem artichokes, has also long been recognized as a potential source of sugar for the ethanol fermentation. Another material which is used is the waste liquor from the sulphite wood-pulping process, which contains 20–30 $g\,l^{-1}$ fermentable sugar. In fact almost any cheap source of sugar which the yeast can convert to fructose or glucose is suitable.

The actual yield of ethanol from glucose in the fermentation is high. The theoretical yield is 0.51 g per g of glucose, and this is equivalent to the recovery of about 95% of the heat of combustion of the glucose. In practice, where the input sugar is derived from sugar cane, 76% of the energy of the sugar is recovered in the distilled ethanol (Table 9.3). However, this hides the energy needed to distil the ethanol from the fermentation liquors. This is provided by bagasse, the dried waste cane which is used to fuel boilers, a surplus of which remains at the end of the process (Table 9.3). Even allowing for all the other energy inputs to the process, the overall balance is good.[5] In fact the process could be viewed as one which converts the low-grade energy of the bagasse into a useful liquid fuel. Since the sugar input is frequently in the form of molasses, itself a by-product, it represents an effective use of waste materials. Other sources of agricultural sugar do not have such a favourable energy balance because no similar waste is available.

Doubts have always chased the expectations that this mixture of fermentation and distillation could provide a route to a plentiful and assured supply of ethanol for use as a fuel, particularly for the internal combustion engine. As long ago as 1921 Sir Charles Bedford, the Chairman of the Empire Motor Fuels Committee, commented:

> For many years past the long-suffering public in this and other countries has been tantalized by visions of cheap and abundant alcohol as an alternative to petrol and other light fuels. The newspapers from time to time announce some new means of obtaining alcohol for a mere song, but unluckily for the public, these dreams do not materialize.

(1) Absorption of ethylene in concentrated sulphuric acid to form mono- and di-ethyl sulphates:

$$CH_2{=}CH_2 + H_2SO_4 \rightarrow CH_3CH_2OSO_3H$$
ethyl hydrogen sulphate
(monoethyl sulphate)

$$2CH_2{=}CH_2 + H_2SO_4 \rightarrow (CH_3CH_2O)_2SO_2$$
diethyl sulphate

(2) Hydrolysis of ethyl sulphate to ethanol:

$$CH_3CH_2OSO_3H + H_2O \rightarrow CH_3CH_2OH + H_2SO_4$$
$$(CH_3CH_2O)_2SO_2 + 2H_2O \rightarrow 2CH_3CH_2OH + H_2SO_4$$

(3) By-product formation:

$$(CH_3CH_2O)_2SO_2 + CH_3CH_2OH \rightarrow CH_3CH_2OSO_3H + (CH_3CH_2)_2O$$
diethyl ether

Figure 9.3 The indirect hydration of ethylene with ethyl sulphates as intermediates. This process was the first to compete with fermentation for the manufacture of ethanol. From Sherman, P.D. and Kavasmaneck, P.R. (1980), in *Encyclopedia of Chemical Technology*, 3rd edn, Vol. 9, p. 342.

Then, as now, petrol distilled from oil was thought unlikely to supply the burgeoning need for a liquid fuel; the known reserves of crude oil were too small. Benzol distilled from coal would not fill the gap, but ethanol produced by fermentation seemed, if not an alternative, at least a supplement which would conserve petrol supplies. Such a scheme was considered particularly attractive where the input sugar could be derived from waste cellulose. To divert ever-increasing areas of good agricultural land to the supply of starch for ethanol manufacture was not thought practical.

In the event, new reserves of oil were discovered, and the following years saw the development of two chemical routes for the manufacture of ethanol. In the earlier of these processes ethylene derived from oil was sulphated and then the organic sulphate was hydrolysed to ethanol (Figure 9.3). Later this route was replaced by a direct catalytic hydration of ethylene (see also Vol. 1, section 11.7.1). These processes came to dominate the manufacture of ethanol (Table 9.4). In the last few years, as industrial societies such as the USA and Europe seek to use agricultural surpluses, or others, such as Brazil, face a shortage of oil, the biological route has regained its former importance (Table 9.5). Yet the justification for this use of good agricultural land remains as doubtful as in 1921. Moreover, the reserves of fossil fuel as a whole, if not of oil in particular, are large (Vol. 1, Figure 2.1), although the cost of their recovery is likely to increase.

The breakthrough in the biological production of ethanol still awaits the efficient use of cellulose and hemicellulose as a source of input sugar. These two polymers of glucose and xylose respectively, together with lignin, a complex aromatic polymer, account for most of the organic content of wood and straw (Table 9.6). However, they are not easily degraded. Mineral acids, which hydrolyse both cellulose and hemicellulose, if rather inefficiently, can

Table 9.4   Estimated proportion of ethanol manufactured by fermentation in the USA between 1920 and 1990

Year	1920	1935	1954	1963	1977	1982	1991
% manufactured	100	90	30	9	6.5	55	94

Table 9.5   Annual production of chemical ethanol

	Production rate (million tonnes per annum)	
	Fermentation	Petrochemical
USA	3.6	0.25
Europe	0.13	0.62
Brazil	9.5	—

Laluce, C. (1991) *Crit. Rev. Biotechnol.*, **11**, 149; Boddey, R. (1993) *Chem. Ind.* **355**

Table 9.6   Composition of wood and straw (%)

	Wood	Straw (rice)	Bagasse
Carbohydrate	65–75	56–63	62–88
Cellulose	40–50	27–35	42–58
Hemicellulose		20–30	20–30
Lignin	18–35	14–22	13–22
Extractives (resins etc.)	5–20		1–3
Ash	1.5–3	12	4
Bulk density		40–256 kg m^{-3}	

provide an input for the yeast fermentation. Indeed, this is the process through which sulphite liquor wastes from the paper industry are used to make ethanol.

The biological hydrolysis of cellulose is also promising but the reaction is a complex one catalysed by the sequential action of the enzymes in the cellulase complex (Figure 9.4).[6] The cellulose needs some pretreatment to soften it if the biological hydrolysis is to be rapid, and the natural organisms which carry out this stage of the process may not efficiently produce ethanol from the liberated glucose. Moreover, the yeasts which are normally used to produce ethanol from glucose cannot metabolize xylose. Some bacterial fermentations will carry out the complete process,[7] and a mixed culture of clostridia will produce 0.2 kg of ethanol from 1 kg cellulose. Clearly on a weight-for-weight basis, this is very much less efficient than the synthesis of ethanol from starch. The energy would be better conserved by simply burning the cellulose, another time-honoured process which still accounts for at least 65% of the energy consumed in non-industrial societies, and which in the USA accounts for almost as much energy generation as does nuclear fission (Table 9.7). The process only begins to look interesting if the input is waste cellulose, and if the output of furfural, in almost equal weight to the ethanol, and of other organic by-products are also

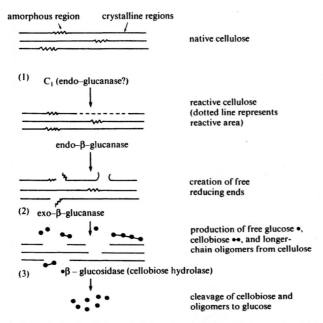

Figure 9.4   The hydrolysis of cellulose. Cellulose is a polymer of glucose in which the glucose units are linked β-1,4. The polymer is partly crystalline and is degraded stepwise to glucose and cellobiose (glucose-β-1,4-glucose). After Kosaric *et al.* (1983), in Dellweg, H., ed., *Biotechnology* 3, 257.

Table 9.7   Energy consumption in the USA (1981)

Source	PJ (1 PJ = 10^{15} J)
Petroleum	33.9
Natural gas	21.3
Coal	17.0
Hydrogen	2.9
Nuclear	3.1
Biomass	
Wood and wood waste (direct combustion	2.5
Municipal and industrial solid waste	0.2
Sewage and agricultural waste	0.1
Alcohol fuels	0.015
Total	81.0

Chem. Eng. News (1983, Mar. 14.) 28.

considered. Already about half of the 4 200 tonnes of vanillin produced each year comes from this source; but it could also provide large amounts of methyl aryl ethers as antiknock components for automobile fuels.[8]

The genes that control the synthesis of the cellulose complex can now be moved between a limited range of micro-organisms. This could help both in the isolation of large amounts of cellulase to catalyse the hydrolysis of cellulose before the fermentation begins, and in the development of new organisms capable of making ethanol direct from cellulose. The former would have the advantage of linking directly with the processes for making ethanol from starch. The way forward is as uncertain as it was in 1921, but any process which makes use of some of the waste we generate from the 2×10^{11} tonnes of carbon fixed each year by photosynthesis must be of interest.

9.2.1.2 Glycerol, acetone and butanol. In 1858 Pasteur noticed that yeasts produced glycerol as a minor component of sugar fermentations. About 3 g was formed for every 100 g input sugar. This observation prompted some research which by the outbreak of World War I, had pushed the amount of glycerol formed up to 7 or 8 g per 100 g sugar. At the time glycerol was manufactured as a by-product of fat saponification, but the war made that supply insufficient in Germany and the yeast fermentation was re-examined.

When *Saccharomyces cerevisiae* is grown in a medium buffered to an alkaline pH with bicarbonate or phosphate, the yield of glycerol rises to about 10 g per 100 g sugar. Unfortunately the alkaline medium also allows the active growth of bacteria in what are, traditionally, non-sterile fermentations. However if the medium is buffered with sodium sulphite (about $60 \, g \, l^{-1}$), which is a salt with well-known antiseptic properties, the bacterial contamination can be prevented and the yield of glycerol rises to about 25 g per 100 g sugar.

The biochemistry of this redirection of the ethanol fermentation is quite simple. The sulphite forms a condensation product with acetaldehyde, which is normally the immediate precursor of ethanol. This cannot be reduced by alcohol dehydrogenase, and the cell must find an alternative route to re-oxidize NADH. It does this by reducing dihydroxyacetone phosphate, and after hydrolysis of the phosphate ester, glycerol is excreted (Figure 9.2). This manipulation of the yeast's metabolism, the details of which were not then understood, allowed a minor fermentation product to become the major one.

Notwithstanding the difficulty of recovering pure glycerol from the fermentation, by 1917 the German fermentation industry produced about 1000 tonnes of glycerol per month.[9] Today the process has little or no commercial significance, the bulk of the world's glycerol (0.5×10^6 tonnes annum^{-1}) being isolated in equal amounts either as a by-product of saponification of animal or vegetable oils and fats or by the chemical oxidation of propylene. However, the fermentation process must represent the first occasion on which the metabolism of an organism was diverted from its natural path into the synthesis of an important chemical product.

Of greater significance is the fermentation for acetone and butanol. Again it was World War I that stimulated development of the process. Until that time acetone was prepared from calcium acetate (Figure 9.5), derived from the pyrolysis of wood and imported into Britain from Austria or the USA. Some years before these supplies were interrupted, Fernbach, working in Paris, showed that some bacteria could produce acetone and butanol from starch,

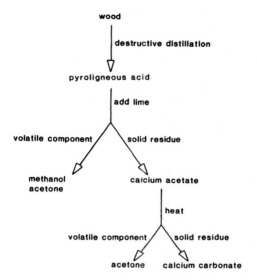

Figure 9.5 Manufacture of acetone from wood. Mixtures of higher ketones are produced if the calcium acetate contains other fatty acid salts. Acetaldehyde will also be formed if any calcium formate is present.

and in 1913 Bayer & Co. took out a German patent for the production of acetone and alcohol by *Bacillus macerans*. Then during the war, Chaim Weizmann, who was originally interested in the butanol, actively promoted the organism, which he reclassified as *Clostridium acetobutylicum*, as a source of acetone.[10]

The process was given official support and was developed to the point where the organism would ferment starch (up to $38\,\mathrm{g\,l^{-1}}$) to a mixture of butanol, acetone and ethanol in a ratio of about 6:3:1 by weight, the total solvent yield being about 35% of the weight of the starch. A strictly anaerobic fermentation was necessary for the solvent production, which was accompanied by a vigorous evolution of carbon dioxide and hydrogen. The fermentation was over in about 36 h.

The biochemistry of this process represents another variation on the metabolism of glucose under anaerobic conditions. *C. acetobutylicum* can decarboxylate pyruvate and synthesize a thioester of acetate, acetyl coenzyme A. From this process it can derive more energy as ATP with the eventual release of acid. This phase of the process, which can begin aerobically, lowers the pH of the medium from around 5.8 to between 5.1 and 5.0, a further drop being prevented by adding calcium carbonate (as chalk granules). In the strictly anaerobic phase which follows, a complex set of reactions enables the acetate to absorb some of the reducing power of the NADH, and the solvents are formed (Figure 9.2) as the pH rises back to 5.8. The surplus reducing potential is converted to the hydrogen which, along with the carbon dioxide, accompanies the synthesis of the solvents.

The fermentation industry, which at that time had no experience of large-scale sterile fermentations, found the scale-up of this process difficult. The process differs radically from the anaerobic ethanol fermentation where the strict exclusion of air is not necessary, and where the growth of competing organisms is limited by the acidity of the medium (pH 4–5), and the vigorous growth, at the start of the fermentation, of the large inoculum of yeast cells. Under the less acid conditions preferred by *C. acetobutylicum* (pH 5.8) many other bacteria will grow on the medium of maize mash, ammonia and inorganic salts, particularly before the anaerobic conditions are established.

In his review of this process, John Hastings points out that all the problems associated with large-scale sterile fermentations had to be solved: how to manufacture a vessel of $125 \, m^3$ capacity with suitable seals and gaskets to allow its sterilization with steam at 121 °C; how to lay out pipework to prevent the dead-legs where contaminating organisms could accumulate; how to design valves and pumps which could provide a sterile seal when transferring inocula; how to manage the foaming, not to speak of the fire hazard, which the gases cause. The latter problem was probably responsible for the explosion which damaged the plant at Kings Lynn shortly after it started production. These were only a few of the new problems which faced the microbiologists and engineers who built the first plant. Clearly the development of this process must have laid the foundations of the modern fermentation industry.

The inoculum which was added to these large fermenters was developed in stages from spores stored in sterile sand or soil (see Figure 9.15). During this process the growing bacteria may become infected with one of several phages. These are viruses which infect bacteria. If this happens at a late stage in the transfers, or in the productive fermentation, it will result in a costly failure. To overcome this problem the bacteria were cultured on a small scale in the presence of the phage, and the spores of those colonies which were resistant to it were selected for the large-scale fermentations. Since the phage was also able to adapt itself to attack these newly-isolated resistant strains of *C. acetobutylicum*, this classic selection technique was continually repeated.

In contrast to the problems of the fermentation, the recovery of the solvents was straightforward. Although the necessary distillation was an important cost in the process, the boiling points of the three main products are sufficiently different to make their separation easy. Certainly the acetone was of high quality and all too suitable for the manufacture of cordite.

Over the years the butanol or the other minor solvents became more important than the acetone, and the process continued to develop. Different species of clostridia were used to produce these other solvents (e.g. *C. butylicum* for butanol and isopropanol, and *C. butyricum* for butyric and acetic acids). Strains were isolated which made a very efficient use of molasses rather than starch. The pH of the medium was controlled by aqueous ammonia rather than by chalk, giving a completely soluble medium suitable for in-line sterilization. The spent cells recovered at the end of the fermentation were

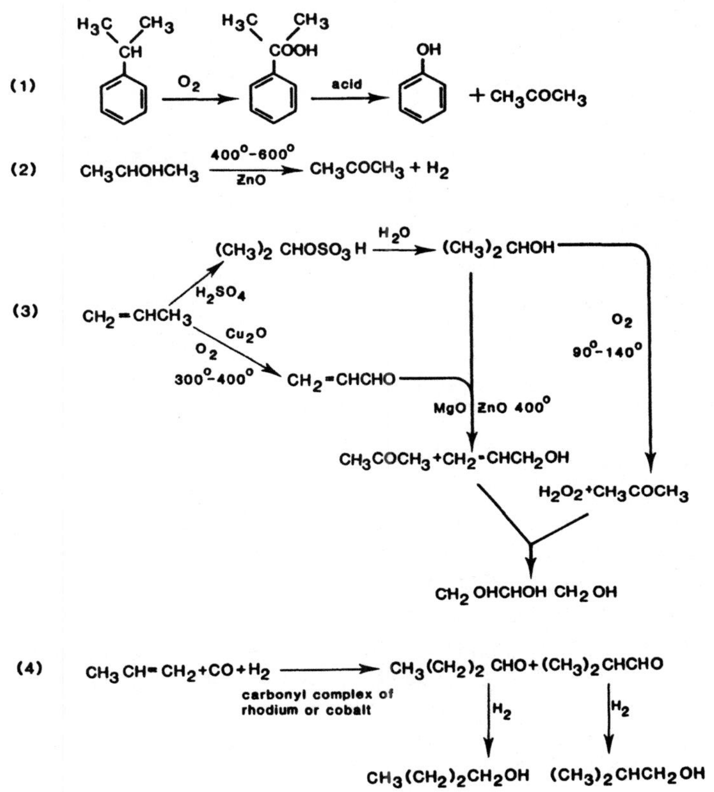

Figure 9.6 Manufacture of acetone (1, 2 and 3) and butanol (4). The cumene hydroperoxide process (1) and the dehydrogenation of propan-2-ol are more important routes to acetone than the Shell glycerol process (3). The oxo-process for butanol manufacture (4) can be controlled so that butan-1-al is the main product of the hydroformylation of propylene.

found to be a good source of riboflavin (Vitamin B_2), and, after drying, were used as a supplement to animal feeds. The carbon dioxide and hydrogen were collected and sold, and continuous distillation was developed for the solvent recovery.

This continual modification of the process allowed it to compete with the petrochemical synthesis of butanol and acetone (see sections 11.8.1 and 11.9.7 of Vol. 1) (Figure 9.6), until molasses found a competing use as an animal feed and its price began to rise. British production stopped in 1957, but the process survives in a few countries, particularly Brazil and Taiwan, where oil supplies are limited. A process which made use of waste cellulose might still be economic. However, its main legacy to biotechnology lies in that early phase of its development which laid the foundations of the modern fermentation industry.

Table 9.8 Manufacture of carboxylic acids. From *Encyclopedia of Chemical Technology*, 3rd and 4th edns; Mattey, M. (1992) *Crit. Rev. Biotechnol.*, **12**, 87.

Carboxylic acid	Methods of manufacture	Annual production (tonnes)
Monocarboxylic acids		
Formic	Chemical	31 000 (USA)
Acetic		1.5×10^6 (USA)
Propionic		54 000 (USA)
Butyric/isobutyric		24 000 (USA)
Stearic (all grades)	Extraction of fats	0.24×10^6 (USA)
Hydroxycarboxylic acids		
Lactic	Fermentation	35 000
Dicarboxylic acids		
Adipic	Chemical	1.7×10^6 (USA)
Tartaric	Fermentation	50 000
Tricarboxylic acids		
Citric	Fermentation	0.55×10^6

9.2.2 Carboxylic acids

Several carboxylic acids are manufactured by fermentation (Table 9.8). The production of acetic and lactic acids underlies the important traditional processes of vinegar and of milk fermentation, but, of these two, only lactic acid is commercially manufactured in this way. From time to time other carboxylic acids which are used in large amounts, notably adipic acid (Table 9.8), are proposed as suitable candidates for a fermentative process. However, where suitable and more economic chemical routes are available from petrochemical feedstocks, there is no real incentive for change.

9.2.2.1 Lactic acid. Lactic acid was the first of the organic acids to be manufactured by fermentation. The process was first used in 1880, but even by 1920 the nature of the organism was not well defined. Today several species of lactobacilli are used, and presumably they were responsible for the earlier fermentations which were inoculated with small amounts of decomposing meat or cheese.

Underlying the synthesis is yet another microbial adaptation to anaerobic growth. In this instance NADH is oxidized by pyruvate, the reaction being catalysed by lactate dehydrogenase (Figure 9.2). Where lactic acid is manufactured from lactose in whey the organism which is used is *Lactobacillus bulgaricus*, but others which convert glucose (*L. delbrueckii*) or sulphite liquor waste (*L. pentosus*) are also important.

Although organisms are available for the synthesis of either enantiomer only the L(+) isomer* is produced commercially. The conversion of sugar to

---

* 'D' and 'L' sterochemical nomenclature is a relative system based on the arbitrary standard D-glyceraldehyde. This is now being replaced by the absolute R, S system as far as chemists are concerned, but biochemists, biologists and some natural product chemists continue to use the D, L system. The L(+) enantiomer of lactic acid has an S configuration.

$$CH_3CHO + HCN \rightarrow CH_3CHOHCN$$
$$CH_3CHOHCN + 2H_2O + HCl \rightarrow CH_3CHOHCOOH + NH_4Cl$$

Figure 9.7 The chemical synthesis of lactic acid. The D- and L-forms are resolved by distillation after the racemate has been esterified with a suitable alcohol. From Van Ness, J.H. (1978) in *Encyclopedia of Chemical Technology*, Vol. 13, p. 85.

lactate is very efficient, about 90% of the theoretical yield being achieved. The acid is isolated as its insoluble calcium salt, and the acid is regenerated with sulphuric acid. About 35 000 tonnes annum^{-1} are produced by the fermentation process, which competes with a chemical synthesis. The latter is based on lactonitrile, with the lactic acid being purified as its methyl ester. Virtually all of the small amount of lactic acid produced in the United States (about 20 tonnes annum^{-1}) is of chemical origin (Figure 9.7), whereas European production is dominated by fermentation.

9.2.2.2 Citric acid. Citric acid is largely responsible for the acidic properties of lemon juice, and Scheele crystallized it from this source in 1784. For many years fruit juices remained the only commerical source of citric acid, and, even today, some is manufactured from lemons and waste pineapple pulp.

The industrial use of citric acid is not confined to the flavouring of food and drink. As a tricarboxylic acid it is an effective metal chelator, and this makes it useful not only as an antioxidant in foods but also as a cleaning and pickling agent in the metals industry. Its chelation of iron in particular is also important in the blood processing industry where it is used as an anticoagulant. It prevents the swelling of clays, reducing their viscosity in suspensions, and it controls the setting of cements. It also keeps metals in solution in processes as different as electroplating and the supply of trace nutrients for agriculture. Finally, it is used as a replacement for polyphosphate in the detergent industry. With such a wide range of uses it is not surprising that nearly 0.55×10^6 tonnes are manufactured each year, over 99% being supplied by fermentation. It is also important to recognize that these varied industrial uses became possible because the fermentation provides a cheap and satisfactory source of material which was able to expand to meet the demand.

Although there are numerous patented processes for the chemical synthesis of citric acid (e.g. Figure 9.8) none can compete with the biological process.

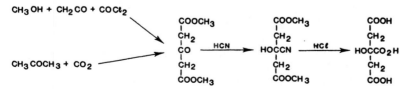

Figure 9.8 Two synthetic routes to citric acid from readily available reagents.

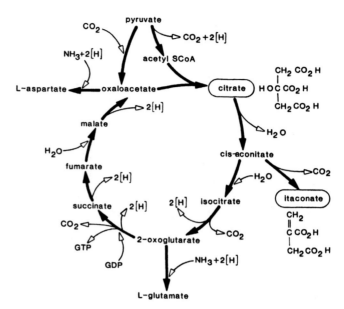

Figure 9.9 An outline of the reaction of the Krebs cycle showing the synthesis of citrate and itaconate. GDP = guanosine-5'-diphosphate; GTP = guanosine-5'-triphosphate.

This was begun by Pfizer Inc. in 1923, and was based on work which James Currie published in 1917.[11] He showed that when the strain and growth conditions were carefully controlled, the mould *Aspergillus niger* would produce citric acid from sucrose. Up to 50 g of citric acid was synthesized from every 100 g sucrose consumed. This is an aerobic process in which one of the normal intermediates of glucose oxidation is excreted from the organism in large amounts. Under aerobic conditions cells decarboxylate pyruvate and condense the acetate with oxaloacetate to form citrate. Two of the carbon atoms in the citrate are normally then oxidized to carbon dioxide, the other product being oxaloacetate. This cyclic process (Figure 9.9), often known as the Krebs cycle in recognition of Sir Hans Krebs who first described it, is central to the oxidation of carbohydrates, but clearly it cannot continue unless the citrate removed from the cycle is replaced. Special reactions known as anapleurotic (filling) reactions feed oxaloacetate and malate back into the cycle from pyruvate and its precursor phosphoenolpyruvate (Figure 9.9). The biological manufacture of citrate is therefore yet another process in which the organism's ability to metabolize glucose is put to good use.

In the original process *A. niger* is grown in shallow pans on the surface of a solution of sucrose or molasses (150–200 g carbohydrate 1^{-1}), ammonium nitrate and metal salts. The spores of *A. niger* are blown over the pans, where they germinate within 24 h. Fans circulate air and the pH falls from about 5 to about 3 as the growing cells use the ammonium salt. Citric acid is produced in large amounts after the growth has stopped, provided the pH and the iron content of the medium remain low. If either is too high, oxalic instead of citric acid will be formed.

The levels of the trace metals in the medium all have a crucial effect on the yield of citric acid, and the concentrations of copper, manganese, magnesium, iron, zinc and molybdenum all need to be controlled. For this reason the fermenter pans are constructed of aluminium or stainless steel to prevent the corrosive solution of citric acid leaching metals into solution. The metals are also removed from the molasses. They are either adsorbed on to ion-exchangers or precipitated with calcium hexacyanoferrate. The latter compound has an interesting secondary effect because the production of citric acid increases if it is present in small excess. Alternatively, the surface cultures may use wheat bran or sweet-potato starch as a solid substrate which replaces the soluble sugars. Citric acid production is then not so dependent on the levels of trace metals.

This technique of surface culture is now largely replaced by submerged culture. The scale (up to $500 \, m^3$) is amongst the largest of the batch fermentation processes. It requires particular control of the inoculum of A. niger which is grown from spores as a pelleted mycelium in the presence of cyanide. Without this treatment the inoculum would grow well in the production phase, but would subsequently give a low yield of citric acid. The same constraints of pH and metal-ion concentration which affect the surface culture also apply to the submerged culture. If the pH is not low enough, oxalic and gluconic acids are produced. Although this is usually undesirable, a modified fermentation for the gluconic acid is commercially important.

The fermentation lasts about 8 days and the yield of citric acid is nearly 70% of the theoretical 1 mole per mole of glucose. For this high yield the dissolved oxygen levels must remain at about 20% of saturation. The air, which is blown through the fermenter at a rate of $0.2–1.0 \, m^3 m^{-3}$ volume min^{-1}, causes the contents to foam. This is controlled with suitable antifoams but even so the foam may take up 30% of the volume of the fermenter. The oxidation of the carbohydrate is exothermic and the necessary aeration and cooling requires an energy input of about $12 \, MJ \, m^{-3}$ fermentation volume h^{-1}. Much less ($2 \, MJ \, m^{-3} \, h^{-1}$) is required for the surface culture where more evaporative cooling occurs.

An alternative to this fermentation with A. niger is the submerged fermentation by the yeasts Candida guilliermondi and C. lipolytica. While the fermentation with A. niger is limited to the oxidation of carbohydrates the yeasts allow a greater range of feedstocks. These include the lower alcohols and, with C. lipolytica, low-molecular-weight normal alkanes. The yeast fermentation is also more rapid. Despite these advantages the conservative regulation of the processes for making food-grade materials will restrict their use. One other variation in the process is the use of A. terreus or A. itaconicus in place of A. niger in which case the fermentation can be directed towards itaconic acid production (Figure 9.9).

Table 9.9   Increasing production of citric acid

Year	1929	1943	1950	1976	1983	1991
Production (tonnes)	5×10^3	12×10^3	50×10^3	0.2×10^6	0.37×10^6	0.55×10^6

When the fermentation is complete the citric acid is isolated by fractional precipitation with calcium. The mycelium of *A. niger* is filtered off and the filtrate is treated with calcium at low pH to remove any oxalate by precipitation, and then at neutral pH to recover precipitated calcium citrate. Calcium is removed from the solid by treatment with sulphuric acid, and the released citric acid is purified through charcoal and an ion-exchange resin. An alternative process is available in which the citric acid is extracted from the filtered fermentation broth into a mixture of C_8 to C_{11} alkanes (kerosene) containing a strongly basic amine (tridodecylamine). The citric acid is re-extracted back into water at a higher temperature. Finally, in either process the citric acid is crystallized from water.

The manufacture of citric acid carried the development of the large-scale processes beyond that reached with the acetone and butanol fermentation. To obtain a high yield of product, not only was the strain of *A. niger* important, but its growth conditions must be closely controlled. The need to control metal-ion concentrations and the corrosive nature of citric acid made stainless steel the material of choice for the construction of the plant, and particularly of the fermenters containing *A. niger* in submerged culture. The success of the process is obvious from the increasing amounts of citric acid manufactured since it was introduced (Table 9.9) and these figures may significantly underestimate the total world production.

9.2.3   Amino acids

In the living organism amino acids are not only the basic units from which proteins are made, but they are also important intermediates in the metabolism of nitrogen. To the extent that the breakdown of sugar to alcohols and acids might be described as a process of degradation yielding only energy, the fixing of ammonia into amino acids, and their subsequent metabolism, might be described as the synthesis of metabolic intermediates important in their own right. However, the distinction is not clearcut, because, while it may be convenient to set down pathways of metabolism leading directly from one compound to another, the interrelations between the pathways create a metabolic network in which the process of degradation must influence, not to say drive, that of synthesis.

It is possible, for example, to set down pathways from glucose to the L-isomers of glutamate, aspartate and lysine (Figures 9.2, 9.9 and 9.10) and to portray their synthesis in the same manner as that of citrate or butanol.

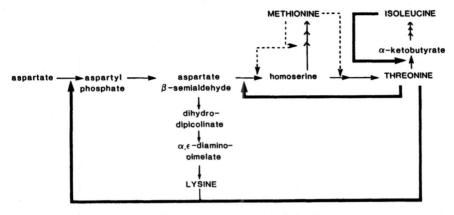

Figure 9.10 The regulation of lysine biosynthesis in *Corynebacterium glutamicum.* ⟶, biosynthetic pathway; −−→, repression; ⟶, feedback inhibition. From Soda, K. *et al.*, (1983) in Dellweg, H., ed., *Biotechnoology*, **3**, 479.

However, the use of micro-organisms to produce amino acids is dependent on a knowledge of the interactions that exist within the metabolic pathways which catalyse their synthesis. This fact sets the development of these fermentations apart from those of the acids and the organic solvents.

Although most of the biologically important amino acids are available commercially in tonne quantities, only a handful (Table 9.10) are manufactured in amounts exceeding 500–1 000 tonnes annum^{-1}.[12] Annual sales of Aspartame (*N*-aspartyl-phenylalanine methyl ester) have now risen to over 2 000 tonnes, and this has increased the production of L-aspartic acid and of L-phenylalanine. The methods of synthesis for these products encompass all possible routes. Some are synthesized chemically, some are extracted from

Table 9.10 Production of amino acids (only those manufactured on a scale of 500 tonnes per annum or more are included)

Amino acid	Method of manufacture	Annual production (tonnes) 1979	1987
L-Glutamic acid	Microbiological	270 000	340 000
DL-Methionine	Chemical	180 000	250 000
L-Methionine	Enzymatic resolution	150	150
L-Lysine	Microbiological	32 000	70 000
Glycine	Chemical	6 000	6 000
L-Aspartic acid	Enzymatic		4 000
L-Phenylalanine	Microbiological		3 000
β-Alanine	Chemical	1 800	2 000
L-Cysteine	Extraction and enzymatic	700	1 000
DL-Alanine	Chemical	700	1 500
L-Arginine	Microbiological	500	1 000
L-Glutamine	Microbiological	500	850

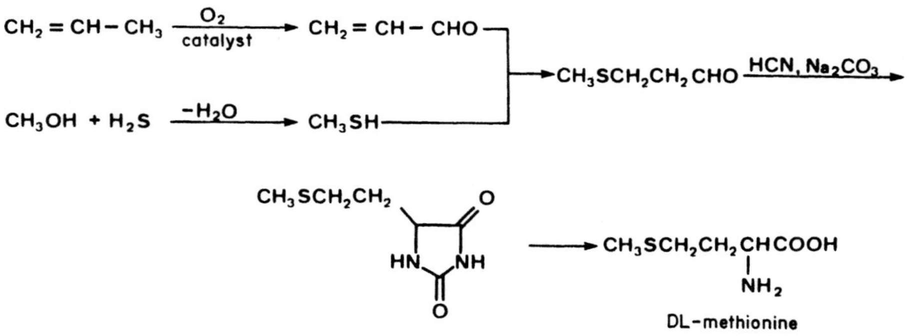

$$CH_2 = CH - CH_3 \xrightarrow[catalyst]{O_2} CH_2 = CH - CHO$$

$$CH_3OH + H_2S \xrightarrow{-H_2O} CH_3SH$$

$$CH_3SCH_2CH_2CHO \xrightarrow{HCN, Na_2CO_3}$$

$$CH_3SCH_2CH_2CH COOH$$
$$NH_2$$
DL-methionine

Figure 9.11   The chemical synthesis of DL-methionine. From Yamamoto, A. (1978), in *Encyclopedia of Chemical Technology*, 3rd edn, Vol. 2, p. 403.

natural sources, others are fermented, while a few are the product of enzymatic catalysis (see section 9.6.3).

The extraction of amino acids from protein is the oldest method of manufacture. Some L-cysteine is still obtained in this way, from hair and feathers, but the process is smelly and the wastes are difficult to treat. A chemical route is often cleaner, and is effective where an achiral molecule (glycine, β-alanine) or the racemate (DL-methionine (Figure 9.11), DL-alanine) is the required product. There are also effective methods which link a chemical synthesis with an enzymatic step (see section 9.3.3), but the manufactures of L-glutamate and L-lysine are good examples of processes based on microbial synthesis.

9.2.3.1   L-Glutamic acid.   Even though it has only a faint taste of its own, monosodium L-glutamate enhances the flavour of foods. This property, which is not shared by D-glutamate, was originally discovered by Ikeda in 1908 when L-glutamate was identified as an important component of vegetable stocks made from kelp seaweeds. Until 1964 it was extracted from acid hydrolysates of wheat gluten. A chemical synthesis is also possible, and one Japanese company (Ajinomoto) did establish and use a process for the enzymatic resolution of the racemate (see section 9.6.3). However, neither of these methods can compete with microbiological synthesis.

In 1957, Kinoshita isolated *Corynebacterium glutamicum*, the first species of bacterium known to excrete L-glutamic acid.[13] Usually important cellular metabolites are held within the cell and their synthesis is kept under careful control. However, the cell membrane of *C. glutamicum* is permeable to glutamate, particularly if it is grown in the presence of limiting levels of biotin. This vitamin has a role in the fixing of the carbon dioxide in the synthesis of fatty acids. Low levels ($2.5-5.0\,\mu g\,l^{-1}$) are sufficient to allow the organism to grow but are too low to allow it to synthesize the fatty acids it requires for its cellular membranes. They therefore become permeable, and L-glutamate escapes from the cell. At higher levels of biotin, succinate and lactate are excreted instead of glutamate.

Other compounds which interfere with the synthesis of the cellular membranes, for example some antibiotics such as penicillin and some surfactants, have a similar effect even when the concentration of biotin is not limiting. This manipulation of the conditions under which the organism grows to ensure the efficient excretion of the L-glutamic acid is reminiscent of the control of citric acid synthesis in *A. niger* (section 9.2.2.2).

C. glutamicum will grow on sucrose or on a wide range of monosaccharides, but starch hydrolysates (see section 9.6.1) or molasses are preferred for industrial purposes. Where the organism is grown on molasses, which naturally contains high levels of biotin ($0.2-1.2\ mg\ kg^{-1}$), penicillins or surfactants are added to stimulate the production of the L-glutamic acid. The amount of oxygen consumed by the organism also needs to be controlled. If it is too high, lactate and 2-oxoglutarate are formed, while if it is too low the products are lactate and succinate (cf. high biotin levels). The titre of L-glutamate can reach over $100\ g\ l^{-1}$, which is 50–60% of the theoretical yield based on the amount of input sugar.

Some organisms can produce L-glutamate from other substrates. *Brevibacterium flavum* will use acetate efficiently if copper(II) ions ($25-250$ $g\ \mu l^{-1}$) are added to the medium, while *Nocardia erythropolis* and *Arthrobacter paraffineus* will use hydrocarbons, although the yields are lower ($60\ g\ l^{-1}$). After the fermentation the L-glutamate is precipitated at acid pH. The cells are separated from the fermentation broth and hydrochloric acid is added to the cleared broth to lower the pH to 3.2. This is close to the isoelectric point of glutamate, that is, the pH at which it has no net charge, and it precipitates as a crude solid. This can be further purified by ion-exchange chromatography.

9.2.3.2 L-Lysine. *Corynebacterium glutamicum* will also synthesize L-lysine. However, this is not a property of the (wild-type) organism as it is isolated from the natural environment, but only of a special strain artificially mutated from it.[13]

The development of the manufacturing process for L-glutamate coincided with advances in our knowledge of the biosynthesis of the amino acids. The synthetic pathways, which frequently branch to produce several amino acids from common intermediates (Figure 9.10), are normally controlled to prevent overproduction of the end-products. If the end-product is supplied in the growth medium, its synthesis may be repressed, that is to say the organism may not produce the enzymes responsible for catalysing its synthesis. Where it does produce these enzymes the simplest alternative form of control is through negative feedback; the end-products of the pathways inhibit one or more of the enzymes catalysing the intermediate stages. Often the main target for this inhibition is the first enzyme in the pathway. In the synthesis of lysine this enzyme is aspartate kinase (Figure 9.10).

Aspartate kinase is effectively inhibited by lysine and threonine acting together, and the pathway shuts down when their concentration in the cell rises above a preset level. That level depends on the actual sensitivity of the aspartate kinase to each of the amino acids, and that is a feature of the amino acid composition of the enzyme, just as is its catalytic activity.

Clearly if the cell cannot synthesize threonine, the aspartate kinase will not be completely inhibited whatever the intracellular level of lysine. It was just such an effect that was observed in mutants of *C. glutamicum* which were found to excrete lysine. The mutation altered the nucleic acid coding for the sequence of amino acids in the enzyme that catalyses that synthesis of L-homocysteine from aspartic semialdehyde (Figure 9.10). It became an inactive catalyst, and the loss is critical to the cell. It will not grow unless L-homoserine is added to enable it to make at least small amounts of the methionine, threonine and isoleucine which are derived from it. It becomes an 'auxotroph' requiring homoserine for growth. If the amounts of homoserine are limited, the intracellular concentration of threonine never becomes high enough to shut down aspartate kinase, whatever the intracellular concentration of lysine.

The cells could produce even more lysine if the kinase were to remain active but insensitive to a high lysine concentration. Mutants of the homoserine auxotrophs of *C. glutamicum* were isolated with this property by growing the organism in the presence of toxic isosteres (close structural analogues) of lysine (e.g. *S*-(2-aminoethyl)-L-cysteine). One way in which the cells can become resistant to the toxic isostere is to overproduce lysine. This is likely to occur in cells where the mutation alters aspartate kinase in such a way as to make it insensitive to inhibition by lysine, while allowing it to retain its full catalytic activity.

Mutant strains of *C. glutamicum* altered in these two ways overproduce lysine in media supplemented with small amounts of homoserine. The production of lysine requires higher levels of biotin ($30 \, \text{mg} \, \text{dm}^{-3}$) than are necessary for glutamic acid production by the wild-type organism, and the amino acid supplement is actually added as soya protein hydrolysate. Under these conditions *C. glutamicum* will produce about $45 \, \text{g} \, \text{l}^{-1}$ L-lysine from $200 \, \text{g} \, \text{l}^{-1}$ molasses. This represents a yield of 30–40% based on the weight of input sugar. However, the manipulation of other organisms such as *Brevibacterium flavum* can lead to lysine titres of at least $75 \, \text{g} \, \text{l}^{-1}$.

The lysine is recovered from the clarified fermentation broth by adsorbing it on to strongly acidic ion-exchange resins. The lysine is eluted with ammonium hydroxide, and is subsequently crystallized as the hydrochloride salt.

These techniques, in which mutant strains are selected for their ability to overproduce metabolites, represent an important advance in the industrial development of microbial synthesis. Their use to improve amino acid manufacture was not new; they had already been used to improve the titres of antibiotics but the nature of the changes introduced into the metabolism of the mutated organisms could not be interpreted in the way that was possible for amino acid synthesis. What is, perhaps, also apparent is that the technique of interfering with the metabolic pathway between aspartate and lysine is, in

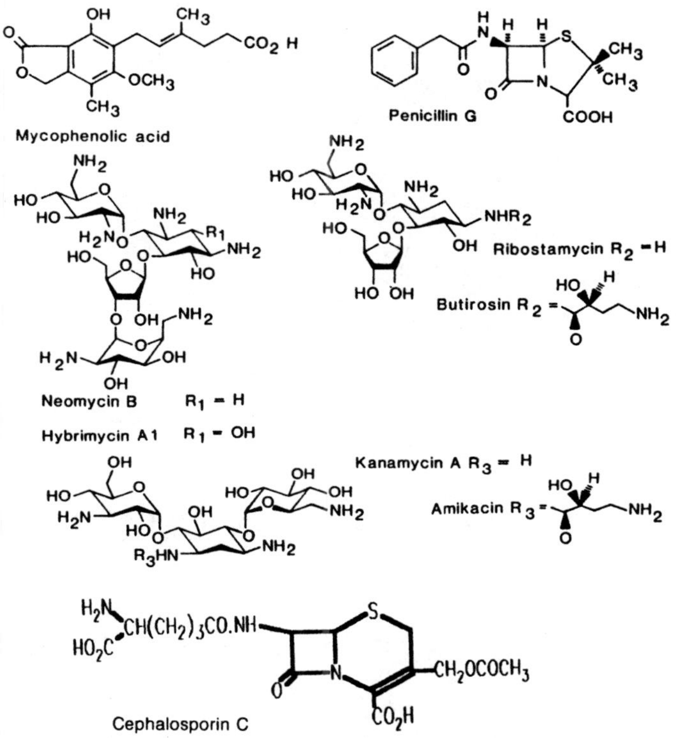

Figure 9.12   The structure of some secondary metabolites.

principle, no different from the use of sulphite to inhibit the synthesis of ethanol (section 9.2.1.2). In one case *C. glutamicum* overproduces lysine, while in the other *S. cerevisiae* will produce glycerol.

To concentrate on the manufacture of primary metabolites is to give an unrepresentative picture of a micro-organism's synthetic capacity. The pathways of primary metabolism broadly encompass all the processes necessary for the continuing life of the cell and its progeny, and exclude those which lead to the odd excreted end-products of metabolism that are the secondary metabolites. Their classification as 'secondary' is often a misnomer because the organisms are capable of producing them in large quantities, sometimes even forming a significant part of their total biomass. Furthermore, their chemical complexity is nothing if not surprising (Figure 9.12).

Animals and plants also produce secondary metabolites (Figure 9.13). Their study in animals is in its infancy, although some are well known for their pharmacological activity or as perfumes. The importance of the materials from plants (Figure 9.13) cannot be underestimated; they are a crucial component of

**9.3
Microbial synthesis of
secondary metabolites**

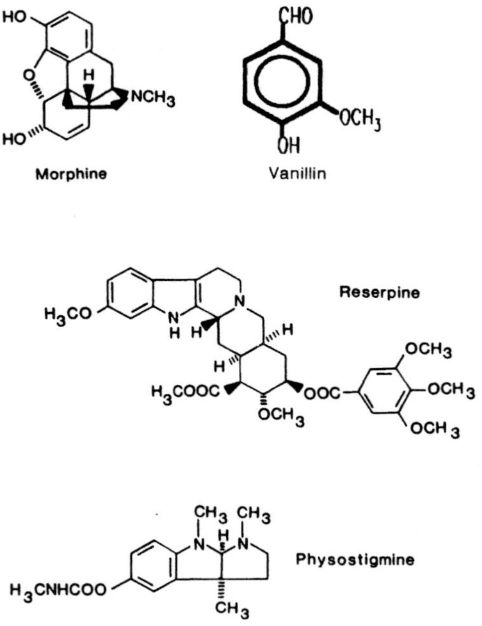

Figure 9.13   A few secondary metabolites of animal and plant origin.

many valuable crops. Where they are isolated as pure compounds their production is usually based either on agriculture or on chemical synthesis. D-Ephedrine and ascorbic acid are two exceptions (sections 9.5.1 and 9.5.2) which use biological catalysis as one stage in their industrial manufacture. Other materials, like the plant steroids (section 9.5.3), are important raw materials for pharmaceutical manufacture.

If micro-organisms are now seen as the most prolific source of new secondary metabolites, this is because the techniques for selecting and growing these organisms under a variety of conditions are particularly well advanced. The higher plants may come to rival the micro-organisms as a new source when, over the next few years, their growth as high densities of single cells in fermenters becomes possible.

Many secondary metabolities are antibiotics, that is, they are toxic to some micro-organism other than the one that synthesizes them. Over 10 000 such compounds are now known[14] in this single class of the secondary metabolites.

When Fleming isolated his organism (*Penicillium notatum*) in 1929, the history of bacterial antibiosis was already well known.[15] Vuilleman introduced the term 'antibiosis' in 1889 and Gioso isolated mycophenolic acid (Figure 9.12) from *P. brevicompactum* in 1896. In 1904 Frost described techniques for detecting antibiosis which are still in use, and which are similar to those used by Fleming in 1929. Against the background of the advances in

Table 9.11 The structures of penicillins isolated from *Penicillium chrysogenum*

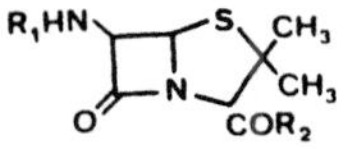

Name	$R_1 (R_2 = OH)$
6-Aminopenicillanic acid	H
Penicillin BT	$CH_3(CH_2)_3SCH_2CO$
Penicillin F	$CH_3CH_2CH{=}CHCH_2CO$
Penicillin G	see Figure 9.14
Isopenicillin N	see Figure 9.14
Penicillin O	$CH_2{=}CHCH_2SCH_2CO$
Pencillin V	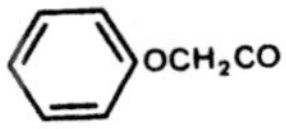
Pencillin X	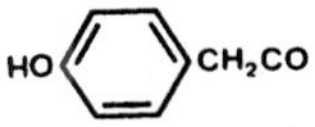

large-scale fermentation between 1900 and 1930, the lapse of ten years or more between the detection of penicillin and its isolation may seem surprising. The biography of Fleming by MacFarlane goes far to explain the lapse.[16] However, it is worth remembering that at the time of Fleming's discovery micro-organisms were usually assumed to play a degradative role in metabolism. Their synthesis of the solvents and the acids could be viewed as the decomposition of sugar. That they might also synthesize complex organic molecules was not appreciated until Raistrick, working at ICI in the 1930s, analysed the structures of some of the natural pigments produced by micro-organisms.[17] Fifty years on, their synthetic ability is taken for granted.

The microbial synthesis of penicillin represents a classic picture of the development of an antibiotic. It is a characteristic of secondary metabolites that they are usually excreted from the cell as a closely related group of compounds rather than as a unique chemical entity. Yet the range of structures found amongst the natural penicillins (Table 9.11) is perhaps no more significant than the range of related di- and tri-carboxylic acids that *Apergillus niger* can excrete instead of citric acid. Each range is a reflection of the biosynthetic pathway up to and beyond the required product. For the penicillins, as for most secondary metabolites, this pathway is complex, requiring as direct precursors three different amino acids as well as one of a number of possible carboxylic acids (Figure 9.14).

In the natural isolates (wild types) of *Penicillium notatum* or *P. chrysogenum* the synthesis of the amino acids is regulated, and this limits their uptake into the pathway that is responsible for penicillin biosynthesis (Figure 9.14). The acidic half of the 6-amino group which replaces that 2-aminoadipyl function is not always synthesized by the cell but may be derived directly from constituents in the fermentation medium. It is therefore not surprising that all these

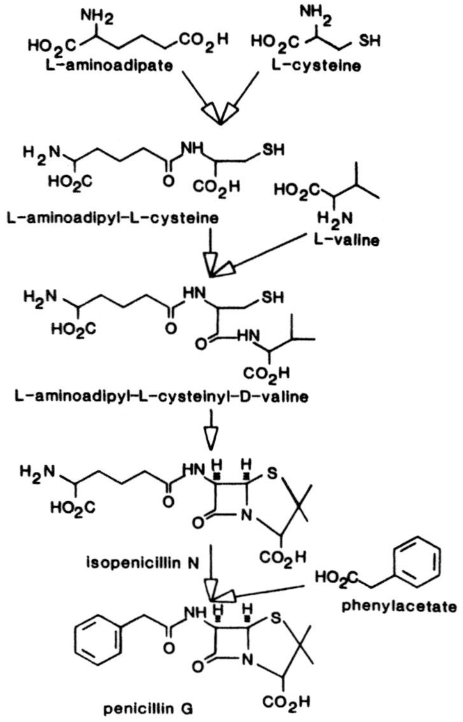

Figure 9.14 Outline of the final stages in the synthesis of penicillin G in *Penicillium chrysogenum*.

constraints limit the productivity of the wild-type organisms to but a low titre of penicillin. Moreover, the product may not be the penicillin G or V which is synthesized in the commercially important fermentations.

Fleming's original strain of *P. notatum* grew as a surface culture and probably produced about 1 mg penicillin l^{-1}. A modern production strain of *P. chrysogenum* grows in submerged culture and produces in excess of $30 \text{ g} l^{-1}$. The difference between the two fermentations is the result of a long period of mutation and strain selection accompanied by careful development of the fermentation medium and conditions. An easy test for the increasing titre relies on the same action that Frost described in 1904. A uniform lawn of a bacterium that is sensitive to penicillin is grown on an agar plate in the presence of samples from the fermentation broth of a producing organism. If these samples contain penicillin the sensitive bacterium will not grow. The diameter of the zone of no growth around the fermentation sample is a function of the titre of penicillin in the sample. The process of mutating the organism, of picking strains with an improved titre from the survivors, and of then repeating the process with the strain showing the best production of penicillin on a large scale underlies the genealogy which connects the original isolate with the commercially important strains. The selected mutants either may be spontaneous natural variants of the original strain, or they may be induced artificially with UV or X-irradiation, or with chemicals such as nitrogen mustards. In one such programme, a strain (NRRL-1951), which produced about $60 \, \text{mg} \, l^{-1}$ of penicillin, passed through four different laboratories whose microbiologists carried out over twenty rounds of mutation and selection. At the end Lilly Industries had a new strain (E15.1) producing $7 \, \text{g} \, l^{-1}$, a titre over 100-fold greater than the original.

Beyond this, it is now possible to cross separate strains of penicillia and to select from the hybrids strains with a greater titre potential than either parent. All these techniques are essentially the same as those which transformed a strain of *C. glutamicum* producing glutamate into another producing lysine. However, the strains used to produce penicillin are not stable and their ability to produce high titres eventually fails. This reflects the large number of mutations on which their high titres depend. Any one of them can revert to some other form with an accompanying loss of titre, and a continual reselection of improved strains is needed to maintain the high production of the penicillin.

These strains of *P. chrysogenum* are fermented in submerged culture under aerobic conditions at a scale between 40 and $200 \, \text{m}^3$. The fermentation conditions that are used depend on the strain of the organism, but the substrates available for the fermentation may determine the strain selected. This is similar to the relation between substrate and organism in other processes (sections 9.2.2.1, 9.2.2.2 and 9.2.3.1). The inoculum is developed in stages (Figure 9.15), and the progress of the final production stage may be complex, with a phase of growth (tropophase) followed by a phase of secondary metabolite production (idiophase) (Figure 9.16). These phases are related to the sequential uptake of substrates such as carbohydrates, phosphate, ammonia, oxygen and the carboxylic acid precursor of the 6-amido side chain. Indeed, some of these substrates need to be fed to the organism during the fermentation (see Figure 9.15).

Under well-controlled conditions *P. chrysogenum* synthesizes essentially one penicillin. Different strains are used to prepare penicillin G and penicillin V, with either phenylacetate or phenyoxyacetate being added to the medium to provide a plentiful source of the side chain. Even so, some penicillin X (Table 9.11) may be produced as an oxidation product of penicillin G, for example. Other related compounds released into the medium include the cyclic lactam of 2-aminoadipate, and small amounts of isopenicillin N and 6-aminopenicillanic acid.

A maximum theoretical yield of about $0.12 \, \text{g}$ penicillin g^{-1} carbohydrate has been calculated for the synthesis,[18] and yields of up to $0.07 \, \text{g}^{-1}$ are achieved in practice. However, it is clear that this is not the yield of a strictly catalytic conversion because the 'catalyst', *P. chrysogenum*, is changed by the process. In recent years some attempts have been made to hold the organism in a state of active penicillin synthesis, and to feed it substrates while it is

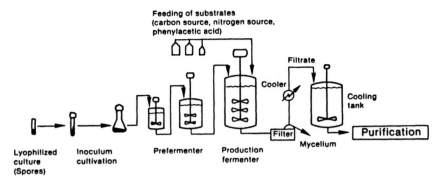

Figure 9.15 Flow chart for the development of a large-scale industrial fermentation. The organism, in this case *Penicillium chrysogenum*, is grown up in stages from the stored organism, in this case lyophilized spores, until there is enough growth to inoculate the production fermenter. Only at this stage is the production of the metabolite, here penicillin, stimulated.

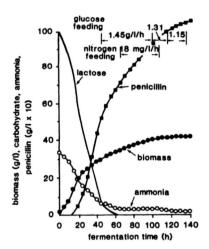

Figure 9.16 The accumulation of penicillin during a fermentation with *P. chrysogenum*. An initial phase of rapid growth (0–20 h, tropophase) is followed by a longer period of slower growth during which penicillin accumulates in the medium (20–140 h, idiophase). From Swartz, R.W. (1979) *Ann. Rep. Ferm. Proc.* **3**, 75.

immobilized in a polymeric matrix. So far these attempts have failed to produce anything other than very low yields of penicillin. Whether they will eventually be successful depends on whether or not the process of the change in the *P. chrysogenum* is itself responsible for the rapid synthesis of the penicillin.

Titres of penicillin G as high as $30\,\mathrm{g\,l^{-1}}$ are not infrequently reported, but the economics of the fermentation may be less dependent on the final titre than on the time taken to achieve it or on the cost of the materials fed to the organism. Short fermentations allow a more intensive use of the

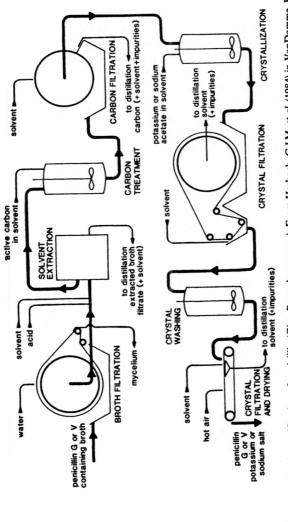

Figure 9.17 Flow chart for the purification of penicillin (Gist–Brocades process). From Herbach. G.J.M. *et al.* (1984) in VanDamme, E.J. (ed.), *Biotechnology of Industrial Antibiotics*, Marcel Dekker, p. 45.

fermenter and should increase the rate of production of penicillin, while the ability to adapt the organism to a variety of substrates allows the most economical feedstock to be used. Moreover, it is important that the final condition of the fermentation broth should be suitable for the extraction of the penicillin. This involves filtration under standard conditions followed by solvent extraction at an acid pH. The penicillin is extracted back into water buffered to a near neutral pH, and then re-extracted into the solvent. After treating the solvent with charcoal the penicillin is precipitated from the solvent as a sodium or potassium salt. This process (Figure 9.17) is dependent only on the fact that penicillin is a weak acid, but strains of *P. chrysogenum* which produce high titres must not produce other materials which interfere with any of the stages. For example they must not affect the filtration, or cause emulsions in the solvent extraction, or co-precipitate with the penicillin.

Despite the nature of the process, penicillin is a cheap bulk intermediate for the pharmaceutical industry. The annual world production of penicillin G and V together is probably about 25 000 tonnes.

The natural penicillins are useful antibiotics, but their full potential was realized only after their further transformation into 'semi-synthetic' products (Table 9.12). Many of these are more active as antibiotics than the natural compounds, or are active against a wider range of organisms, or are more favourably absorbed or retained by the body. It is important to recognize that although the large scale of the fermenters with their relatively low aqueous concentrations of penicillin may dominate the manufacturing sites where they

Table 9.12 The structures of some semi-synthetic penicillins

Name	R_1	R_2
Ampicillin	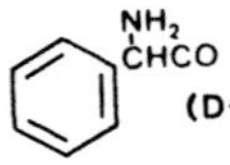 phenyl-CHCO with NH₂ (D-isomer)	OH
Methicillin	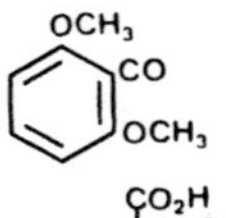 phenyl(OCH₃)(OCH₃)-CO	OH
Ticarcillin	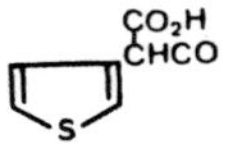 thienyl-CHCO with CO₂H	OH
Floxacillin	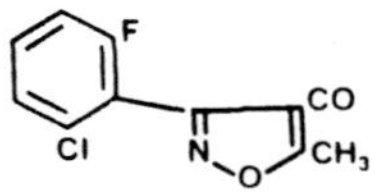 (F, Cl)phenyl-isoxazole-CO with CH₃	OH
Penamecillin	phenyl-CH₂CO	OCH₂OCOCH₃

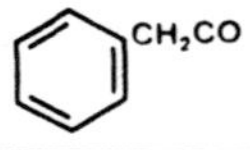

Refer to Table 9.11 for R_1 and R_2.

are made it is the subsequent chemical processes at higher concentrations, and frequently comprising many stages, which are responsible for the wide range of the manufactured products. In this respect the fermentation of penicillins should now be seen as providing a raw-material input to the fine chemicals industry rather than being an industry in its own right.

9.4.1   Biologically active proteins

The microbial synthesis of proteins is the aspect of biotechnology which, more than any other, has attracted the public imagination. It is easy to see why this should have happened; the use of micro-organisms to make large amounts of complex macromolecules which have high pharmacological activity and which were previously only synthesized in small amounts by specialized animal tissues must represent a breakthrough. The advance is real enough, and it has extended the catalytic potential of micro-organisms, but it ought not to be taken out of context.

The information which specifies the amino acid sequence of a protein is stored in the nucleotide sequence of the double helix of deoxyribonucleic acid (DNA). The transcription of sections of this information into ribonucleic acid (RNA) is catalysed by RNA polymerases. These enzymes not only control the synthesis of RNA but also recognize stop and start signals on the DNA. The start signals are complex and may be blocked by repressor molecules which inhibit the transcription process. Once synthesized, the (messenger) RNA is processed and exported to ribosomes where its nucleotide sequence is translated into protein. Triplets of three nucleotides (codons) in the messenger RNA each specify (encode) one amino acid. The linear sequence of nucleotides in the messenger RNA thus specifies the sequence of amino acids in the protein whose primary structure will therefore correspond directly to the sequence of nucleotides in the DNA.

The protein is synthesized starting from the amino acid which will carry the free amino group of the completed peptide chain. As this chain builds it folds and is further modified. In particular, sulphydryl bridges may form where two thiols are brought together by the folding process and are oxidized; segments of the peptide chain may be removed; sugars, both singly and in groups, may glycosylate the free hydroxyl groups of serine and threonine and the protein may eventually carry a large burden of sugars; and the hydroxyl groups may be phosphorylated. All these reactions are themselves catalysed by other proteins, and the modifications affect the activity and stability of the newly synthesized protein (Figure 9.18).

These processes underlie the effects brought about by mutations which change the nucleotide sequence in the DNA itself, and by gene transfer or cell fusion, each of which eventually adds extra DNA to the cell. The significance of the effect depends on the position of the change within the DNA. For example, a change in a single nucleotide could affect the activity of an enzyme, or the

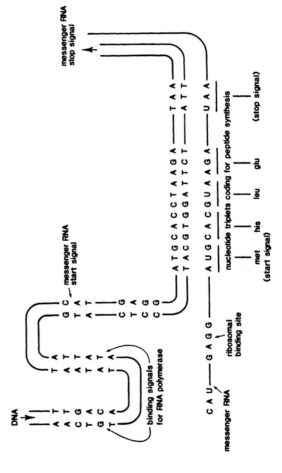

Figure 9.18 The genetic information in most organisms is stored in DNA as a sequence of nucleotide base pairs (A = adenine, C = cytosine, G = guanine, T = thymine, U = uracil). The information which corresponds to the primary sequence of amino acids in a protein is encoded in triplets of the base pairs. It is transcribed into (messenger) RNA before being translated into a peptide sequence on the ribosomes (see Figure 9.19). The DNA also carries special sequences (promoters, and initiator and terminator regions) which the enzymes responsible for RNA synthesis (RNA polymerases) recognize as binding sites, and as start and stop signals. These are not transcribed into RNA. However, other signals, which control the binding of the RNA to the ribosomes, and the start and stop signals at the beginning and the end of the peptide chain, are transcribed.

Table 9.13 The microbial production of enzymes (calculated as the weight of the pure protein) in 1979

Enzyme	Tonnes per annum
Bacterial protease	500
Amyloglucosidase	300
Bacterial amylase	300
Glucose isomerase	50
Rennet (chymosin) substitutes	10
Fungal amylase	10
Pectinase	10

response of DNA to the repressors. In the latter case the activity of RNA polymerase would be affected and the amount of enzyme synthesized could change dramatically.

An organism can synthesize only the proteins already encoded in its DNA. The synthesis of the 'recombinant' protein requires the insertion of new DNA; the genome is cut and then 'recombined' to include the new insert of DNA. Where the insertion is limited to a section which codes for a particular protein, it should allow the recipient cell to make that protein. The development of the techniques which made these transfers possible is responsible for our ability to manufacture a wide range of recombinant proteins in micro-organisms. However, this advance should be seen in the context of an established industry for the manufacture of bacterial proteins.

The commercial production of proteins from micro-organisms began in the USA around 1890, when Takamine introduced a traditional Japanese fermentation process for takadiastase. This product, which was derived from *Aspergillus niger* (cf. section 9.2.2.2), was a mixture of enzymes which catalysed the hydrolysis of starches and proteins. Some years later, in 1913, Boidin and Effront discovered that *Bacillus subtilis* produces a heat-stable α-amylase. This enzyme also catalyses the hydrolysis of starches, and was used in the textile industry for desizing cloth.

The enzymes that catalyse the hydrolysis of proteins and starch remain the major bulk proteins with biological activity which are derived from micro-organisms (Table 9.13).[19] The proteases have many uses (Table 9.14),[20] but the list is little different to that which Webb recorded in his review of biochemical engineering in 1964. The enzymes, such as α-amylase and amyloglucosidase, which catalyse the hydrolysis of starch, are important for the manufacture of sugar (section 9.6.1). Since this is one stage in the synthesis of ethanol from starch (see section 9.2.1.1), their level of production has doubled. However, improved fermentation processes now yield greater amounts of enzyme selected for better catalytic performance. The result is that the total weight of enzyme manufactured has risen less quickly since 1979 than their increased use would suggest. This will also be true of the bacterial proteases used in detergents, which now often include lipases.

Table 9.14   Major industrial uses of proteases

Industry	Process
Food	Cheese manufacture
	Flour modification
	Stabilization of emulsions
	Extraction of fish and vegetable oils
	Tenderizing of meat
	Chillproofing of beer
Textile	Cleaning and laundering
	Dehairing of leather
Photographic	Recovery of silver from gelatin

Table 9.15   Some clinically useful human polypeptides which are potentially attractive for biosynthesis in micro-organisms

Polypeptide	No. of amino acids
Growth hormone	191
Insulin	51
Corticotrophin (ACTH)	39
Calcitonin	32
Glucagon	29
Secretin	27
Active fragment of ACTH	24

Other speciality proteases, such as chymosin, which were once only available from the stomach of calves, are now produced by fermentation following the transfer of the encoding DNA into *E. coli* and other organisms. The recovered product contains about 60% chymosin, compared with only about 2% in the traditional rennet obtained from calves. The annual consumption of chymosin for cheese manufacture is about 26 tonnes, and the recombinant enzyme already accounts for 20% of the market in the USA.[21] This will have increased the manufacture of bacterial rennet well beyond 10 tonnes annum^{-1}, of which approximately 2 tonnes is from recombinant *E. coli*.

At the other end of the scale of production are the pharmacologically active proteins. In 1982 the United States Office of Technology Assessment published a list of small human proteins (mol. wt. < 25 000) which seemed attractive targets for microbial synthesis (Table 9.15). Some of these, such as insulin and growth hormone, are available as microbial products.

Their synthesis is not always straightforward. The transfer of the necessary segments of DNA which code for insulin's natural precursor proinsulin (Figure 9.19) did not ensure that the human genetic information would be expressed in the new bacterial host. When the protein was synthesized very large amounts were produced, but in an inactive state. Like many genetically-engineered proteins the proinsulin formed dense granules of insoluble protein inside the bacterial cell. These appear to be a matrix of incorrectly folded protein which results from the methods which bacterial and animal cells use to

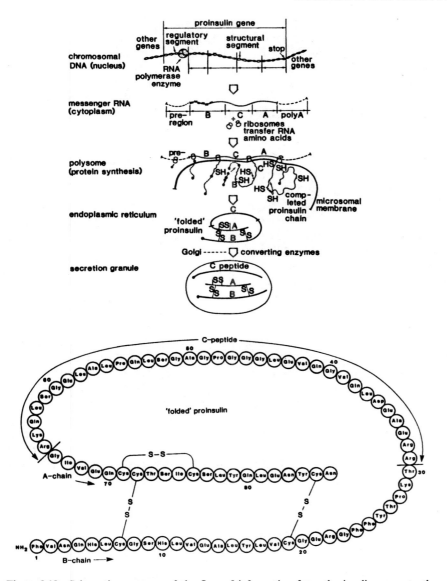

Figure 9.19 Schematic summary of the flow of information from the insulin genes to the biosynthetic machinery of the animal pancreas. The proinsulin gene is represented schematically in the upper panel. RNA polymerase is necessary for the transcription of preproinsulin messenger RNA from the gene, and this then guides the synthesis of preproinsulin on the polysomes. Preproinsulin is discharged and cleaved to proinsulin. The proinsulin is then passed through the endoplasmic reticulum to the Golgi region where conversion to insulin and storage in secretion granules begin. When the proinsulin gene (with the pre-region removed) is transferred to *E. coli,* proinsulin is synthesized on the bacterial polysomes but its subsequent folding is not the same as in the animal pancreas, and the protein is inactive. After Steiner, D.F. (1976) *Diabetes* **26,** 322.

process their proteins after they are synthesized. Whatever causes them to form, their chemical conversion to active protein is difficult. The fact that genetically engineered insulin is now available for human use is no small triumph (see Vol. 1, section 2.2.3).

Growth hormone is also manufactured in *E. coli* or in *S. cerevisiae*, but erythropoeitin, which is of a similar size (166 amino acids) must be prepared with animal cells in tissue culture. Only they are able to add the correct range of carbohydrates, which account for about half of its molecular weight (35 kDa), and which are essential for its activity in stimulating the production of erythrocytes (precursors of red blood cells). The larger protein, tissue plasminogen activator (527 amino acids, mol. wt. 70 kDa) is also made in animal cells, even though glycosylation is not essential for its activation of the proteolytic enzymes that digest fibrin clots in blood. When related plasminogen activators are made in *E. coli* they are incorrectly folded and are inactive. The animal cell route for synthesis is quite practical, if more expensive than the use of microorganisms such as *E. coli*.[22]

The impact of this technique does not lie in the scale of the synthesis. The annual demand for insulin is about 2 tonnes, while that for erythropoeitin and tissue plasminogen activator is much less. With fermentation titres from recombinant *E. coli* near to $1 g l^{-1}$, this could all be supplied in a very few fermenters. More important is the range of proteins which the biochemist now has to hand in sufficient quantities to study their activity. Some proteins, such as growth hormone, are available only in minute amounts from their natural source, and genetic engineering is now a viable alternative technique for their isolation in large quantities. This has made a significant impact on our understanding of their function, and has turned some into major pharmaceutical targets. Erythropoeitin and tissue plasminogen activators are now used in human medicine, and the former is among the world's fifty most valuable pharmaceutical products. The combined annual sales of these two proteins are now estimated to be about $750 million. When sales of recombinant insulin of $500 million are also included, the economic importance of this process for recombinant protein manufacture is clear, even allowing for the hyperbole which often accompanies such market estimates. It is also now reasonable to assume that any protein or enzyme could be made available in quantity, whatever its original source and abundance.

9.5 Microbial transformations

Microbial synthesis as a process is not catalytic. The excretion of the product into the fermentation broth is accompanied by, if not dependent on, the growth of the micro-organism, and, strictly, if the organism is changing during the process it cannot be a catalyst. In microbial syntheses many stages are necessary to convert the substrates fed to the organism into a product, and enzymes catalyse each of these stages. Individually, and together, these enzymes function as true catalysts and they may be inhibited (poisoned), degraded, or resynthesized like any other.

Occasionally the synthesis of a microbial product, for example that of ethanol from glucose, is catalysed by non-viable cells (section 9.2.1.1). Then the process is properly catalytic, because the *Saccharomyces cerevisiae* cells do not change, for a time at least. However, there are some industrially important reactions in which micro-organisms are first grown to a high biomass and are then added to a substrate which is almost quantitatively converted to a product. These are effectively catalytic processes in which one or a few enzymes in the organism transform an added substrate into a useful product. These transformations are divorced from cell growth, in contrast to syntheses such as those in which carbohydrates are converted into citric acid or complex feedstocks into secondary metabolites.

The division between the non-catalytic microbial syntheses and the catalytic microbial transformations is perhaps not very sharp. The oxidation of ethanol to acetic acid in the manufacture of vinegar by the traditional microbial route uses a permanent culture of bacteria growing on wooden slats. The bacteria oxidize the ethanol in the wine as it flows over the slats, and, at the same time, they draw sufficient nutrients from the wine to maintain a population which is both constant and healthy. Individual organisms divide and die, but the overall bacterial culture is unchanged, and is catalytic in its action. Here, in a real sense, the distinction between a catalytic and a non-catalytic process depends on the scale on which the process is described.

9.5.1 *L-Ascorbate (vitamin C)*

About 50 000 tonnes of L-ascorbate is synthesized each year from D-glucose. The multi-step reaction sequence (Figure 9.20) contains, as a central feature, a microbial transformation first described by Brown in 1886.[23] He showed that a bacterium, then called *Bacterium aceti*, would oxidize (D-)mannitol to laevulose (D-fructose). The work was extended by Bertrand between 1896 and 1904, and he showed, that, as a general rule, *Acetobacter xylinum* would oxidize the secondary hydroxyl group of a polyol to a ketone if that group itself lay between a primary and a secondary hydroxyl group, the two secondary alcohols being in a *cis* configuration. As an example Bertrand showed that D-sorbitol was oxidized to L-sorbose (see Figure 9.20).

Reichstein devised the first efficient synthesis of L-ascorbate in 1934, making use of this reaction. In it the combined action of a chemical reduction and the biological oxidation switches the sugar's reducing group from one end of the carbon chain in glucose to the other in sorbose. The result is that the sugar is inverted from D- to L-, allowing the synthesis of L-ascorbate from D-glucose. Such regiospecificity is not unusual in the reactions which enzymes catalyse, but it would be difficult to achieve in a chemical process.

The important oxidation step is catalysed by a single enzyme, sorbitol dehydrogenase. However, *A. xylinum* is not now used as a source of this enzyme because its growth is inhibited by small amounts of nickel ions which

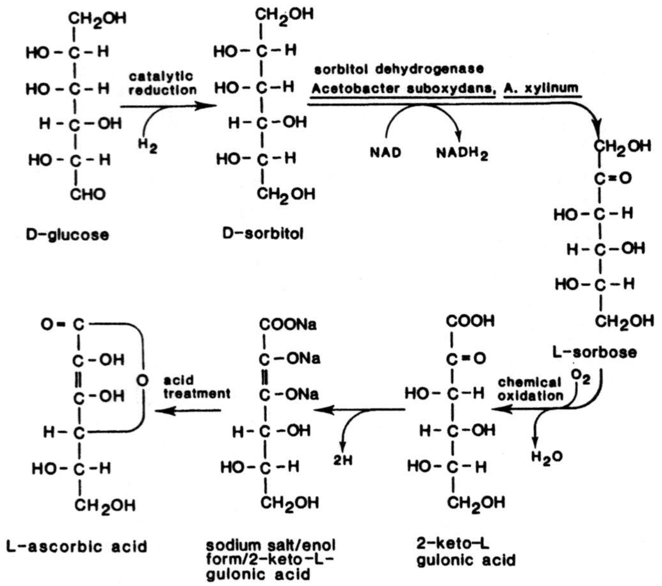

Figure 9.20   Stages in the synthesis of L-ascorbate.

the catalytic reduction of glucose with Raney nickel leaves in the sorbitol. A more resistant organism, *Acetobacter suboxydans*, will tolerate higher levels of nickel and it is grown in a medium containing 20% (w/v) sorbitol. Other nutrients for the growth are usually provided by adding a dried yeast extract (0.5%) to the medium. The fermentation is vigorously aerated to allow the organism to grow quickly while it consumes the nutrients and a small amount of the sorbitol. During this growth some of the sorbitol is also oxidized to sorbose without being further degraded, but aeration is continued after growth has stopped to complete this oxidation. Oxygen may even be used in place of air to speed up the process.

The oxidation in such a batch process would be complete in 24 h, but it can be extended by adding additional sorbitol (up to a final concentration of 28%) during the fermentation. However, some modern processes use continuous fermentation so that a continuous input of D-sorbitol is matched by the output of L-sorbose, and the fermenter itself becomes a large catalytic converter. The molar yield of L-sorbose in these processes is usually greater than 85%.

9.5.2   D-Ephedrine

In 1921 Carl Neuberg discovered that actively growing yeast cells would convert added benzaldehyde into (R)-1-phenyl-1-hydroxy-propan-2-one. The

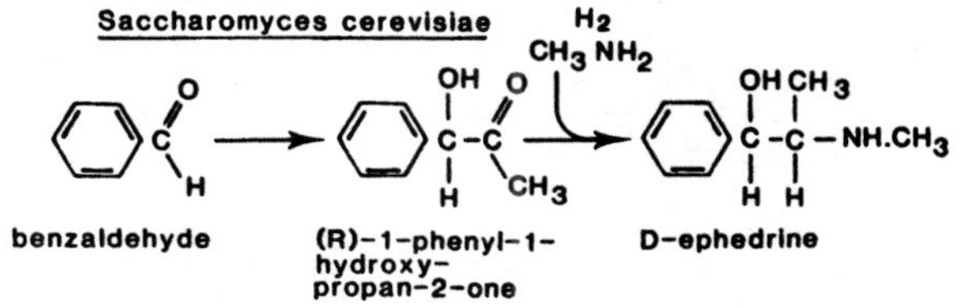

Figure 9.21 The stages in the first commercial synthesis of D-ephedrine.

other substrate in this asymmetric condensation reaction is acetaldehyde, which the yeast derives from its degradation of carbohydrate (see section 9.2.1.1). In 1930 this reaction was developed by Hildebrandt and Klavehn as a key stage in the synthesis of D-ephedrine, (1R, 2S)-1-phenyl-1-hydroxy-2-methylaminopropane (Figure 9.21).[24]

This process, which is still used, illustrates the other important feature of microbial transformations, which is their stereoselectivity. This together with the regioselectivity apparent in the oxidation of sorbitol, is an important argument in favour of biological catalysis.

9.5.3 Steroid transformations

The raw materials from which steroid hormones are manufactured (Figure 9.22) are derived either from plants (e.g. stigmasterol, β-sitosterol) or from the bile acids of oxen (e.g. deoxycholate). When the anti-inflammatory properties of cortisone were first recognized in 1949 the chemical route from deoxycholate to cortisone involved over thirty stages. One of the most difficult stages in this synthesis was the introduction of the 11-keto group of cortisone. For this reason some cortisone is manufactured from plant steroids such as hecogenin in which a 12-keto group is already present, which can easily be converted to an 11-keto steroid. However, some 11 β-hydroxy steroids are also pharmacologically active and may be preferred to cortisone. These can be synthesized from hecogenin but six steps are needed to convert the carbonyl at C-12 into an 11β-hydroxyl. These difficulties led to the development of a microbial hydroxylation, whose 11α-product is easily inverted.

In 1952 Peterson and Murray[25] observed that the mould *Rhizopus arrhizus* oxidized progesterone to 11α-hydroxyprogesterone. Since that time microorganisms have been observed to hydroxylate separately virtually every carbon atom on the steroid molecule, and these reactions have become key stages in the manufacture of steroids.

The organism is first grown in a suitable fermentation medium for about 24 h before the steroid is added. Since the steroid is almost insoluble in water it is added in a way which will present a finely divided suspension to the growing organism. One technique is to suspend a finely divided powder in a non-ionic

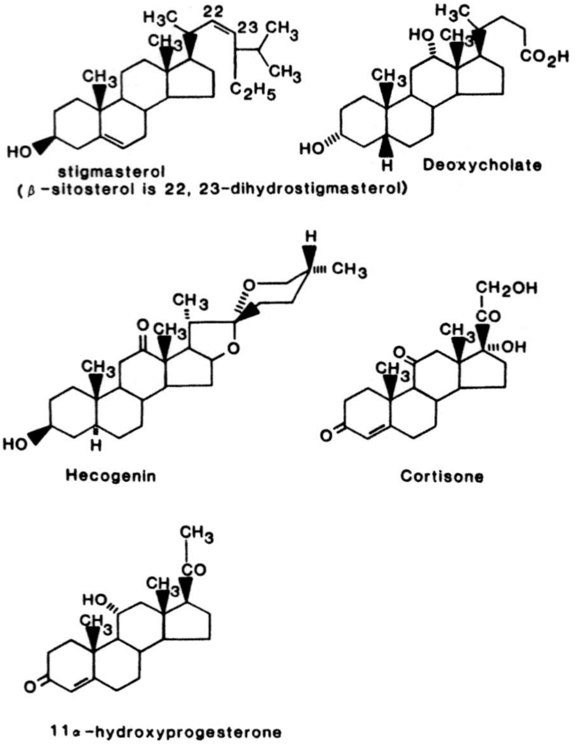

stigmasterol
(β-sitosterol is 22, 23-dihydrostigmasterol)

Deoxycholate

Hecogenin

Cortisone

11α-hydroxyprogesterone

Figure 9.22 Structures of steroids referred to in this chapter.

detergent. Enough is then mixed with the fermentation to give a final steroid concentration between 0.1 and 1% (w/v). Under these conditions the steroid is sufficiently well dispersed for it to be hydroxylated in good yield. Enough is transported from the solid into the aqueous phase of the fermentation for it to enter the cells, where it is hydroxylated. The 11α-hydroxysteroid is transported back into the solid phase, so that over a period of three or four days at 28 °C over 90% of the steroid is hydroxylated. It is important that the reaction is not prolonged to the point where all the substrate is consumed, because other unwanted reactions, particularly 6β-hydroxylation, begin to occur.

This process graphically illustrates two of the problems that attend many interesting transformations: that the substrates are frequently insoluble in water, and that the total concentration of substrate in the reaction is very low. The oxidation of a 0.1 to 1% suspension of a steroid substrate in three or four days is very slow when compared to the oxidation of a 20% solution of sorbitol in 24 h. Attempts are being made to overcome this problem, principally by incorporating the steroid in a second liquid phase, such as carbon tetrachlor-

Table 9.16 Commercial transformations of steroids catalysed by micro-organisms

Reaction	Substrate	Product
Hydroxylation		
11α-	Progesterone	11α-Hydroxyprogesterone
11β-	11-Deoxycortisol	Cortisol
16α-	9α-Fluorocortisol	9α-Fluoro-16α-hydroxycortisol
Dehydrogenation		
Δ-1, 2	Cortisol	Prednisolone
Side-chain removal with 3β-oxidation	β-Sitosterol	Androstenedione

ide or hexane, in which it is more soluble. The transformation is then carried out in an emulsion with water containing the cells as the continuous phase, and the solvent as the dispersed phase. Unfortunately, organic solvents extract essential components from the cells, and particularly from the membranes which contain the enzymes responsible for steroid hydroxylation. Enzymes themselves are not necessarily incompatible with organic solvents (see section 9.8), but the complex systems responsible for hydroxylation frequently contain several components whose spatial arrangement is crucial for their activity. Solvents which disrupt the membranes in which they are embedded will usually reduce, if not destroy, their activity.

Despite these problems, micro-organisms play an important role in the manufacture of steroids, and they now catalyse several different transformations (Table 9.16). The reactions which degrade the 17α-side chain of the steroid are interesting because of the interrelating parts played by chemistry and biology. The chemical process is dependent on the $\Delta 22$-desaturation already present in the aliphatic side chain of a natural product such as stigmasterol. For some years β-sitosterol, which has a saturated side chain and which was extracted along with stigmasterol from plants, accumulated as a by-product in larger amounts than the stigmasterol. The Upjohn Company took this β-sitosterol, formed it into bricks, and buried it.

Subsequent attempts were made to make use of this waste. A number of micro-organisms were known to grow on steroids. They obtain carbon for their growth by degrading the steroid, and an early stage in this process involved the oxidation and removal of the 17α-side chain, even where it was saturated. The attack on the steroid nucleus, starting with a $\Delta 1(2)$-desaturation and a 9β-hydroxylation, led to its complete breakdown to carbon dioxide and water. A number of mutant strains of mycobacteria were isolated which lacked the later enzymes in the pathway. These strains were only able to degrade the 17α-side chain, the nucleus being left intact as a 17-keto steroid. Upjohn now found their waste β-sitosterol to be a valuable raw material, literally a 'sitosterol mine' as Charney and Herzog described it in 1967.[25]

Considering the wide range of catalytic activities that go on inside the living organism it is surprising that just one of them can be harnessed to a

9.6 Enzymatic processes

manufacturing process, independently of the others which are operating concurrently. Even so the complete separation of growth and catalytic activity is only possible where the process uses a single isolated enzyme.

Such processes have existed for many thousands of years in the manufacture of cheese from milk and in the brewing of beer from barley. In the former process the enzyme chymosin catalyses the hydrolysis of one peptide bond in casein, causing the milk to clot, and in the latter the amylases from malted barely (the diastases of Peyen and Persoz) catalyse the hydrolysis of starch. While these remain important processes, the application of enzymes such as these to industrial chemistry is quite recent, dating back only to about 1960.

It is true that first Takamine and then Boidin and Effront manufactured enzymes for commercial use much earlier (see section 9.4), but these were not used as catalysts for chemical manufacture. Since 1960, new techniques have been introduced for handling enzymes when they are used in a chemical reaction, and since then they have made a significant contribution to a number of manufacturing processes. Probably the most important advance has been the linking of soluble enzymes to solid supports in ways that do not destroy their catalytic activity. This work was pioneered particularly by Katchalski, and it is now commonplace for enzymes to be prepared as solid-phase catalysts.

9.6.1   The synthesis of fructose from starch

Starches are polymers of glucose. The sugars are linked together both α-1,4 and α-1,6 with the former predominating to give a highly branched structure. The α- and β-amylases from barley, which degrade starch in the brewing of beer, attack only the α-1,4 links. The α-amylase acts internally, throughout the branched structure, except where it is rather compact at its central core. The β-amylase acts only on the ends of the chains, releasing only maltose (glucose-α-1,4-glucose). These two enzymes therefore do not completely degrade the starch. They release a mixture of glucose, maltose, and some small oligosaccharides containing both α-1,4 and α-1,6 links, and leave a small glucose polymer (a limit dextrin) derived from the core of the starch molecule.

Although the mixture provides a suitable substrate for the brewing process, it is not suitable for the manufacture of glucose or fructose. Complete hydrolysis of the starch is necessary if this process is to succeed, and glucose must be the only product. This complete hydrolysis is catalysed by two fungal enzymes. One is amyloglucosidase, which is isolated from various strains of *Aspergillus* spp. (including *A. niger*) and of *Rhizopus* spp. This enzyme catalyses a hydrolysis in which glucose units are released, one at a time, from the non-reducing end of the chains. Both α-1,4 and α-1,6 bonds are hydrolysed, although the latter are attacked more slowly than the former. For this reason a second enzyme is added which catalyses the hydrolysis of the

α-1,6 links only. This enzyme, which is called pullulanase, is isolated from the bacterium *Aerobacter aerogenes*.

The large-scale process for the production of glucose from corn starch uses one other enzyme, which is a heat-stable α-amylase. Most enzymes are stable in the temperature range between 25 and 45 °C. As the temperature rises their catalytic activity increases, but at the same time, the increased thermal energy disrupts the three-dimensional polypeptide structure on which their activity depends. This process becomes very rapid at higher temperatures, and most enzymes do not survive heating for long periods above 60 °C.

For starch to be effectively degraded it must first be heated above 65 °C. The starch grains then burst and they form a viscous gel. This must be thinned so that the viscosity is low enough for the dispersed starch to be handled. The α-amylase from barley is fairly stable to heat, and it is active at 65 °C. However, some α-amylases from the organisms which grow at temperatures up to 80 or 90 °C (e.g. *Bacillus stearothermophilus*) are stable for short periods at temperatures above 100 °C, and these heat-stable enzymes are very useful for the controlled thinning of starch gels.

In the process[26] a starch suspension containing the heat-stable α-amylase is heated briefly to 140 °C so that it forms a gel. This is sufficiently hydrolysed before the α-amylase is destroyed to allow it to be pumped to a vessel where more α-amylase is added, and the hydrolysis continues for about 30 min at 100 °C. At the end of this period about 10% of the α-1,4 links in the starch are hydrolysed and the gel is thin enough to be cooled to 55 °C without setting solid. Amyloglucosidase and pullulanase are stable at this lower temperature, and enough of these two enzymes is added to catalyse the hydrolysis of the starch to glucose over a two- or three-day period. Only some 2 or 3% of the links between the glucose units remain, and this, in fact, represents the equilibrium position of the hydrolysis.

The glucose syrup can be used as an input for the ethanol fermentation (see section 9.2.1.1), but where a purer product is required it is filtered and decolorized with charcoal. Metal ions, including those added to stabilize the α-amylases, are removed with ion-exchange resins. If it is to be used in foods as confectioner's syrup, it must be partly converted to fructose, which is a more intense sweetener than glucose. This is done with the enzyme glucose isomerase which catalyses the interconversion of glucose and fructose. The final proportions of glucose and fructose are dependent on the thermodynamic equilibrium between these two sugars. At 61 °C this tends to be in favour of glucose (55:45) and it moves further towards glucose at lower temperatures.

There are several sources of the isomerase, one being *Bacillus coagulans*. The cells of this organism are harvested and treated with glutaraldehyde (1,5-pentandial). This kills the cells and cross-links them into a matrix without destroying their isomerase activity. The solid is formed into pellets which are packed into a column. This column behaves as a catalytic reactor, and isomerizes the glucose syrup which passes through it. The product from the

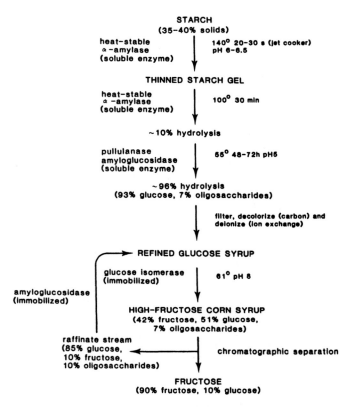

Figure 9.23   A flow sheet for the manufacture of fructose and of high-fructose corn syrup. The degree of hydrolysis of the starch is calculated from the release of reducing sugar.

column, which has a dry solids content of between 40 and 45%, contains about 42% of its dry weight as fructose, 51% as glucose and about 7% as small oligosaccharides which remain from the starch. This syrup is concentrated to a dry solids content of about 70% before being used in the food industry as high-fructose corn syrup.

If necessary, pure fructose can be separated from the syrup chromatographically. The glucose is returned to the isomerization process, but only after being treated again with amyloglucosidase which is chemically bonded to a supporting matrix. This preparation hydrolyses the residual oligosaccharides, which would otherwise build up in the process.

This versatile process (Figure 9.23) can be adapted to produce a variety of grades of starch hydrolysate. These range from crude mixtures of partly hydrolysed starch, through maltose or glucose for ethanol fermentations, to high-fructose corn syrup and pure fructose. Non-enzymatic processes can replace some of the stages, particularly the thinning of the starch gels, but the acids used generate colour and off-flavours which contaminate the final products. It is unlikely therefore that chemical catalysts would replace the enzymes used in this process. Its scale is very large. The market for high-fructose syrup is about 7.8 million tonnes per year, and a further amount of about 4 million tonnes per year of glucose is needed to supply the input for ethanol fermentations in the United States (section 9.2.1.1). It is the largest of the processes in which enzymes catalyse the manufacture of a product whose chemical structure is well defined.

9.6.2 Hydrolysis of penicillin G to 6-aminopenicillanic acid (6-APA)

Of the natural penicillins, only two, G and V (section 9.3), are produced commercially. They are useful antibiotics in their own right, but these two compounds do not do justice to the full potential of the antibiotic action of the penicillin structure. This can only be realized through a much wider range of compounds in which different groups replace those naturally present on the 6-amino group (see Tables 9.11 and 9.12). These changes can improve the stability and uptake of the penicillins in the stomach, where the pH is low. They can also increase their ability to kill bacteria, and make them more resistant to microbial degradation. The manufacture of these compounds only became possible when an enzyme was discovered that would catalyse the hydrolysis of the 6-amido group.

The β-lactam nucleus of the penicillins is stable between about pH 3 and pH 8, but this is too narrow a range to allow a chemical route to 6-APA to be easily developed. However, between 1956 and 1960 it was discovered that some micro-organisms contained an enzyme that could efficiently hydrolyse the 6-amido group. More importantly, amongst these organisms were some, including E. coli, which did not also catalyse the hydrolysis of the β-lactam ring. The enzyme in E. coli catalyses a much more rapid hydrolysis of penicillin G than penicillin V, the ratio of the two rates being about 10:1.

At first whole E. coli cells were harvested and this cell paste was added to a solution of penicillin G (about 50 mg cm^{-3}). Alkali was added to neutralize the released acid, the amount added being a good measure of the degree of hydrolysis. When the hydrolysis was complete the cells were removed and the 6-APA was isolated from the filtrate. The 6-APA was subsequently re-acylated with a new side chain by treating it with a suitable acyl chloride.

Such a crude process was likely to add impurities to the 6-APA. At least some of these could be high-molecular-weight fragments from degraded E. coli cells. If these were not carefully removed during the subsequent purification of the 6-APA they could have the undesirable effect of inducing sensitivity to the penicillins in patients given the antibiotics made from the 6-APA. This was one reason why attempts were made to produce a cleaner catalytic process, but another, more obvious, reason was a desire to re-use the catalyst.

The enzyme, which is an amidohydrolase usually known as penicillin acylase, is intracellular. A much cleaner process can be operated if the enzyme is purified from the *E. coli* cells.[27] At least this limits the amount of high-molecular-weight material which is added to the hydrolysis medium along with the enzyme. To do this the cells are first broken by forcing them through a narrow annulus at high pressure (about 340 atm) in a homogenizer. The machine which is used is similar to those used in the dairy industry for homogenizing milk. The rapid decompression releases the enzyme into solution from which it is purified by any one of a number of the classical techniques used to purify enzymes in the laboratory.

The purified enzyme is soluble, and so would be difficult to recover from a reaction mixture. It must therefore be made insoluble again, that is to say immobilized, before it is used. A variety of methods are possible, ranging from its adsorption on to bentonite or on to a resin such as a polymethacrylate where it may be cross-linked with glutaraldehyde to form an insoluble matrix (cf. glucose isomerase, section 9.6.1), to its incorporation into fibres of cellulose triacetate. In any of these immobilized forms the amidohydrolase retains much, if not all, of its catalytic activity. The nature of the catalytic reactor depends largely on the form of the immobilized enzyme, but essentially the enzyme replaces the whole *E. coli* cells of the original process. When the hydrolysis is complete it can be easily recovered from the reaction mixture and re-used in tens, if not hundreds, of reactions.

The reaction is very efficient. If the hydrolysis is continuously titrated to pH 7.8 with sodium hydroxide about 96% of the penicillin G is hydrolysed to 6-APA. This is close to the equilibrium position for a dilute solution ($33 \, g \, dm^{-3}$) of penicillin G. After the hydrolysis the enzyme is removed and the solution is acidified (pH 2.5). The phenylacetic acid released is extracted into a solvent (methyl isobutyl ketone) and is recycled to the penicillin G fermentation (section 9.3.1). The crude solution of 6-APA which remains is then neutralized and concentrated until its concentration is between 120 and 150 $g \, dm^{-3}$. At its isoelectric point (pH 4.3) it precipitates from this concentrate and the crystals are washed and dried. The overall yield of 6-APA is about 90% of theory.

The process is now well established, although after it was introduced a chemical process was developed which used phosphorus pentachloride to break the amide bond. This has some advantages over the enzymatic hydrolysis, not least of which is the chemist's, as distinct from the biochemist's, general familiarity with phosphorus pentachloride as a reagent. It will also operate at a higher penicillion G concentration, and it does not require the back-up of a fermentation plant to provide the catalyst. Partly for these reasons, and partly because of patent coverage, the two processes operated on an almost equal footing. However, of the 10 000 tonnes of 6-APA which are now produced each year, the bulk is the product of enzymatic hydrolysis. The low process intensity of the hydrolysis is not quite such a problem as it is in the steroid hydroxylations (section 9.5.3), but it does illustrate again this disadvantage of enzymatic catalysis. Even so, the cost, at approximately $75 per kg, is remarkably low for such a complex organic intermediate.

Since the hydrolysis which the enzyme catalyses is an equilibrium reaction it is theoretically possible to use the enzyme to synthesize the semisynthetic penicillins from 6-APA and a suitable carboxylic acid. At pH 5 the equilibrium of the reaction lies in favour of the penicillin, and there are numerous patented processes which make us of this. However, its value is limited because the 6-APA and the new carboxylic acid side-chain are too valuable to waste in a reaction which does not go to completion. For this reason alone, it is most unlikely that the amidohydrolase will ever find a use as a synthetic, as distinct from an hydrolytic, catalyst in the reaction.

9.6.3   The synthesis of L-amino acids

The virtue of the enzymes in the manufacture of fructose and glucose, and of 6-APA, is the mild conditions under which they act. This prevents the development of colours and off-flavours in the sugars, and the loss of the β-lactam nucleus from the penicillins. One of their other virtues is their stereochemical specificity and selectivity. This is well illustrated by the processes used to manufacture the amino acids L-methionine, L-cysteine and L-aspartic acid.

Amino acylases catalyse the stereospecific hydrolysis of N-acyl-L-amino acids (Figure 9.24) but not of the corresponding D-isomers. This reaction is comparable to that catalysed by penicillin acylase, which in fact shows a similar specificity when hydrolysing amides other than penicillin G. If the

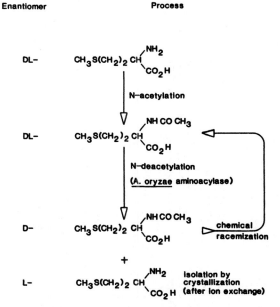

Figure 9.24   Process for the resolution of DL-methionine with amino acid acylase.

racemic mixture of an amino acid is first acylated and is then treated with the acylase, the product will be a mixture of optically pure L-amino acid and pre-dominantly D-acyl amino acid. These are easily separated on the basis of their solubilities at the isoelectric point of the amino acid. The pure L-amino acid can be crystallized from the mixture, leaving the residual acyl amino acid to be racemized and recycled through the process.

The enzyme for such a process is usually isolated form *Aspergillus oryzae*. In the period between 1953 and 1969 this mould was used to carry out the hydrolysis in batch reactors. The process was similar to that originally used to manufacture 6-APA; however, soluble enzyme is now extracted from the cells and is adsorbed on to a weakly basic resin (e.g. diethylaminoethyl cellulose) to provide an immobilised catalyst. This is packed into 1 m³ columns which become the catalytic reactors at the heart of the process (Figure 9.25). The reactor maintains at least 60% of its activity over five weeks of continuous operation. When the activity has fallen too far, the residual enzyme can be stripped from the resin, which is then recharged with a fresh batch.[28]

When this process was introduced in 1969 it was the first to use an immobilized enzyme. The scale of the process is not large (see Table 9.10), but where the chemical synthesis of DL-amino acids is cheap it provides an efficient route through to the L-isomer.

The enzymatic synthesis of L-cysteine also couples a stereospecific hydrolysis with a racemization, but it is performed on a much larger scale. This route is progressively replacing the rather dirty extraction from keratin (hair) hy-

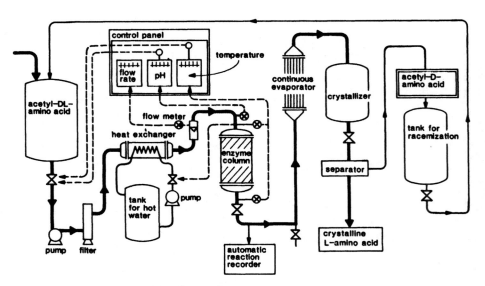

Figure 9.25 Flow diagram for the continuous production of L-amino acids by immobilized aminoacylase. From Chibata, I. (1979) *Immobilized Enzymes: Research and Development*, Halsted, p. 172.

drolysates. The synthesis starts with methyl 2-chloroacrylate which is converted to DL-amino-Δ^2-thiazoline-4-carboxylate. An enzyme from a strain of *Pseudomonas* or from *Sarcina lutea* will hydrolyse both enantiomers directly to L-cysteine; the necessary racemization occurs spontaneously during the reaction (Figure 9.26).[29]

The stereoselective synthesis of the required enantiomer from an achiral precursor would be a preferred route for any chiral product. Even if a racemization, such as that which accompanies the synthesis of L-methionine and L-cysteine, is practical, there will be fewer steps in the selective synthesis. The manufacture of L-aspartic acid from fumaric acid exemplifies such a process. The reaction is the reverse of one first discovered in 1926, in which resting cells of *E. coli* deaminate L-aspartate to fumarate. In the 1950s it was shown that growing cells of *E. coli* would accumulate L-aspartate when their growth medium was fed fumarate and ammonium ions. The enzyme responsible for this reversible reaction is L-aspartate ammonia lyase (Figure 9.26).

The manufacture is very efficient. It is catalysed either by whole *E. coli* cells or by enzyme extracted from them. In either case the preparation is immobilized, for example in polyacrylamide gel, and the reagents percolate through the enzyme column, as in the process for the manufacture of L-methionine (Figure 9.24). The L-aspartic acid formed is recovered from the effluent at pH 2.8. The yield of product is over 90%.[30]

The established industrial processes for the enzymatic synthesis of amino acids are simple. A few reactions of greater complexity, such as the synthesis of

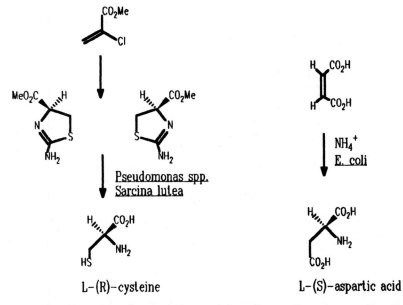

Figure 9.26 Routes for the enzymatic manufacture of L-cysteine and L-aspartic acid.

L-tryptophan from indole and L-serine, are described in patents, but these are not commercial processes. A few complex reactions have been taken to the scale of a large pilot plant, for example the stereoselective synthesis of amino acids from racemic α-hydroxyacids. This must compete with other processes for the manufacture of amino acids, but it does already operate at a tonne scale.

The α-hydroxyacids are first oxidized to ketoacids with a mixture of L- and D-α-hydroxyacid dehydrogenases. The dehydrogenases use NAD^+ as co-substrate, and this is reduced to NADH. In the second stage of the reaction this NADH is used to synthesize an L-amino acid from ammonium ions and the ketoacid in a reaction catalysed by an amino acid dehydrogenase. For the synthesis of leucine the L- and D-hydroxyacid dehydrogenases are obtained from *Lactobacillus confusus* and *L. casei* respectively, while the L-amino acid dehydrogenase is obtained from *Bacillus sphaericus* (Figure 9.27).[31]

The coupling of an oxidation and a reduction stage through NAD^+ recycles the cofactor, just as does *Saccharomyces cerevisiae* in the synthesis of ethanol (see section 9.2.1.1). Without such coupling each half of the reaction would be limited by the concentration of NAD^+, and this is a general problem which much be faced in the development of many synthetic reactions.

Coupled reactions of this type also need to retain the low-molecular-weight cofactor in the system. At best it may be possible to bind the cofactor covalently to the enzyme which uses it, but the techiques for doing this, while allowing sufficient flexibility for the cofactor to shuttle between two enzymes in an oxidation and reduction cycle, are in their infancy. In the synthesis of the amino acids the problem was solved by covalently attaching the NAD^+ to a soluble polymer. Its molecular weight was then sufficiently increased for it to be retained behind an ultrafiltration membrane with the enzymes which require it (Figure 9.27). Although this reduces the catalytic potential of the NAD^+, enough remains for it to function efficiently. Moreover, the chemical stability of this coupled NAD^+ is greater than that of NAD^+ itself, and this is an important advantage.[31]

The catalytic reactor thus consists of an ultrafiltration cell containing the enzymes and their cofactor. Substrates diffuse into the cell across the membrane, and the amino acids diffuse back. The substrate stream would eventually reach an equilibrium containing a mixture of hydroxyacids, ketoacids and amino acids were it not for the removal of the amino acids with a cation exchange resin.

The integrated balance of oxidation and reduction which this process makes possible cannot always be achieved. Yet the recycling of the cofactor is obviously crucial. A second, quite separate reaction can be introduced, for example the oxidation of ethanol to acetaldehyde with an alcohol dehydrogenase. However, it is possible that an electrolytic reduction of the NAD^+ could be developed. This is illustrated by the manufacture of glutathione, which is an imporant thiol-containing peptide. The dimeric disulphide is isolated from yeast cells as a copper complex. This is treated with sulphide to remove the copper, and the free peptide is

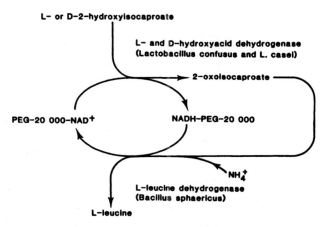

Figure 9.27   A process for the synthesis of an L-amino acid (leucine) from a racemic hydroxyacid (2-hydroxyisocaproate, i.e. 2-hydroxy-4-methylpentanoate). The reaction is carried out in an ultrafiltration reactor with the cofactor NAD^+ immobilized to a soluble polymer, polyethylene glycol 20 000, so that it remains behind the ultrafiltration membrane with the enzymes.

then reduced at acid pH in the catholyte of an electrolytic half cell. The oxidation of the thiol at the anode is prevented by an ionic membrane which separates the two halves of the cell.

The reduced glutathione could indirectly provide the reducing power for many enzymatic reductions, but similar processes for the direct electrochemical reduction of NAD^+ are now being developed.[32] If they are successful, some powerful catalysts for chemical reduction in aqueous media will become available.

Biological catalysis is often considered friendly to the environment, as somehow more 'natural' than chemistry. The reasons why this is an unrealistic assumption are economic as well as practical. The article which Sir Charles Bedford wrote in 1921 (section 9.2.1.1) hints at some of the relevent issues, and they really require a separate analysis. Nevertheless, like any other branch of chemistry, biological catalysis does offer improvements to existing processes which can reduce the burden which our use of modern technology imposes on the environment. In short, it offers some cleaner chemistry, of which the resolution of 2-chloropropionate and the hydrolysis of acrylonitrile are two more examples to set alongside the new process for the manufacture of cysteine.

The herbicides based on 2-phenoxypropionate are synthesized from 2-chloropropionate. Dichlorprop and Fluazifop (Figure 9.28) are two examples, which, like other members of the group, are chiral. Although only the R-enantiomer is active, they were originally marketed as racemates because of the added cost of resolving the mixture (see also section 7.5.8).

**9.7
Clean chemistry**

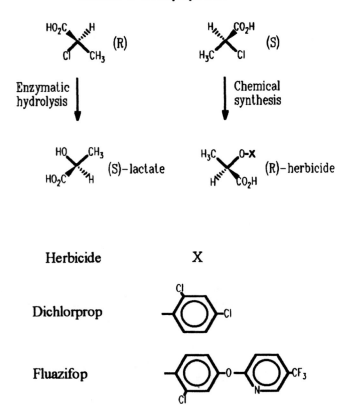

Figure 9.28   Synthesis of herbicides based on 2-phenoxypropionate.

Recently the need to reduce the evironmental burden of unnecessary chemicals placed a premium on a product containing only the active isomer. Moreover, if the resolution is performed at an early enough stage in the synthesis there is a large cost to be saved in not wasting materials to synthesize the inactive isomer. This is important where the aromatic portion of the molecule is complex, as it is in Fluazifop.

The herbicides are synthesized from (S)-2-chloropropionate, since the chiral centre is inverted when the aromatic portion is added. ICI isolated a number of micro-organisms which were able to grow on chloropropionate. Amongst these were several strains of *Pseudomonas* that contained separate enzymes for hydrolysing the R- and the S-enantiomers. The product of the hydrolysis, which cannot be recycled, is lactate with an inverted chirality. Since the (S)-2-chloropropionate is required, one strain was genetically manipulated to remove the enzyme which specifically hydrolysed this enantiomer, and to enhance the yield of the enzyme with the opposite specificity. The enzyme in this organism is now the basis for resolving (R,S)-chloropropionate at a scale of about 2000 tonnes annum^{-1}.[33]

The enzymatic hydrolysis of nitriles is useful not only in the manufacture of the important intermediate acrylamide, but also in the treatment of wastes containing high concentrations of organic nitriles and inorganic cyanide. Micro-organisms contain a variety of enzymes that hydrolyse organic nitriles, often distinguished by their attack on aliphatic and aromatic substrates (Figure 9.29).

This distinction is now accepted as too superficial, but it does illustrate the different hydrolytic processes, one of which yields on amide as an intermediate. The Nitto Chemical Company in Japan now uses a nitrile hydratase from *Rhodococcus rhodochrous* to hydrolyse acrylonitrile to acrylamide. The established chemical process for this hydrolysis is catalysed by copper at 100 °C. It is difficult not only to prepare and recover the catalyst, but also to separate and purify the acrylamide. The enzymatic process operates at 10° C and so requires less energy, and there is less chance of either the acrylonitrile or the acrylamide polymerizing during the reaction. The recovered product is also of greater purity than the material from the chemical process.[34]

This purer acrylamide can be polymerized to a higher molecular weight than is possible with the chemical product. Since much of the polyacrylamide is used as a flocculant whose effectiveness increases with its molecular weight, smaller amounts of polymer prepared from the enzymatic acrylamide are needed in precipitation processes for which it is the preferred flocculant. Some of these processes are in the water treatment industry, where a reduced input of a purer reagent is preferred. It therefore seems likely that the cleaner enzymatic process, which currently produces approximately 20 000 tonnes of acrylamide

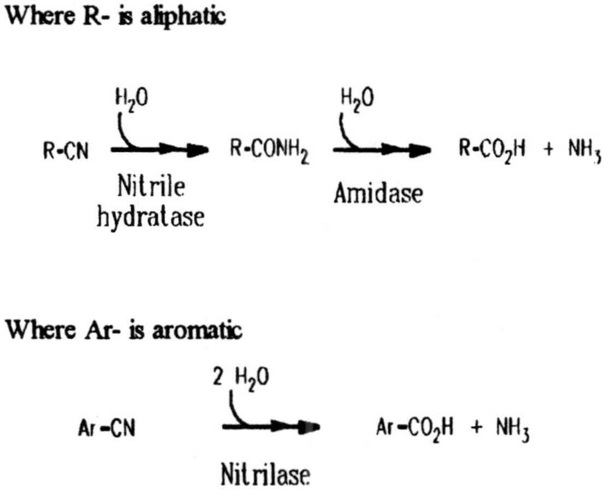

Where R- is aliphatic

$$R\text{-}CN \xrightarrow[\text{Nitrile hydratase}]{H_2O} R\text{-}CONH_2 \xrightarrow[\text{Amidase}]{H_2O} R\text{-}CO_2H + NH_3$$

Where Ar- is aromatic

$$Ar\text{-}CN \xrightarrow[\text{Nitrilase}]{2\ H_2O} Ar\text{-}CO_2H + NH_3$$

Figure 9.29 Enzymatic hydrolysis of nitriles.

annually, will take a progressively larger share of total annual production of about 200 000 tonnes.

9.8
Enzymes in unusual reaction conditions

The structure and stability of proteins is dependent on the presence of water, and enzymes are, quite reasonably, associated with catalysis in water. Indeed, this is one of their advantages, since they catalyse reactions over the range of temperatures at which water is liquid at normal atmospheric pressures, or even that at which organisms live. Nevertheless there are exceptions, for example in the high-temperature digestion of starch. Nor does the aqueous environment in which enzymes usually act preclude their use catalysts in organic solvents.

There are good reasons for wanting to use enzymes outside their normal range. High temperatures may favour an increased solubility of the substrate, or at least a suitable colloidal state in the case of starch. They may prevent a process from becoming infected with micro-organisms and thus are an aid to hygiene, if not sterility. Finally, they may affect the equilibria of reactions in favour of the required products. Organic solvents are useful for the same reasons, but they may have an added advantage in partitioning out of the aqueous phase both substrates and products which can inhibit the catalysis.

However, proteins become less stable or unfold as temperatures increase, or when they are used in organic solvents. The chemistry of this process is sufficiently understood to allow it to be influenced by deliberate redesign of their polypeptide chain, by chemical modification of their existing structure or by careful choice of their reaction conditions. For example, the stability of proteins in organic solvents is dependent on the $\log P$ value of the solvent (Figure 9.30), so that solvents with a low $\log P$ are considered less suitable for the catalysis than those with a higher value. Moreover, some thermophilic micro-organisms, which naturally grow at high temperatures ($> 60\,°C$), contain particularly stable enzymes which are useful as catalysts.

Often the enzymes are used in mixtures where the water and the solvent are present as two distinct phases. This is useful where the substrate is poorly soluble in water. If the water content of reactions involving hydrolytic enzymes is further reduced then the solvent not only affects the overall concentrations of substrates and products at equilibrium, but it can also trap an unstable intermediate which might otherwise be hydrolysed. The effect is to 'reverse' the action of the hydrolytic enzyme which now catalyses a synthetic reaction eliminating water from the substrates.

Already there are a few examples of large-scale reactions under such conditions (Table 9.17), although it is likely that some are demonstration projects at a pilot scale. The control of water activity is a significant technical problem where the enzyme is in nearly dry solvent. Where a lipase catalyses ester synthesis, stoichiometric amounts of water are released. Although some water is essential for enzyme activity, too much will affect the balance between ester synthesis and hydrolysis. Effective scale-up will require better methods of control than are presently available.

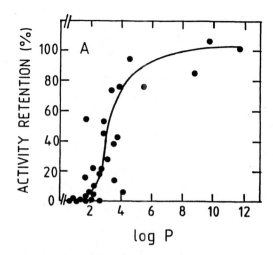

Figure 9.30 The degree to which organic solvents inhibit enzymes is dependent on their log P value. P is the partition coefficient of the solvent measured between octanol and water (i.e. P = solubility in octanol/solubility in water). Each point on the graph represents the activity of the enzyme in a reaction mixture containing water saturated with a solvent having the log P value shown. (After Laane, C. *et al.* (1987) Biocatalysis in organic media (Laane, C. *et al.* eds.). *Studies in Organic Chemistry,* **29**, 73, Elsevier.)

Table 9.17 Large-scale processes catalysed by enzymes in organic solvents

Catalyst	Reaction	Solvent
Nocardia corillina	Epoxidation	
	1-Octene	n-Hexadecane
	Styrene	n-Hexadecane
	1-Tetradecene	1-Tetradecene (substrate)
Horseradish peroxidase	Phenol polymerization	Ethyl acetate
Subtilisin (protease)	Ester hydrolysis	Dioxan, chloroform, etc.
Mucor miehei	Transesterification of	Triglycerides (substrate)
Aspergillus niger	triglycerides	
Arthrobacter simplex	Steroid dehydrogenation	Toluene

Adapted from Lilly, M.D. *et al.* (1990), 'Opportunities in Biotransformations' (Copping, L.G. *et al.* eds) Elsevier, p. 13.

This chapter has concentrated on the use of biological catalysis at an industrial scale for chemical synthesis. At a laboratory scale some sections of the industry make much wider use of the technology. Several excellent books have reviewed this topic (see Bibliography), and the chemical and biochemical literature is full of examples.

Many pharmaceutical products are mixtures of enantiomers, of which only one is likely to have the required pharmacological activity. Occasionally both are active, but with different pharmacological effects. Thalidomide fell into this category with disastrous consequences. The authorities who license pharmaceuticals have begun to argue that where the products are chiral

**9.9
Laboratory scale
synthesis (the chiral
switch)**

then the isomers should be separated and each rigorously tested on its own. Moreover, they believe that it is undesirable that an active chiral drug should be sold as a mixture with its inactive enantiomer.

These arguments have their parallel in agrochemical manufacture (see section 9.7), and the issue is complex. The debate has already affected programmes of chemical synthesis in the pharmaceutical industry, with some groups resolving to synthesize only achiral compounds. Others will concentrate on the need for synthetic methods that are targeted at only one enantiomer. The drive towards chiral purity, known as the 'chiral switch,' has emphasized the need for more effective methods of asymmetric synthesis so that the separation of enantiomers is avoided.[35]

Some fine-chemical manufacturers who supply the pharmaceutical industry now specialize in the use of enzymes for the synthesis of chiral synthons. Chiral cyanohydrins, α-hydroxycarboxylic acids, and α- and β-amino acids are just a few of the products now available, often produced directly from achiral precursors. Chiral glycerol-phosphate, which is used in phospholipid manufacture, is synthesized from glycerol and ATP with the enzyme glycerol kinase as catalyst, the ADP being recyled with a separate enzyme (Figure 9.31). These

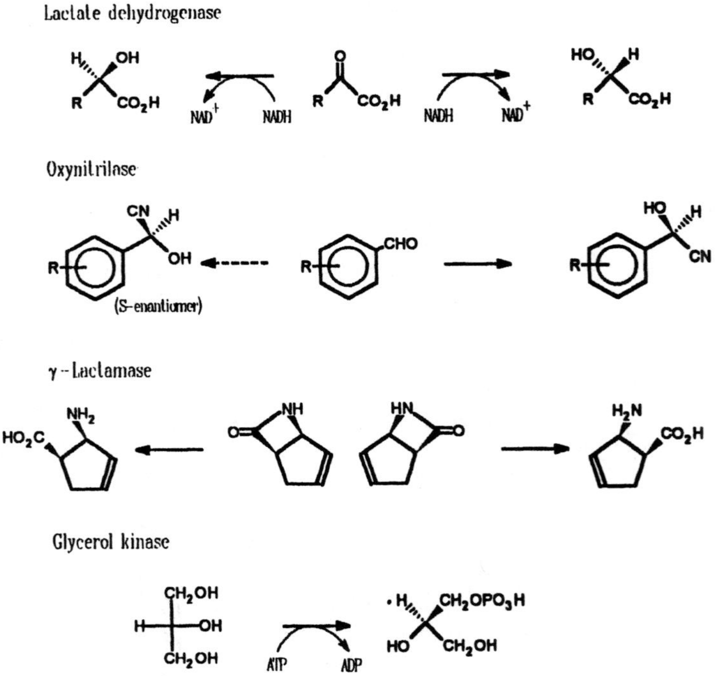

Figure 9.31 Some chiral synthons currently available as industrial intermediates. Note that three of the group are prepared from achiral precursors. Where two enantiomers are available they are prepared with different enzymes. The oxynitrilase which synthesizes the S-cyanohydrin is only available in small quantities.

are initially produced in kilogram amounts on the assumption that scale-up to about 1 tonne annum^{-1} is practical. The pharmaceutical industry almost certainly produces other synthons on this scale from its own resources.

The outcome is likely to be an increasing use of enzymes in chemical manufacture. As enzymes become more entrenched in synthesis at the laboratory and pilot scale so will their use in manufacture increase. This is not to say that every process that uses an enzyme to make an effective product in the laboratory will have an enzyme in its manufacturing process. Manufacture often involves development that radically alters the route of the synthesis. Nevertheless, the constraints in bringing materials to the market in the pharmaceutical industry leave little time for the complete reworking of a process so that it is likely that many enzyme steps will remain. This is already presenting a challenge for the industry, which is finding some engineering problems in the way of the scale-up of the technology.

When Berzelius identified his concept of catalysis as a special power of decomposition he compared the acidic digestion of starch with its biological equivalents, and he drew a general analogy between catalysis and the chemistry of living organisms. At the end of the 20th century this conceptual linkage has become a practical one.

9.10
A new catalysis

9.10.1 Metabolic pathway engineering

The metabolic pathways which micro-organisms use to manufacture amino acids and antibiotics were once only manipulated through mutation and selection (section 9.2.3.2). These are blunt techniques which act on the organism as a whole, and have little control over the molecular events that finally yield useful new strains of micro-organisms. It is the selection test itself, which is able to recover the few useful isolates from the very large number of mutant organisms, which makes this a practical method. However, it leaves the nature of the useful mutation undefined at the molecular level.

The practice of genetic engineering, through which a micro-organism is enabled to synthesize a foreign protein such as insulin (see section 9.4), is now sufficiently refined to allow its use in making precise changes to the organism. A microbiologist can change the base sequence of the DNA in a micro-organism, just as would a chance mutation, but knowing precisely what change is made, and being able to monitor its effect on the organism's activity. No longer is the change itself a random event; the relation between the cause and the effect can be precisely determined, if not yet entirely predicted.

The primary sequence of the amino acids in a microbiologically important enzyme may be altered, with a potential change in its catalytic activity or its stability. The amount of enzyme which is present in the cell may be increased or reduced, so affecting the flow of metabolites through a pathway. The enzyme could be deleted altogether, so blocking the synthesis of a range of end products from the pathway. It is true that all of these changes are similar to those introduced into the pathways affecting lysine synthesis (section 9.2.3.2), but now they are directed changes, not random ones.

Genetic engineering makes one more change practical. A cell can be altered to produce not only a hormone such as insulin, but also a new enzyme which is metabolically active in the cell itself. This raises the possibility of extending a metabolic pathway by one or two desirable steps, an action which is clearly impossible through the normal process of mutation and selection, which can only act on the complement of enzymes which the cell already contains.

One of the first uses of this technique enabled *E. coli* to produce the purple dye, indigo. *E. coli* contains tryptophanase, an enzyme which will degrade the amino acid tryptophan to indole (Figure 9.32). The enzyme naphthalene dioxygenase, which is present in another organism, *Pseudomonas putida*, will oxidize indole to indoxyl, and the latter is spontaneously oxidized to indigo in air. The transfer of the dioxygenase into *E. coli* allowed it to make indigo directly from tryptophan, a process neither it nor *P. putida* was able to catalyse on its own. With a range of tryptophan analogues as input the recombinant *E. coli* was able to make a series of dyes of different colours all related to indigo.[36]

A more targeted use has arisen in the synthesis of ascorbic acid (section 9.5.1). A number of attempts were made to simplify Reichstein's synthesis. They were mostly based on the oxidation of glucose to 5-keto-D-gluconate, a reaction which, like the oxidation of D-sorbitol to L-sorbose, is catalysed by *A. xylinum*. Other organisms, notably other *Acetobacter* spp. and *Erwinia* spp. will take the oxidation further to 2,5-diketo-D-gluconate (Figure 9.32), a compound which *Corynebacterium* spp. can reduce to 2-keto-L-gluonate. In the Reichstein synthesis this is easily converted to L-ascorbate (Figure 9.32).

These two steps, of oxidation and reduction, were demonstrated, as a mixed fermentation of *Acetobacter* spp. and *Corynebacterium* spp., to convert glucose

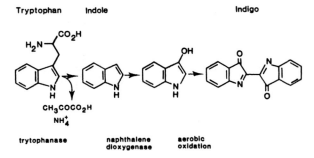

Figure 9.32   The synthesis of indigo by genetically engineered strains of *E. coli*. *E. coli*. which naturally contains tryptophanase, oxidizes tryptophan to indole, pyruvate and ammonia. When the cells are genetically engineered so that they also contain the naphthalene dioxygenase derived from *Pseudomonas putida*, the indole is further oxidized to indigo.

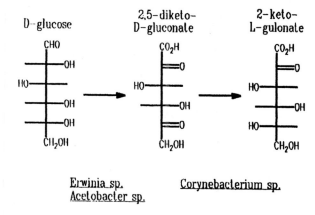

Figure 9.33   An alternative short synthesis of 2-keto-L-gulonate.

directly to 2-keto-L-gulonate. As a final development the reductase from *Corynebacterium* spp. was transferred into *Erwinia* spp. and this single organism alone was now able to perform both the oxidation and the reduction (Figure 9.33).[37] In the proceess the carbon skeleton of glucose enters the ascorbate with its orientation reversed compared to its entry in the Reichstein synthesis, a nice demonstration of the power of biological catalysis.

Further potential of the technology is seen in a recent manipulation of the synthesis of cephalosporin C. This β-lactam antibiotic, which is manufactured with *Acremonium chrysogenum*, is related to penicillin G, and its synthesis is derived from the same metabolic intermediate, isopenicillin N (Figure 9.34; see Figure 9.14). Unlike the penicillins there is no therapeutic use for the natural product itself, and the manufacture of all useful cephalosporins requires the removal of the 7-aminoadipyl side chain to provide the 7-aminocephalosporanic acid (7-ACA).

Metabolic pathway engineering has addressed two problems in the synthesis of the cephalosporins. The first is the tendency of strains of *A. chrysogenum* to produce not only cephalosporin C, but also penicillin N (Figure 9.34). The latter is a wasteful fermentation product which the cell should be capable of converting to cephalosporin C were the catalytic activity of the intervening enzymes sufficiently active. The second is the lack of an enzymatic route which could compete with the chemical cleavage of the 7-amido group.

The first problem has been attacked by improving the expression of the intermediate enzymes in the pathway, particularly the oxygenase which converts penicillin N into desacetyl cephalosporin C. This appears to have been successful, and it seems likely that further manipulation of the pathway will lie further back in the metabolic process, at the level of the supply of the three constituent amino acids (see Figure 9.14).

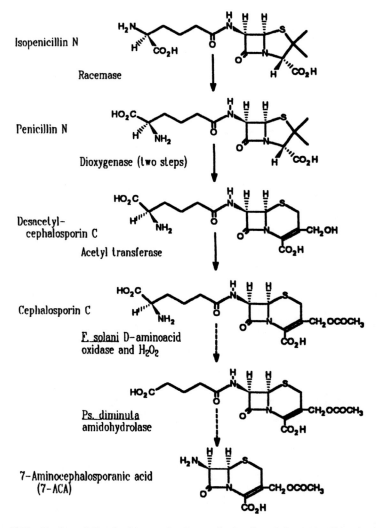

Figure 9.34 Outline of the final stages in the synthesis of cephalosporin C in *Acremonium chrysogenum* (cf. Figure 9.14) and its conversion to 7-ACA. Dotted arrows show the stages added by genetic engineering.

The enzymatic preparation of 7-ACA is technically feasible, but it is a curious fact that no single enzyme has shown more than a low activity in hydrolysing this particular amide link. An effective process requires two steps. If the 7-aminoadipyl function is first oxidized as far as 7-glutaryl, and this is a reaction which the D-amino acid oxidase in *Fusarium solani* will catalyse, then an enzyme in *Pseudomanas diminuta* will hydrolyse this amide to yield 7-ACA.

If both enzymes are transferred to *A. chrysogenum*, the engineered strain will produce 7-ACA directly.[38]

With such catalysts to hand, it is interesting to reflect on why none is used in manufacture. Ultimately the answers are economic rather than technical. While there is an occasional vogue for 'natural' dyes like indigo, the fine-chemical industry is the more efficient producer, and costs are important (see section 5.5.3.4). The new route for the synthesis of ascorbate omits the synthesis of sorbitol, which is an important product in its own right. It also yields 2-keto-L-gulonate as an intermediate, which is not easy to isolate, and the yields and the process intensity are low compared to the Reichstein synthesis. Similar problems would affect the manufacture of 7-ACA by the new route where a new fermentation and recovery process would need to be developed. In essence, each competes with an existing and well-developed manufacturing process which it is difficult to displace.

At some point, a new product will arise out of these methods which would not have been available in any other way. It could well be soon given the speed with which this technology is advancing.

9.10.2 Artificial enzymes

An important development for the future will be the creation of new catalysts that mimic enzymes. As the catalytic activity of enzymes becomes better understood so the chemist will take a greater hand in the synthesis of organic reagents that can extend the range of the catalysis into new reactions.

The activity of an enzyme is dependent on its three-dimensional structure, and the folding and the forces that create this shape are essentially delineated in the primary sequence of amino acids in the peptide chain. The specificity and selectivity of the catalysis result from the interaction of the substrate with the side chains of a few nicely positioned amino acids in the protein.

The analysis of these interactions usually relies on X-ray crystallography, either of the protein itself, or of one closely related to it. Some rules are beginning to emerge about the way in which proteins fold. Their overall shape is often conserved amongst related enzymes despite considerable changes in their primary amino acid sequence. This allows the gross features of the catalytic centre to be predicted without an X-ray structure, and the general nature of the catalysis can often be described when this structural information is combined with other data from the reaction kinetics, and from amino acid substitutions which genetic engineering can introduce. Unfortunately, some important features of the catalysis seem to depend on the precise location of interactions which are below the limit of resolution of X-rays.

Despite this, chemical changes can be made at the catalytic centre which affect the catalysis in a predictable fashion. The proteolytic enzyme papain was modified by covalently attaching a flavin to a thiol group at the active centre of the enzyme. The flavins are cofactors for oxidation and reduction processes, and they have a redox potential somewhat less negative than does NAD^+. Some enzymes which carry flavins are able to oxidize NADH to NAD^+ and hydrogen peroxide if molecular oxygen is also supplied. This is a typical oxidase reaction. The modified papain will also catalyse this reaction at a neutral pH at rates which are several orders of magnitude faster than the reaction between NADH and the flavin in the absence of papain. If an aliphatic group is attached to the NADH the rates of reaction become even faster because its association with the normal substrate binding site of the papain becomes firmer. In this simple way a proteolytic enzyme is transformed into an oxidase.[39]

Other attempts to modify enzymes have concentrated on those whose substrate specificity was not ideal; that is to attempt to retain the catalytic activity, but to apply it more widely. Enzymes frequently catalyse useful reactions, but not with substrates that are of commercial interest. In the past, the high specificity of these enzymes has been an insurmountable disadvantage, and a search for an alternative enzyme with a different specificity was the only way forward. Now genetic engineering offers an alternative approach, although it has yet to prove useful in practice.

One of the enzymes which take part in the activation of amino acids before their incorporation into proteins was successfully modified to increase its specificity with full retention of its catalytic activity.[40] However, an attempt to extend the substrate specificity of an amidohydrolase to make it capable of hydrolysing a wider range of penicillins and related antibiotics (section 9.6.2) has so far only succeeded at the expense of a considerable loss of catalytic activity, and the results with other enzymes are rather variable. This does not seem to be a useful method for extending the substrate range of enzymes, although it is important for a study of their action.

There has been greater success with modifying the rates of the various reactions that they catalyse. Some proteases will hydrolyse esters and amides as well as peptides, and will often catalyse a transamidation. When the enzymes are modified, the relative rates of these reactions may change, and under the right equilibrium conditions the proteases will reverse the hydrolysis and will synthesize peptide bonds. One method for the manufacture of human insulin from porcine insulin takes advantage of this. The two hormones differ at the carboxyterminal end of the B chain. In the human hormone threonine replaces the alanine of the porcine hormone (see amino acid 30 in Figure 9.19). If the enzyme carboxypeptidase Y is treated with mercuric chloride its normal action in removing the terminal alanine of the porcine hormone is modified; it will also catalyse a transpeptidation from threonine amides allowing the replacement of the alanine with threonine.[41]

Another important feature of the enzyme that the genetic engineer can change is its stability, both inside the cell where it is synthesized and in the catalytic reactor where it operates in a manufacturing process. Enzymes can operate as catalysts for only short periods at high temperature (section 9.6.1), and some of the causes of this instability are dependent on the primary amino acid sequence. New disulphide bridges within the folded structure have been introduced into one protein to give it greater strength to resist thermal denaturation, and the chemical stability of the peptide chain has been increased in another. The peptide links which involve the carboxyl group of aspartate are particularly prone to hydrolysis at high temperature, and in some instances the aspartate has been replaced by another amino acid, with an accompanying increase in the stability of the protein.[42] In a similar way it should be possible to identify and alter those peptide bonds which are susceptible to proteolytic attack inside the cells where they are made.

Thus the genetic engineer and the chemist can alter the protein to make it a more effective manufacturing catalyst. In the end the total design of an enzyme may become a possible target. The notion of molecular recognition in biology and chemistry has established a concept of a host–guest relationship into the binding of a substrate to an enzyme. The substrate fits within a pocket of the enzyme which pulls it into a transition state, so catalysing its transformation into products. Artificial ways of inducing this change include, in biology, the raising of antibodies that will recognize the transition state, and these proteins do have predictable catalytic activities.[43] However, chemical methods for polymer synthesis have made the idea of entirely synthetic peptides with catalytic activity a reality. Nor would such structures be confined to single-chain polypeptides containing just the twenty natural amino acids.

The influence on chemical manufacture of such a radical redesign of polypeptide catalysis will not be quickly felt. Indeed as the basis for new enzymatic catalysts it may be unnecessary. There is, after all, a considerable pool of natural catalytic activities for us still to exploit. It is more likely that the genetic engineer will help to throw light on to the actual catalytic mechanisms themselves, so that they can be reinterpreted within the framework of organic chemistry.

What has been described as biomimetic control of chemical selectivity is already possible. When the steroid 3-α-cholestanol is esterified with 4-iodo-phenylacetic acid and treated with chlorine in the dark, the iododichloride can generate free radicals which attack the C-17 of the steroid. The shorter ester derived from benzoic acid attacks C-9 (Figure 9.35). In another context metal porphyrin complexes have been devised which can hydroxylate hydrocarbons (Figure 9.36).[44] Large synthetic molecules with holes, such as those formed naturally in some bacteria from rings of linked glucose molecules (cyclodextrins), or synthesized by organic chemists, such as crown ethers or linked porphyrin rings, will all act as hosts to smaller molecular guests. The host will orientate the guest, and can control the stereoselectivity in reactions with the latter. A recently prepared catalyst for a Diels–Alder reaction is only one example of many now in the chemical literature (Figure 9.37). With these host–guest relationships chemists are even beginning to explore the concept of molecular self-assembly which is a feature of the living organism.[45]

These are exciting new ideas for organic synthesis, and many more are likely

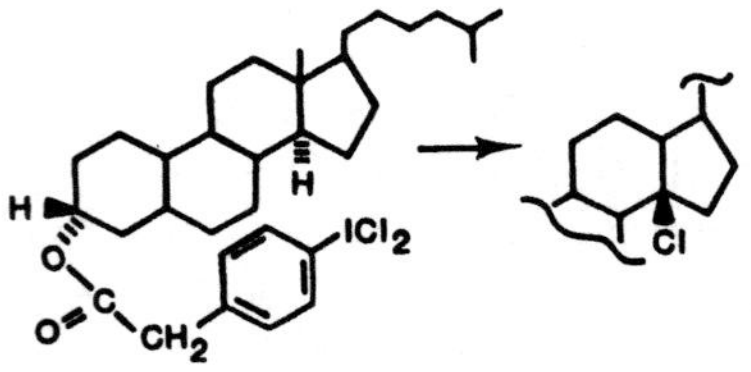

Figure 9.35   The biomimetic chlorination of steroids. The dichloride of 3α-(4-iodophenylacetyl)-cholestanol (1) activates the C-14 of the steroid nucleus, while the dichloride of the shorter 3-iodobenzoate ester (2) activates C-9. A chloride radical is delivered to these carbon atoms from a second molecule of the respective iodochlorides, and for this reason the chlorine atom derived from (1) lies above the plane of the steroid nucleus, while that derived from (2) lies below it. After Breslow, R. (1980) *Acc. Chem. Res.* **13**, 170.

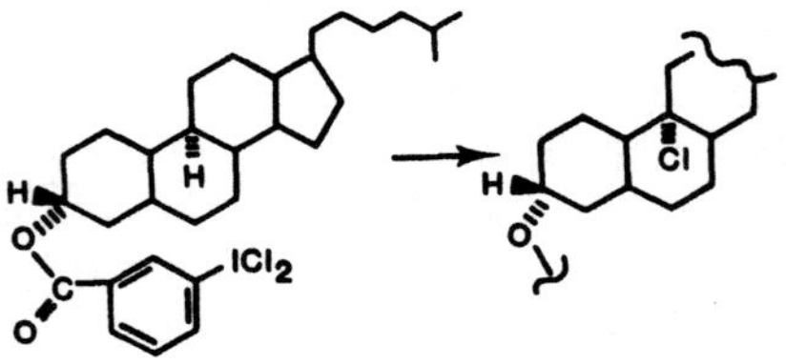

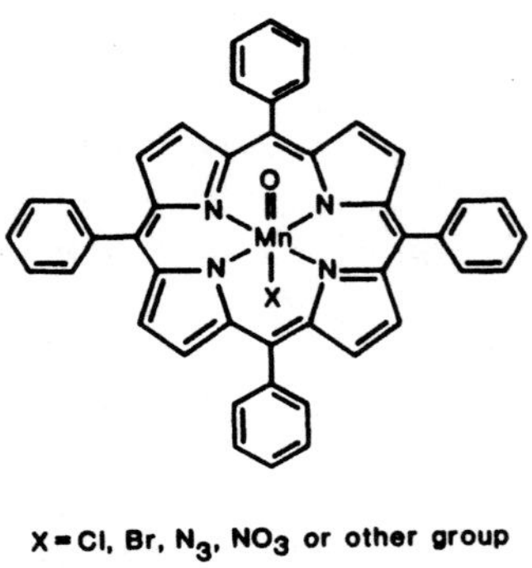

X = Cl, Br, N₃, NO₃ or other group

Figure 9.36   A synthetic porphyrin containing manganese(V) which can hydroxylate saturated hydrocarbons through a free-radical mechanism at room temperature. Manganese(III) is oxidized to manganese(V) with iodosylbenzene.

to follow a better understanding of the mechanisms of enzyme catalysis. It would be easy to interpret this as heralding a revolutionary new phase in the development of catalysts brought about by the impact of biotechnology. It is more rationally viewed as the synthesis of two strands in the development of catalysis, one vitalist and the other mechanistic, which were separated in the nineteenth century. Those who take the revolutionary view are likely, once again, to miss the antecedents, and may have to rediscover what is already

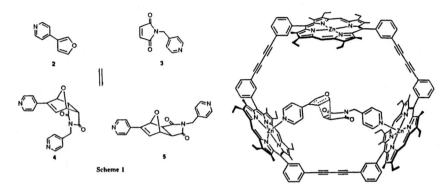

Figure 9.37 A trimeric porphyrin host which catalyses the Diels–Alder reaction between reactants **2** and **3** to form product **5**. (After Walter, C.J. *et al.* (1993) *J. Chem. Soc. Chem. Commun.*, 459).

known. It is to be hoped that this brief outline of large-scale biological catalysis as it is applied to chemical manufacture has placed the recent advances in context so that this aspect of biotechnology can be seen for what it is; a useful and evolving technique with an established place in manufacture, but, like any other, one with its own peculiar limitations.

References

1. 'Biotechnology,' Report of a Joint Working Party under the chairmanship of Dr. A. Spinks, London, HMSO, 1980.
2. Dunnill, P., *Chem. Ind.* 1981, 204.
3. Boyde, T.R.C., 'Foundation Stones of Biochemistry,' Voile et Aviron, 1980.
4. Brown, H.T., *J. Inst. Brewing*, 1916, 13, 265; Porter, J.R., *Science*, 1972, 178, 1249.
5. Chem. Ind., 1984, 425–443.
6. Kosaric, N. *et al.* 'Biotechnology,' vol. 3 (Dellweg, H., ed.), Verlag Chemie, 1983, p. 257; Winkelmann, G., ed., 'Microbial Degradation of Natural Products,' VCH Publishers, 1992, p. 83 and p. 127.
7. Ingram, L.O. *et al.*, *Dev. Ind. Microbiol.*, 1990, 31, 21.
8. Bergeron, P.W. and Hinman, N.D., *Appl. Biochem. Biotechnol.*, 1990, 24/25, 15.
9. Chapman, A.C., *J. Roy. Soc Arts*, 1921, 69, 581, 597, 609.
10. Hastings, J.J.H., 'Economic Microbiology,' vol. 2 (Rose, A.H., ed.), Academic Press, 1978, p. 31.
11. Rohr, M. *et al.*, 'Biotechnology,' vol. 3 (Dellweg, H., ed.), Academic Press, 1983, p. 31.
12. Araki, K. and Ozeki, T., Kirk-Othmer Encyclopedia of Chemical Technology, 4th edn, vol. 2, (Kroschwitz, J.I. and Howe-Grant, M., eds), John Wiley, 1992, p. 504.
13. Soda, K. *et al.*, 'Biotechnology,' vol. 3 (Dellweg, H., ed.), Academic Press, 1983, p. 479.
14. Berdy, J., 'Handbook of Antibiotic Compounds,' CRC Press, 1980—in progress.
15. Florey, H.W. 'Antibiotics,' vol. 1 (Florey, H.W. *et al.*, eds), Oxford University Press, 1949, p. 1.
16. Macfarlane, G., 'Alexander Fleming,' Chatto & Windus, 1984.
17. Raistrick, H. *et al.*, *Proc. Roy. Soc. Ser. B*, 1931, 220, 1.
18. Swartz, R.W., *Ann. Rep. Ferm. Proc.*, 1979, 3, 75.
19. Aunstrup, K., 'Applied Biochemistry and Engineering,' vol. 2, (Wingard, L. *et al.*, eds), Academic Press, 1979, p. 27.
20. Cheetham, P.S.J., 'Handbook of Enzyme Biotechnology,' 2nd edn, (Wiseman, A., ed.), Ellis Horwood, 1985, p. 274.
21. Teuber, M., 'The Release of Genetically Modified Microorganisms,' (Stewart-Tull, D.E.S. and Sussman, M., eds), Plenum Press, 1992, p. 59.

22. Pierard, L. & Bollen, A., *J. Biotechnol.*, 1990, 15, 283.
23. Brown, A.J., *J. Chem. Soc.*, 1886, 49, 172.
24. British Patent 360 334 (1931).
25. Charney, W. and Hertzog, H.L., 'Microbial Transformations of Steroids,' Academic Press, 1967.
26. Reichelt, J.R., 'Industrial Enzymology' (Godfrey, T. & Reichelt, J., eds), Macmillan, 1983, p. 375.
27. Savidge, T.A., 'Biotechnology of Industrial Antibiotics,' (Vandamme, E.J., ed.), Marcel Dekker, 1984, p. 171.
28. Chibata, I., 'Immobilised Enzymes,' Halstead Press, 1978.
29. Sano, K. *et al.*, *Appl. Environ. Microbiol.*, 1977, 34, 806.
30. Chibata, I. *et al.*, *Progr. Ind. Microbiol.*, 1986, 24, 145.
31. Schmidt-Kastner, G. and Egerer, P., 'Biotechnology,' vol. 6a (Keislich, K., ed.), Verlag Chemie, 1984, p. 387.
32. Bowen, R. and Pugh, S., *Chem. Ind.*, 1985, 323; Laval, J.-M. *et al.*, *Ann. N.Y. Acad. Sci.*, 1993, 672, 213.
33. Taylor, S.C., In 'Opportunities for Biotransformations,' (Copping, L.G. *et al.*, eds), Elsevier Applied Science, 1990, p. 170.
34. Kobayashi, M. *et al.*, *Trends Biotechnol.*, 1992, 10, 402.
35. Stinson, S.C. *Chem. Eng. News*, 1992, Sep. 28, 46.
36. Ensley, B.D. *et al.*, *Science*, 1983, 222, 167.
37. Boundrant, J., *Enzyme Microb. Technol.*, 1990, 12, 322.
38. Skatrud, P.L., *Trends Biotechnol.*, 1992, 10, 324.
39. Kaiser, E.T. *et al.*, 'Biomimetic Chemistry,' (Dolphin, D. *et al.*, eds) *Adv. Chem. Ser.*, 1980, 191, 35, Amer. Chem. Soc.
40. Wilkinson, A.J. *et al.*, *Nature*, 1984, 307, 187.
41. Markussen, J., 'Recent Advances in Diabetes,' vol. 1 (Nattrass, M., ed.), Churchill Livingstone, 1984, p. 45.
42. Graycar, T.P., 'Biocatalysts for Industry,' (Dordick, J.S., ed.), Plenum Press, 1991, p. 257.
43. Chadwick, D.J. & Marsh, J., 'Catalytic Antibodies,' *Ciba Foundation Symposium* 159, John Wiley & Sons, 1991.
44. Groves, J.T. *et al.*, J. Amer. Chem. Soc., 1981, 103, 2884; Khenkin, A.M. & Shilov, A.E., New J. Chem., 1989, 13, 659.
45. Stoddart, J.F., 'Host–Guest Molecular Interactions: From Chemistry to Biology,' (Chadwick, D.J. and Widdows, K., eds) *Ciba Foundation Symposium* 158, 5, John Wiley & Sons, 1990.

Bibliography

'Basic Biotechnology,' Bu'lock, J. & Kristiansen, B., Academic Press, 1987.
'Biocatalysis,' Abramowicz, D.A., ed., Van Nostrand Reinhold, 1990.
'Biocatalysts for Industry,' Dordick, D.A., ed., Plenum Press, 1991.
'Biochemistry,' Zubay, G.L., 2nd edn, Macmillan, 1989.
'Biochemical Engineering and Biotechnology Handbook,' Atkinson, B. and Mavituna, F., 2nd edn, Macmillan, 1991.
'Biotechnology. A Textbook of Industrial Microbiology,' Creuger, W. & Creuger, A., 2nd edn, Sinauer Ass. Inc., 1989 (English trans. Brock, T.D., 1990).
'Biotransformations in Organic Chemistry,' Faber, K., Springer-Verlag, 1992.
'Biotransformations in Preparative Organic Chemistry,' Davies, H.G. *et al.*, Academic Press, 1989.
'Immobilised Biocatalysts,' Hartmeier, W., Springer Verlag, 1986.
'Kirk-Othmer Encyclopedia of Chemical Technology,' 3rd edn, 24 vols and supplements (Grayson, M. & Eckroth, D., eds), John Wiley, 1978–84.
'Kirk-Othmer Encyclopedia of Chemical Technology,' 4th edn, vols in progress (Kroschwitz, J.I. & Howe-Grant, M., eds), John Wiley 1992-93.
'Microbiology,' Prescott, L.M. *et al.*, Wm. C. Brown, 1990.
'Opportunities in Biotransformations,' Copping, L.G. *et al.*, eds, Elsevier Applied Science, 1990.
'Principles of Gene Manipulation,' Old, R.W. and Primrose, S.B., 4th edn, Blackwell Scientific, 1989.

The future 10
Alan Heaton

The late 1970s and early 1980s were traumatic years for the major chemical producers with serious problems arising from reduced demand coupled with overcapacity (caused by rapid expansion in the early to mid-1970s), particularly in the petrochemicals and polymer sectors of the industry. Rationalisation of business areas and reductions in the labour force at this time, followed by recovery of the world economy enabled companies to improve their performance and profitability through the mid- to late 1980s.

By the end of the decade and into the start of the 1990s the cyclical nature of the chemical industry was again apparent with a dramatic downturn in activity and a resultant reduction in profitability. This was linked to a major world recession in economic activity, which has proved much longer and more serious in its effects than the one at the end of the 1970s and has seen companies cutting the labour force even more strongly, and also retrenching into areas which have been their traditional strengths, whilst disposing of their weaker and unprofitable businesses. For example the number of employees working in the UK chemical industry fell as follows:

Year	Number of employees
1980	390 000
1985	339 000
1991	303 000

There were signs in 1993 of a slow climb out of this recession.

10.1
Current situation

Attention was drawn in the corresponding section of the first edition of this book to several factors which had profoundly affected the development of the chemical industry during the previous decade, and which would continue to be influential during the next decade. These included developments in electronics and computer technology, developments in chemical engineering, energy costs and public concern for the safety of chemicals and chemical operations.

There is no doubt that the extremely rapid advances in electronics and computer technology—particularly the increasing computer processing power for ever lower costs—has led to major improvements in the efficiency of operations within the chemical industry. This has been the major factor in

10.2
Significant influences

enabling output levels to be maintained (or even increased) whilst reducing manpower.

Although energy costs are still a major focal point, with the exception of a few processes where they are the major cost item, e.g. electrolysis of brine, the drive to reduce them is not at the same intensity as it was at the end of the 1980s. This is due to the reductions already achieved, which obviously limit the scope for future improvements, and the attention devoted to minimising these costs at the design stage for new plants. Advances in chemical engineering, coupled with a reduction in energy costs, have been the major factors in improving process efficiencies.

Public interest in, and concern for, the safety of chemicals—particularly for their effect on the environment— has continued to grow. This has been recognised here by including an entirely new chapter (chapter 3) devoted to environmental issues plus half of another new chapter (chapter 2) to safety. These sorts of issues have also been highlighted in specific parts of the book such as agrochemicals (section 7.2), pharmaceuticals (section 8.4.1) and biological catalysis and biotechnology (section 9.6.4). The consequences of this public pressure for safer and more environmentally acceptable (even friendly!) chemicals are felt in almost all parts of the industry. In general this takes the form of much tighter controls, i.e. lower limits on gaseous emissions and effluent discharges. This in turn increases the cost of: (1) producing any chemical; (2) probably the price for which it sells; and (3) the price that we, as consumers, pay for the end product. This pressure is felt even more acutely in sectors producing biologically active chemicals, e.g. pesticides (plant protection agents) and drugs or medicines. Here the drive to produce new and safer products has led to an enormous increase in the cost of testing and development programmes. This is a result of the increased number of prescribed tests leading to a lower success rate and an increased time-scale. Details were given in section 7.3 (particularly 7.3.2) and section 8.4.1.

There is no doubt that there is a greater effort by the chemical industry to produce safer products and a genuine attempt to be more open in discussing and tackling environmental issues (although some would argue that there is still some way to go to reach an acceptable position). The industry certainly could do a lot more to publicise the benefits which its products bring to us and so improve its still poor public image. There is some evidence that this is starting to happen. The remarkable story of the development of CFC replacements was cited in section 3.5 as a pointer to the way forward on environmental issues for the benefit of all. A unique international agreement by governments was the result of the worldwide outcry against CFCs. The positive response of the chemical industry and its phenomenal achievement in developing environmentally acceptable alternatives over such a short time period should be fully recognised. It is worth stressing here, again, that there is a price to be paid for this environmental improvement—the replacements are considerably more expensive than the CFCs. However, for the consumer this is tempered by the fact that in most applications, e.g. refrigeration, polyurethane foam, relatively small amounts of CFC replacements are used. It does show just how much can be achieved if there is sufficient united concern over an environmental issue coupled with a willingness by companies to overcome this concern by producing acceptable alternatives.

10.3.1 Constraints

The constraints cited in section 7.3.1 of the first edition of this book still apply very much to the chemical industry. These are: increasing the safety of chemicals and chemical processes; control of energy usage and therefore costs; control of emissions to increasingly lower levels; and restricted availability of oil leading to increasing interest in alternative feedstocks such as coal and biomass. The growing influence of the oil-producing countries such as Mexico and Saudi Arabia, and others, like Korea, as suppliers of basic petrochemicals will enhance the tendency of the mature chemical-producing countries (Europe, Japan and North America) to concentrate more on downstream processes where their experience and technological expertise make it much harder for the 'developing' chemical countries to compete. Following the remarkable political changes in the Communist bloc during the last few years these countries offer both a challenge and an opportunity, since they have the potential to become major chemical producers but also represent a vast market because of the size of their populations, particularly in the former Soviet Union and China.

Chapter 2 of this volume has also drawn attention to the growing importance of quality issues. The commitment of chemical companies to implement 'total quality management,' and hence conduct their business more effectively and efficiently to satisfy their customers' needs (and of course reduce their costs to become more competitive), is apparent. Many also require their suppliers to operate in a similar manner and to have received external inspection and certification (such as BS5750) to confirm this.

10.3.2 Prospects

Despite the very severe economic difficulties of the last few years the future prospects for the chemical industry as a whole look good. For the foreseeable future chemical pesticides will continue to be the main weapons in the battle against pests and the agrochemicals sector will continue to develop new, more effective products which are environmentally acceptable. The intensive research effort in the pharmaceutical area to produce novel drugs which are effective against 'newer' diseases such as Aids and Alzheimer's will undoubtedly continue as will the development of a complementary approach using more biochemically oriented methods such as gene therapy. It is however a concern that the rate at which new pesticides and drugs are being introduced has fallen significantly in recent years. The constraints already discussed (relating to safety aspects) have undoubtedly had a major influence on this situation. There seems little chance of things improving over the next few years because research efforts—particularly long-term basic research—have been somewhat curtailed during the recession.

Another concern has been the relatively poor return on the enormous investment over the last decade in advanced materials such as speciality polymers. Although there have been some successes (see section 4.1.3) this is scant reward for the enormous effort expended and the financial investment made.

Returning to the theme of safety and environmental acceptability of chemicals the public pressure to improve these will undoubtedly continue. Already we are seeing the results of different approaches and strategies. Whereas the strategy with existing plants has been to contain, treat and discharge as effluent waste by-products (unless of course they can be utilised), current thinking on new plants is to design out or minimise waste production, i.e. avoid producing it in the first place, by using cleaner technologies. Examples of this include: (1) the use of membrane rather than mercury cells for chloralkali production (section 6.5.4.4); and (2) the development of the use of 'clean chemistry' (section 9.6.4) where, with the aid of biological catalysis using organisms (or extracts thereof) which have been genetically engineered, only a single biologically active enantiomer is produced (see also section 7.5.8). Use of a single stereoisomer rather than a racemic mixture halves the amount of chemical needed and also reduces waste by 50%. Asymmetric synthesis and/or optical resolution are complementary approaches to achieve the same aim. A third example of cleaner technology can be found in the dyestuffs industry where more environmentally acceptable products and more efficient processes to produce more effective dyes are being developed. Modern reactive dyes having a higher fixation to fabrics will minimise waste during cleaning of the dyed garments.

From a business point of view the importance of adopting a worldwide or global approach will continue to grow in importance. A major influence on this is that the developing markets, which are growing rapidly, are in areas like the Pacific Rim of South East Asia and China (with its enormous potential). This might suggest that the trend of the last few decades for the multi-national chemical companies to pursue a policy of acquisitions to become even larger will continue. It will therefore be interesting to observe the performances of the UK's largest chemical company, ICI, over the next few years. In 1993 ICI adopted the opposite approach and split the company into two parts—Zeneca (which includes broadly the agrochemical, pharmaceutical and speciality chemicals businesses) and ICI (which encompasses all the other business areas).

The great dependence of our lifestyle on the products of the chemical industry, plus the world's rapidly growing population, which requires increasing quantities of e.g. food and medicines, will ensure a continuous demand for the industry's products.

Index

Printed in the United States
23994LVS00001B/208

WARNER MEMORIAL LIBRARY
EASTERN UNIVERSITY
ST. DAVIDS, PA 19087-3696